William Jacob Holland

The Butterfly Book

A Popular Guide to a Knowledge of the Butterflies of North America

William Jacob Holland

The Butterfly Book
A Popular Guide to a Knowledge of the Butterflies of North America

ISBN/EAN: 9783742810724

Manufactured in Europe, USA, Canada, Australia, Japa

Cover: Foto ©Klaus-Uwe Gerhardt /pixelio.de

Manufactured and distributed by brebook publishing software
(www.brebook.com)

William Jacob Holland

The Butterfly Book

THE BUTTERFLY BOOK

A POPULAR GUIDE
TO A KNOWLEDGE OF THE BUTTERFLIES OF
NORTH AMERICA

BY

W. J. HOLLAND, Ph. D., D. D., LL. D.

CHANCELLOR OF
THE WESTERN UNIVERSITY OF PENNSYLVANIA; DIRECTOR OF
THE CARNEGIE MUSEUM, PITTSBURGH, PA.; FELLOW OF THE ZOÖLOGICAL AND ENTOMOLOGICAL
SOCIETIES OF LONDON; MEMBER OF THE ENTOMOLOGICAL SOCIETY
OF FRANCE, ETC., ETC.

WITH 48 PLATES IN COLOR-PHOTOGRAPHY, REPRO-
DUCTIONS OF BUTTERFLIES IN THE AUTHOR'S COL-
LECTION, AND MANY TEXT ILLUSTRATIONS PRESENTING
MOST OF THE SPECIES FOUND IN THE UNITED STATES

LONDON
HUTCHINSON & CO.
PATERNOSTER ROW
1902

TO MY GOOD WIFE
AND MY TWO BONNY BOYS,
THE COMPANIONS OF MY LEISURE HOURS
AND MY VACATION RAMBLES,
I DEDICATE THIS BOOK,
WITHOUT ASKING THEIR PERMISSION

PREFACE

AT some time or other in the life of every healthy young person there appears to be developed what has been styled "the collecting mania." Whether this tendency is due to the natural acquisitiveness of the human race, to an innate appreciation of the beautiful and the curious, or to the development of an instinct such as is possessed by the bower-bird, the magpie, and the crow, which have the curious habit of gathering together and storing away trifles which are bright and attractive to the eye, I leave to students of the mind to decide. The fact is patent that there is no village without its youthful enthusiast whose collection of postage-stamps is dear to his heart, and no town in which there are not amateur geologists, archæologists, botanists, and zoölogists, who are eagerly bent upon the formation of collections of such objects as possess an attraction for them.

One of the commonest pursuits of boyhood is the formation of a collection of insects. The career of almost every naturalist of renown has been marked in its early stages by a propensity to collect these lower, yet most interesting and instructive, forms of animal life. Among the insects, because of their beauty, butterflies have always held a foremost place in the regard of the amateur collector. For the lack, however, of suitable instruction in the art of preserving specimens, and, above all, by reason of the almost entire lack of a convenient and well-illustrated manual, enabling the collector to identify, name, and properly classify the collections which he is making, much of the labor expended in this direction in the United States and Canada fails to accomplish more than the furnishing of temporary recreation. It is otherwise in Europe. Manuals, comprehensive in scope, and richly adorned with illustrations of the

v

leading insect forms of Great Britain and the Continent, have been produced in great numbers in recent years in England, France, and Germany. The result is that the youthful collector enters the field in those countries in the possession of a vast advantage over his less fortunate American fellow. It is to meet this want on this side of the Atlantic that this volume has been written. Its aim is to guide the amateur collector in right paths and to prepare him by the intelligent accomplishment of his labors for the enjoyment of still wider and more difficult researches in this and allied fields of human knowledge. The work is confined to the fauna of the continent of North America north of the Rio Grande of Texas. It is essentially popular in its character. Those who seek a more technical treatment must resort to the writings of others.

If I shall succeed in this book in creating a more wide-spread interest in the world of insect life and thereby diverting attention in a measure from the persecuted birds, which I love, but which are in many species threatened with extinction by the too eager attentions which they are receiving from young naturalists, who are going forth in increased numbers with shot-gun in hand, I think I shall render a good service to the country.

I flatter myself that I have possessed peculiar facilities for the successful accomplishment of the undertaking I have proposed to myself, because of the possession of what is admitted to be undoubtedly the largest and most perfect collection of the butterflies of North America in existence, containing the types of W. H. Edwards, and many of those of other authors. I have also enjoyed access to all the other great collections of this country and Europe, and have had at my elbow the entire literature relating to the subject.

The successful development in recent months of the process of reproducing in colors photographic representations of objects has been to a certain degree the argument for the publication of this book at the present time. A few years ago the preparation of such a work as this at the low price at which it is sold would have been an utter impossibility. "The Butterflies of North America," by W. H. Edwards, published in three volumes, is sold at one hundred and fifty dollars, and, as I know, is sold even at this price below the cost of manufacture. "The Butterflies of New England," by Dr. S. H. Scudder, in three volumes, is sold at seventy-five dollars, and likewise represents at this price only a partial return to the learned author for the money, labor, and time expended upon it. The present volume, while not pretending to vie in any respect with the magnificence of the illustrations contained in these beautiful and costly works, nevertheless presents in recognizable form almost every species figured in them, and in addition a multitude of others, many of which have never before been delineated. So far as possible I have employed, in making the illustrations, the original types from which the author of the species drew his descriptions. This fact will no doubt add greatly to the value of the work, as it will not only serve as a popular guide, but have utility also for the scientific student.

I am under obligations to numerous friends and correspondents who have aided me, and take the present opportunity to extend to them all my hearty thanks for the generous manner in which they have assisted me in my pleasant task. I should fail, however, to follow the instincts of a grateful heart did I not render an especial acknowledgment to Mr. W. H. Edwards, of Coalburg, West Virginia, and Dr. Samuel H. Scudder, of Cambridge, Massachusetts. Justly esteemed as the two foremost lepidopterists of America, it is my honor to claim them as personal friends, whose kindness has much aided me in this labor of scientific love which I have undertaken. For the kind permission given me by Dr. Scudder to use various illustrations contained in the "Butterflies of New England" and other works, I am profoundly grateful.

I am under obligations to Messrs. Charles Scribner's Sons for permission to use the cuts numbered 46–49, 51–56, 59, 61, 62, and 73, which are taken from the work entitled "Taxidermy and Zoölogical Collecting," by W. T. Hornaday, and to the authorities of the United States National Museum and the heirs of the late Professor C. V. Riley for other illustrations.

Should this book find the favor which I have reason to think it deserves, I shall endeavor shortly to follow it by the preparation of a similar work upon the moths of the United States and Canada.

OFFICE OF THE CHANCELLOR,   W. J. H.
WESTERN UNIVERSITY OF PENNSYLVANIA,
August 16, 1898.

TABLE OF CONTENTS

INTRODUCTION

CHAP.   PAGE

I. THE LIFE-HISTORY AND ANATOMY OF BUTTERFLIES . . 3–25

The Eggs of Butterflies. Caterpillars: Structure, Form, Color, etc.;
Moults; Food of Caterpillars; Duration of Larval State; Transformation.
The Pupa, or Chrysalis: The Form of Chrysalids; Duration of Pupal
Life; The Transformation from the Chrysalis to the Imago. *Anatomy
of Butterflies:* The Head; The Thorax; The Abdomen; The Legs;
The Wings; Internal Organs; Polymorphism and Dimorphism; Albi-
nism and Melanism; Monstrosities; Mimicry. *The Distribution of But-
terflies.*

II. THE CAPTURE, PREPARATION, AND PRESERVATION OF SPECI-
MENS 26–57

Collecting Apparatus : Nets; Collecting-Jars; Field-Boxes; The Use
of the Net; Baits; Beating. *The Breeding of Specimens :* How to Get
the Eggs of Butterflies; Breeding-Cages; How to Find Caterpillars;
Hibernating Caterpillars. *The Preservation of Specimens :* Papering
Specimens; Mounting Butterflies; Relaxing Specimens; The Preparation and Preservation of Butterfly Eggs; The Preservation of Chrysalids; The Preservation of Caterpillars. *The Preservation and
Arrangement of Collections :* Boxes; Cabinets and Drawers; Labeling; Arrangement of Specimens; Insect Pests; Greasy Specimens;
Mould; Repairing Specimens; Packing and Forwarding Specimens;
Pins; The Forceps.

III. THE CLASSIFICATION OF BUTTERFLIES 58–68

The Place of Butterflies in the Animal Kingdom; The Principles of
Scientific Arrangement; The Species; The Genus; The Family, etc.;
Scientific Names; Synonyms; Popular Names.

IV. BOOKS ABOUT NORTH AMERICAN BUTTERFLIES . . . 69–74

Early Writers; Later Writers; Periodicals.

ix

THE BOOK

PAGE

THE BUTTERFLIES OF NORTH AMERICA NORTH OF MEXICO.

Family I. *Nymphalidæ*, the Brush-footed Butterflies . . 77
    Subfamily *Euplœina*, the Milkweed Butterflies . . . 80
    Subfamily *Ithomiina*, the Long-winged Butterflies . . 85
    Subfamily *Heliconiina*, the Heliconians 91
    Subfamily *Nymphalina*, the Nymphs 93
    Subfamily *Satyrina*, the Satyrs, Meadow-browns, and
      Arctics 197
    Subfamily *Libytheina*, the Snout-butterflies . . . 226

Family II. *Lemoniidæ* 228
    Subfamily *Erycinina*, the Metal-marks 228

Family III. *Lycænidæ* 236
    Subfamily *Lycænina*, the Hair-streaks, the Blues, and
      the Coppers 236

Family IV. *Papilionidæ*, the Swallowtails and Allies . . 272
    Subfamily *Pierinæ*, the Whites, the Sulphurs, the
      Orange-tips 272
    Subfamily *Papilionina*, the Parnassians and Swallowtails 304

Family V. *Hesperiidæ*, the Skippers 318
    Subfamily *Pyrrhopyginæ* 319
    Subfamily *Hesperiina*, the Hesperids 320
    Subfamily *Pamphilinæ* 339
    Subfamily *Megathyminæ*, genus *Megathymus* 367

DIGRESSIONS AND QUOTATIONS

PAGE

Immortality (Sigourney) 57
Hugo's "Flower to Butterfly" (Translated by Eugene Field) 74
Superstitions (Frank Cowan) 90
Luther's Saddest Experience (Yale Literary Magazine, 1852) 100
A Race after a Butterfly 127

Suspicious Conduct 136
Collecting in Japan 149
Faunal Regions 161
Widely Distributed Butterflies 171
The Butterflies' Fad (Ella Wheeler Wilcox) 186
Fossil Insects 195
In the Face of the Cold 224
Uncle Jotham's Boarder (Annie Trumbull Slosson)   . . . 233
Mimicry 235
The Utility of Entomology 256
Size 271
Instinct 280
Red Rain (Frank Cowan) 299
For a Design of a Butterfly Resting on a Skull (Mrs.
    Hemans) 303
The Caterpillar and the Ant (Allan Ramsay)   316
Collections and Collectors 337
Exchanges 344

LIST OF ILLUSTRATIONS IN TEXT

FIG.   PAGE

1. Egg of Basilarchia disippus, magnified 3
2. Egg of Basilarchia disippus, natural size 3
3. Egg of Papilio turnus, enlarged 4
4. Egg of Anosia plexippus, magnified 4
5. Egg of Anosia plexippus, natural size 4
6. Egg of Anthocharis genutia, magnified 4
7. Egg of Lycæna pseudargiolus, magnified 4
8. Egg of Melitæa phaëton, magnified 4
9. Micropyle of egg of Pieris oleracea, magnified . . . 5
10. Eggs of Grapta comma, magnified 5
11. Eggs of Vanessa antiopa, magnified 5
12. Caterpillar of Papilio philenor 6
13. Head of caterpillar of Papilio asterias, magnified . . . 6
14. Head of caterpillar of Anosia plexippus, magnified . . 6
15. Head of caterpillar of Anosia plexippus, side view,
    enlarged 7
16. Caterpillar of Anosia plexippus, natural size 7
17. Fore leg of caterpillar of Vanessa antiopa, enlarged . . 7
18. Anterior segments of caterpillar of A. plexippus . . 7
19. Proleg of caterpillar of Vanessa antiopa, enlarged . . 7
20. Caterpillar of Basilarchia disippus 8
21. Early stages of goatweed butterfly 9
22. Head of caterpillar of Papilio troilus 9
23. Caterpillar of milkweed butterfly changing into
    chrysalis 11
24. Chrysalis of milkweed butterfly 12
25. Chrysalis of Papilio philenor 12
26. Caterpillar and chrysalis of Pieris protodice 12
27. Chrysalis of Pieris oleracea 13

FIG.		PAGE
28.	Butterfly emerging from chrysalis.	13
29.	Head of milkweed butterfly, showing parts	14
30.	Cross-section of sucking-tube of butterfly	15
31.	Longitudinal section of the head of the milkweed butterfly	15
32.	Interior structure of head of milkweed butterfly	16
33.	Labial palpus of butterfly	16
34.	Legs of butterfly	17
35.	Parts of leg of butterfly	17
36.	Scales on wing of butterfly	18
37.	Androconia from wing of butterfly	18
38.	Outline of wing of butterfly	20
39.	Arrangement of scales on the wing of a butterfly	20
40.	Figure of wing, showing names of veins	21
41.	Internal anatomy of caterpillar of milkweed butterfly	22
42.	Internal anatomy of milkweed butterfly	23
43.	Plan for folding net-ring	27
44.	Insect-net	27
45.	Plan for making a cheap net	27
46.	Cyanide-jar	29
47.	Paper cover for cyanide	29
48.	Method of pinching a butterfly	30
49.	Cheap form of breeding-cage	35
50.	Breeding-cage	36
51.	Butterfly in envelope	38
52.	Method of making envelopes	38
53.	Setting-board	39
54.	Setting-block	39
55.	Butterfly on setting-block	39
56.	Setting-needle	40
57.	Setting-board with moth upon it	40
58.	Butterfly pinned on setting-board	41
59.	Drying-box	41
60.	Drying-box	42
61.	Apparatus for inflating larvæ	45
62.	Tip of inflating-tube	46
63.	Drying-oven	46
64.	Drying-oven	47
65.	Detail drawing of book-box	48
66.	Detail drawing of box	48

FIG.   PAGE
67. Detail drawing of box 49
68. Insect-box 49
69. Detail drawing of drawer for cabinet 51
70. Detail drawing for paper bottom of box to take place
    of cork 52
71. Manner of arranging specimens in cabinet or box . 52
72. Naphthaline cone. 53
73. Butterflies packed for shipment 55
74. Forceps 56
75. Forceps 57
76. Antennæ of butterfly. 61
77. Antennæ of moths 62
78. Neuration of genus Anosia 81
79. Swarm of milkweed butterflies, photographed at night 83
80. Neuration of genus Mechanitis 86
81. Neuration of genus Ceratinia 88
82. Neuration of genus Dircenna 89
83. Fore leg of female Dircenna klugi 89
84. Neuration of genus Heliconius 91
85. Young caterpillar of Vanessa antiopa 94
86. Neuration of genus Colænis 95
87. Neuration of genus Dione 96
88. Neuration of genus Euptoieta 98
89. Neuration of genus Argynnis 101
90. Neuration of genus Brenthis 129
91. Neuration of genus Melitæa 138
92. Neuration of genus Phyciodes 151
93. Neuration of genus Eresia 157
94. Neuration of genus Synchloë 159
95. Neuration of genus Grapta 163
96. Neuration of genus Vanessa 167
97. Neuration of genus Pyrameis 170
98. Neuration of genus Junonia 172
99. Neuration of genus Anartia 174
100. Neuration of genus Hypanartia 175
101. Neuration of genus Eunica 176
102. Neuration of genus Cystineura 177
103. Neuration of genus Callicore 178
104. Neuration of genus Timetes 179
105. Neuration of genus Hypolimnas 181

FIG.   PAGE
106. Neuration of genus Basilarchia 182
107. Leaf cut away at end by the caterpillar of Basilarchia . 183
108. Hibernaculum of caterpillar of Basilarchia 183
109. Neuration of genus Adelpha 187
110. Neuration of genus Chlorippe 188
111. Neuration of genus Pyrrhanæa 192
112. Neuration of genus Ageronia 193
113. Neuration of genus Victorina 195
114. Neuration of genus Debis 199
115. Neuration of genus Satyrodes 200
116. Neuration of genus Neonympha 201
117. Neuration of genus Cœnonympha 205
118. Neuration of genus Erebia 208
119. Neuration of genus Geirocheilus 211
120. Neuration of genus Neominois 212
121. Neuration of genus Satyrus 214
122. Neuration of genus Œneis 219
123. Caterpillars of Œneis macouni 221
124. Neuration of genus Libythea 226
125. Neuration of base of hind wing of genus Lemonias . . 228
126. Neuration of genus Lemonias 229
127. Neuration of genus Calephelis 232
128. Neuration of genus Eumæus 237
129. Neuration of Thecla edwardsi 238
130. Neuration of Thecla melinus 242
131. Neuration of Thecla damon 246
132. Neuration of Thecla niphon 249
133. Neuration of Thecla titus 250
134. Neuration of genus Feniseca 251
135. Neuration of genus Chrysophanus 252
136. Neuration of Lycæna pseudargiolus 267
137. Neuration of Lycæna comyntas 268
138. Neuration of genus Dismorphia 273
139. Neuration of genus Neophasia 274
140. Neuration of genus Tachyris 276
141. Neuration of genus Pieris 277
142. Neuration of genus Nathalis 281
143. Neuration of genus Euchloë 282
144. Neuration of genus Catopsilia 236
145. Neuration of genus Kricogonia 287

FIG.   PAGE

146. Neuration of genus Meganostoma 288
147. Neuration of genus Colias 289
148. Neuration of genus Terias 295
149. Neuration of genus Parnassius 305
An Astronomer's Conception of an Entomologist . . 317
150. Head and antenna of genus Pyrrhopyge 319
151. Neuration of genus Pyrrhopyge 319
152. Neuration of genus Eudamus 321
153. Antenna and neuration of genus Plestia 322
154. Neuration of genus Epargyreus 323
155. Neuration of genus Thorybes 324
156. Neuration of genus Achalarus 326
157. Antenna and neuration of genus Hesperia 327
158. Neuration of genus Systasea 329
159. Neuration of genus Pholisora 330
160. Neuration of genus Thanaos 332
161. Neuration of genus Amblyscirtes 340
162. Neuration of genus Pamphila 342
163. Neuration of genus Oarisma 343
164. Neuration of genus Ancyloxypha 345
165. Neuration of genus Copæodes 346
166. Neuration of genus Erynnis 347
167. Neuration of genus Thymelicus 351
168. Neuration of genus Atalopedes 352
169. Neuration of genus Polites 353
170. Neuration of genus Hylephila 354
171. Neuration of genus Prenes 355
172. Neuration of genus Calpodes 355
173. Neuration of genus Lerodea 356
174. Neuration of genus Limochores 357
175. Neuration of genus Euphyes 360
176. Neuration of genus Oligoria 361
177. Neuration of genus Poanes 362
178. Neuration of genus Phycanassa 362
179. Neuration of genus Atrytone 364
180. Neuration of genus Lerema 366
181. Megathymus yuccæ, ♀ 367
182. Larva of Megathymus yuccæ 368
183. Chrysalis of Megathymus yuccæ 368
The Popular Conception of an Entomologist 369
xvii

LIST OF COLORED PLATES

Produced by the color-photographic process of the Chicago Colortype Company, 1205 Roscoe Street, Chicago, Ill.

FACING
PAGE

I. Spring Butterflies . . . . *Frontispiece*
II. Caterpillars of Papilionidæ and Hesperiidæ . . . 6
III. Caterpillars of Nymphalidæ 18
IV. Chrysalids in Color and in Outline — Nymphalidæ 30
V. Chrysalids in Color and in Outline — Nymphalidæ, Lycænidæ, Pierinæ 44
VI. Chrysalids in Color and in Outline — Papiloninæ and Hesperiidæ 58
VII. Anosia and Basilarchia 80
VIII. Ithomiinæ, Heliconius, Dione, Colænis, and Euptoieta 88
IX. Argynnis 100
X. Argynnis 104
XI. Argynnis 108
XII. Argynnis 112
XIII. Argynnis 116
XIV. Argynnis 122
XV. Brenthis 130
XVI. Melitæa 138
XVII. Melitæa, Phyciodes, Eresia 152
XVIII. Argynnis, Brenthis, Melitæa, Phyciodes, Eresia, Synchloë, Debis, Geirocheilus 156
XIX. Grapta, Vanessa 164
XX. Grapta, Vanessa, Junonia, Anartia, Pyrameis . . 168
XXI. Timetes, Hypolimnas, Eunica, Callicore 178
XXII. Basilarchia, Adelpha 184
XXIII. Chlorippe 190
XXIV. Pyrrhanæa, Ageronia, Synchloë, Cystineura, Hypanartia, Victorina 196

XXV. Satyrodes, Cœnonympha, Neonympha, Neominois, Erebia 204
XXVI. Satyrus 214
XXVII. Œneis 220
XXVIII. Libythea, Lemonias, Calephelis, Eumæus, Chrysophanus, Feniseca 228
XXIX. Chrysophanus, Thecla 236
XXX. Thecla, Lycæna 246
XXXI. Lycæna 256
XXXII. Lycæna, Thecla, Nathalis, Euchloë 266
XXXIII. Catopsilia, Pyrameis 272
XXXIV. Euchloë, Neophasia, Pieris, Kricogonia 280
XXXV. Tachyris, Pieris, Colias 288
XXXVI. Meganostoma, Colias 294
XXXVII. Terias, Dismorphia 298
XXXVIII. Papilio. 302
XXXIX. Parnassius 306
XL. Papilio. 310
XLI. Papilio. 314
XLII. Papilio. 316
XLIII. Papilio, Colias, Pyrameis, Epargyreus 318
XLIV. Papilio 322
XLV. Papilio, Pholisora, Eudamus Achalarus, Pyrrhopyge, Plestia, Calpodes, Thanaos 330
XLVI. Hesperiidæ 338
XLVII. Hesperiidæ 350
XLVIII. Hesperiidæ and Colias eurytheme 360

INTRODUCTION

INTRODUCTION

CHAPTER I

THE LIFE-HISTORY AND ANATOMY OF BUTTERFLIES

*" The study of butterflies,—creatures selected as the types of airiness and frivolity,—instead of being despised, will some day be valued as one of the most important branches of biological science."—*BATES, *Naturalist on the Amazons.*

In studying any subject, it is always well, if possible, to commence at the beginning; and in studying the life of animals, or of a group of animals, we should endeavor to obtain a clear idea at the outset of the manner in which they are developed. It is a familiar saying that "all life is from an egg." This statement is scientifically true in wide fields which come under the eye of the naturalist, and butterflies are no exception to the rule.

THE EGGS OF BUTTERFLIES

The eggs of butterflies consist of a membranous shell containing a fluid mass composed of the germ of the future caterpillar and the liquid food which is necessary for its maintenance and development until it escapes from the shell. The forms of these eggs are various. Some are spherical, others hemispherical, conical, and cylindrical. Some are barrel-shaped; others have the shape of a cheese, and still others have the form of a turban. Many of them are angled, some depressed at the ends. Their surface is variously ornamented. Some-

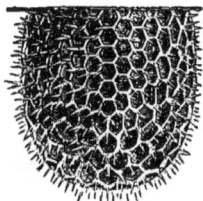

Fig. 1.—Egg of *Basilarchia disippus*, magnified 30 diameters (Riley).

Fig. 2.—Egg of *Basilarchia disippus*, natural size, at the end of under surface of leaf (Riley).

3

times they are ribbed, the ribs running from the center outwardly and downwardly along the sides like the meridian lines upon a globe. Between these ribs there is frequently found a fine network of raised lines variously arranged. Sometimes the surface is covered with minute depressions, sometimes with a series of minute elevations variously disposed. As there is great variety in the form of the eggs, so also there is great variety in their color. Brown, blue, green, red, and yellow eggs occur. Greenish or greenish-white are common tints. The eggs are often ornamented with dots and lines of darker color. Species which are related to one another show their affinity even in the form of their eggs. At the upper end of the eggs of insects there are one or more curious structures, known as micropyles (little doors),

FIG. 3.—Egg of *Papilio turnus*, greatly magnified.

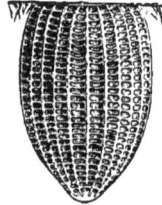

FIG. 4.—Egg of *Anosia plexippus*, magnified 30 diameters (Riley).

FIG. 5.—Egg of *Anosia plexippus*, natural size, on under side of leaf (Riley).

FIG. 6.—Egg of *Anthocharis genutia*, magnified 20 diameters.

FIG. 7.—Turban-shaped egg of *Lycæna pseudargiolus*, greatly magnified.

FIG. 8.—Egg of *Melitæa phaëton*, greatly magnified.

through which the spermatozoa of the male find ingress and they are fertilized. These can only be seen under a good microscope.

The eggs are laid upon the food-plant upon which the caterpillar, after it is hatched, is destined to live, and the female reveals wonderful instinct in selecting plants which are appropriate to the development of the larva. As a rule, the larvæ are restricted in the range of their food-plants to certain genera, or families of plants.

Fig. 9.—Upper end of egg of *Pieris oleracea*, greatly magnified, showing the micropyle.

Fig. 10.—Eggs of *Grapta comma*, laid in string-like clusters on the under side of leaf. (Magnified.)

The eggs are deposited sometimes singly, sometimes in small clusters, sometimes in a mass. Fertile eggs, a few days after they have been deposited, frequently undergo a change of color, and it is often possible with a magnifying-glass to see through the thin shell the form of the minute caterpillar which is being developed within the egg. Unfruitful eggs generally shrivel and dry up after the lapse of a short time.

The period of time requisite for the development of the embryo in the egg varies. Many butterflies are single-brooded; others produce two or three generations during the summer in temperate climates, and even more generations in subtropical or tropical climates. In such cases an interval of only a few days, or weeks at the most, separates the time when the egg was deposited and the time when the larva is hatched. When the period of hatching, or emergence, has arrived, the little caterpillar cuts its way forth from the egg through an opening made either

Fig. 11.—Eggs of *Vanessa antiopa*, laid in a mass on a twig.

at the side or on the top. Many species have eggs which appear to be provided with a lid, a portion of the shell being separated from the remainder by a thin section, which, when the caterpillar has reached the full limit allowed by the egg, breaks under the pressure of the enlarging embryo within, one portion of the egg flying off, the remainder adhering to the leaf or twig upon which it has been deposited.

CATERPILLARS

Structure, Form, Color, etc.—The second stage in which the insects we are studying exist is known as the larval stage. The insect is known as a larva, or a caterpillar. In general cater-

pillars have long, worm-like bodies. Frequently they are
thickest about the middle, tapering before and behind, flattened on the under side. While the
cylindrical shape is most common, there
are some families in which the larvæ
are short, oval, or slug-shaped, sometimes
curiously modified by ridges and prominences. The body of the larvæ of lepidoptera consists normally of thirteen rings,
or segments, the first constituting the
head.

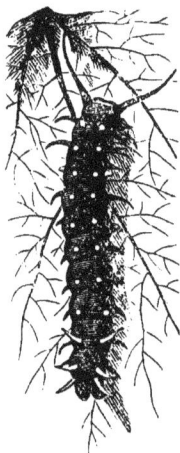

The head is always conspicuous, composed of horny or chitinous material,
but varying exceedingly in form and
size. It is very rarely small and retracted.
It is generally large, hemispherical,
conical, or bilobed. In some families it
is ornamented by horn-like projections.
On the lower side are the mouth-parts,
consisting of the upper lip, the mandibles,
the antennæ, or feelers, the under lip, the
maxillæ, and two sets of palpi, known as
the maxillary and the labial palpi. In
many genera the labium, or under lip, is provided with a
short, horny projection known as the spinneret. through
which the silk secreted by the cater-
pillar is passed. On either side,
just above the man-
dibles, are located the
eyes, or ocelli, which
in the caterpillar are
simple, round, shining
prominences, generally
only to be clearly dis-
tinguished by the aid
of a magnifying-glass.
These ocelli are fre-
quently arranged in series on each side. The palpi are organs
of touch connected with the maxillæ and the labium, or under
lip, and are used in the process of feeding, and also when the

FIG. 12.—Caterpillar of
Papilio philenor (Riley).

FIG. 13.—Head
of caterpillar of
*Papilio aste-
rias,* front view,
enlarged.

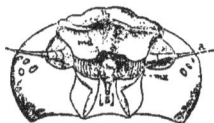

FIG. 14.—Head of caterpillar
of *Anosia plexippus,* lower side,
magnified 10 diameters: *lb,* labrum, or upper lip; *md,* mandibles; *mx,* maxilla, with two
palpi; *lm,* labium, or lower lip,
with one pair of palpi; *s,* spinneret; *a,* antenna; *o,* ocelli.
(After Burgess.)

6

Explanation of Plate II

Reproduced, with the kind permission of Dr. S. H. Scudder, from "The Butterflies of New England," vol. iii, Plate 70.

Caterpillars of Papilionidæ and Hesperiidæ

1. *Colias eurytheme.*
2. *Callidryas eubule.*
3. *Terias lisa.*
4. *Callidryas eubule.*
5. *Euchloë genutia.*
6. *Terias nicippe.*
7. *Pieris protodice.*
8. *Pieris napi,* var. *oleracea.*
9. *Pieris napi,* var. *oleracea.*
10. *Colias philodice.*
11. *Pieris rapæ.*
12. *Pieris rapæ.*
13. *Papilio philenor.*
14. *Papilio ajax.*
15. *Papilio turnus.* Just before pupation.
16. *Papilio cresphontes.*
17. *Papilio asterias.* In second stage.
18. *Papilio troilus.*
19. *Papilio troilus.* In third stage; plain.
20. *Papilio philenor.*
21. *Papilio philenor.* In third stage; dorsal view.
22. *Papilio troilus.* In third stage, dorsal view.
23. *Achalarus lycidas.* Dorsal view.
24. *Papilio asterias.* In fourth stage; dorsal view.
25. *Thorybes pylades.*
26. *Papilio turnus.* Dorsal view.
27. *Papilio asterias.*
28. *Papilio turnus.*
29. *Thorybes pylades.*
30. *Epargyreus tityrus.*
31. *Epargyreus tityrus.*
32. *Thorybes bathyllus.*
33. *Epargyreus tityrus.*
34. *Eudamus proteus.*
35. *Epargyreus tityrus.* In third stage.

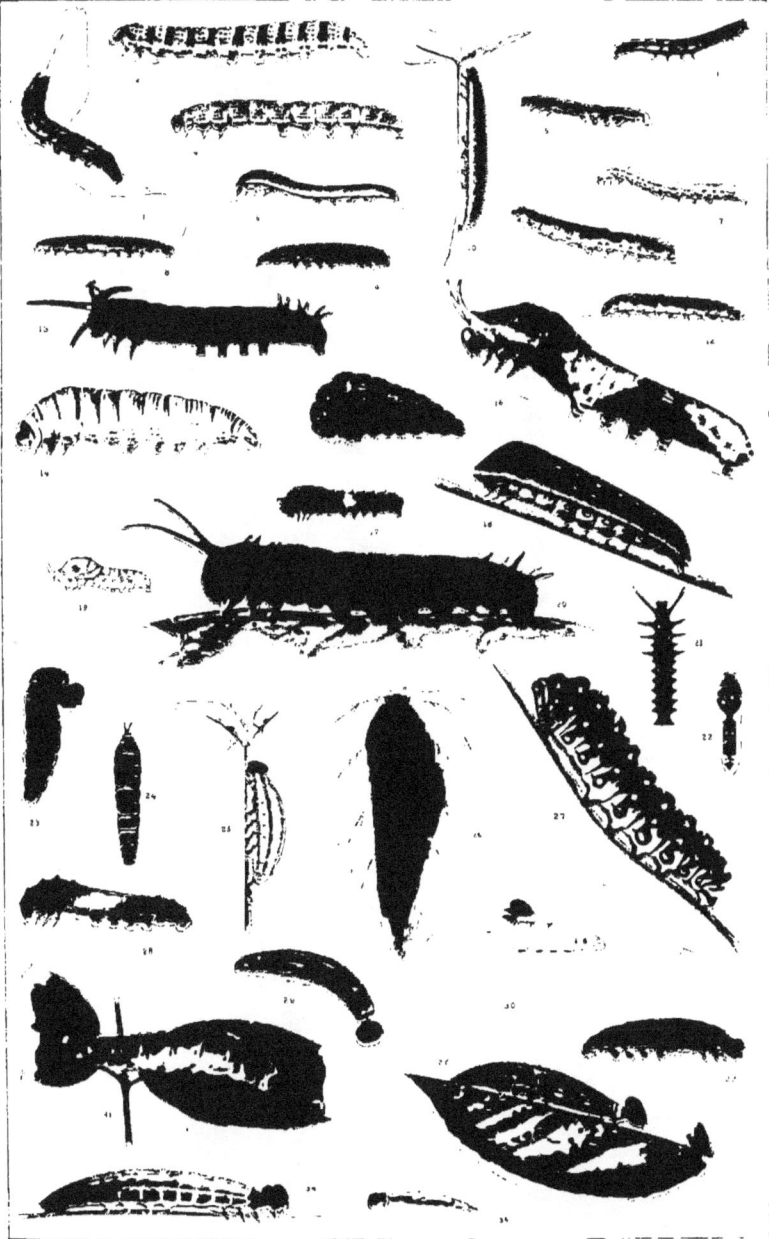

caterpillar is crawling about from place to place. The larva appears to guide itself in great part by means of the palpi.

The body of the caterpillar is covered by a thin skin, which often lies in wrinkled folds, admitting of great freedom of motion. The body is composed, as we have seen, of rings, or segments, the first three of which, back of the head, correspond

FIG. 15.—Head of caterpillar of *Anosia plexippus*, side view, showing ocelli.

FIG. 16.—Caterpillar of *Anosia plexippus*, milkweed butterfly (Riley).

to the thorax of the perfect insect, and the last nine to the abdomen of the butterfly. On each ring, with the exception of the second, the third, and the last, there is found on either side a small oval opening known as a spiracle, through which the creature breathes. As a rule, the spiracles of the first and eleventh rings are larger in size than the others.

Every caterpillar has on each of the first three segments a pair of legs, which are organs composed of three somewhat horny parts covered and bound together with skin, and armed at their extremities by a sharp claw (Fig. 17). These three pairs of feet in the caterpillar are always known as the fore legs, and corre-

FIG. 17.—Fore leg of caterpillar of *Vanessa antiopa*, enlarged.

FIG. 18.—Anterior segments of caterpillar of milkweed butterfly, showing thoracic or true legs (Riley).

FIG. 19.—Proleg of caterpillar of *Vanessa antiopa*, enlarged.

spond to the six which are found in the butterfly or the moth. In addition, in most cases, we find four pairs of prolegs on the under side of the segments from the sixth to the ninth, and another pair on the last segment, which latter pair are

7

called the anal prolegs. These organs, which are necessary to the life of the caterpillar, do not reappear in the perfect insect, but are lost when the transformation from the caterpillar to the chrysalis takes place. There are various modifications of this scheme of foot-like appendages, only the larger and more highly developed forms of lepidoptera having as many pairs of prolegs as have been enumerated.

The bodies of caterpillars are variously ornamented: many of them are quite smooth; many are provided with horny projections, spines, and eminences. The coloration of caterpillars is as remarkable in the variety which it displays as is the ornamentation by means of the prominences of which we have just spoken. As caterpillars, for the most part, feed upon growing vegetation, multitudes of them are green in color, being thus adapted to their surroundings and securing a measure of protection. Many are brown, and exactly mimic the color of the twigs and branches upon which they rest when not engaged in feeding. Not a few are very gaily colored, but in almost every case this gay coloring is found to bear some relation to the color of the objects upon which they rest.

Fig. 20.—Caterpillar of *Basilarchia disippus*, the viceroy, natural size (Riley).

Caterpillars vary in their social habits. Some species are gregarious, and are found in colonies. These frequently build for themselves defenses, weaving webs of silk among the branches, in which they are in part protected from their enemies and also from the inclemencies of the weather. Most caterpillars are, however, solitary, and no community life is maintained by the vast majority of species. Many species have the habit of drawing together the edges of a leaf, in which way they form a covering for themselves. The caterpillars of some butterflies are wood-boring, and construct tunnels in the pith, or in the soft layers of growing plants. In these cases, being protected and concealed from view, the caterpillars are generally white in their coloration, resembling in this respect the larvæ of wood-boring beetles. A most curious phenomenon has

8

within comparatively recent years been discovered in connection with the larval stage of certain small butterflies belonging to the family *Lycænidæ*. The caterpillars are carnivorous, or rather aphidivorous; they live upon aphids, or plant-lice, and scale-insects, and cover themselves with the white exudations or mealy secretions of the latter. This trait is characteristic of only one of our North American species, the Harvester (*Feniseca tarquinius*).

In addition to being protected from enemies by having colors which enable them to elude observation, as has been already stated, some caterpillars are provided with other means of defense. The caterpillars of the swallowtail butter-

Fig. 21.—Early stages of the goatweed butterfly: *a*, caterpillar; *b*, chrysalis; *c*, leaf drawn together at edges to form a nest. (Natural size.) (Riley.)

flies are provided with a bifurcate or forked organ, generally yellow in color, which is protruded from an opening in the skin

Fig. 22.—Head of caterpillar of *Papilio troilus*, with scent-organs, or *osmateria*, protruded.

back of the head, and which emits a powerful odor (Fig. 22). This protrusive organ evidently exists only for purposes of defense, and the secretion of the odor is analogous to the secretion of evil odors by some of the vertebrate animals, as the skunk. The majority of caterpillars, when attacked by insect or other enemies, defend themselves by quickly hurling the anterior part of the body from side to side.

Moults.—Caterpillars in the process of growth and development from time to time shed their skins. This process is called *moulting.* Moulting takes place, as a rule, at regular intervals, though there are exceptions to this rule. The young larva, having emerged from the egg, grows for a number of days, until the epidermis, or true skin, has become too small. It then ceases feeding, attaches itself firmly to some point, and remains quiet for a time. During this period certain changes are taking place, and then the skin splits along the middle line from the head to the extremity of the last segment, and the caterpillar crawls forth from the skin, which is left behind it, attached to the leaf or branch to which it was fastened. The skin of the head sometimes remains attached to the head of the caterpillar for a time after it has moulted, and then falls off to the ground. Ordinarily not more than five, and frequently only four, moults take place between hatching from the egg and the change into the chrysalis. In cases where caterpillars hibernate, or pass the winter in inaction, a long interval necessarily elapses between moults. Some arctic species are known in which the development from the egg to the perfect insect covers a period of two or three years, long periods of hibernation under the arctic snows taking place. The manner in which the caterpillar withdraws itself from its exuviæ, or old skin, is highly interesting. Every little spine or rough prominence is withdrawn from its covering, and the skin is left as a perfect cast of the creature which has emerged from it, even the hairs and spines attached to the skin being left behind and replaced by others.

The Food of the Caterpillar.—The vast majority of the caterpillars of butterflies subsist upon vegetable food, the only exceptions being the singular one already noted in which the larvæ feed upon scale-insects. Some of the *Hesperiidæ*, a group in which the relationship between butterflies and moths is shown, have larvæ which burrow in the roots and stems of vegetation.

Duration of the Larval State.—The duration of the larval state varies greatly. In temperate climates the majority of species exist in the caterpillar state for from two to three months, and where hibernation takes place, for ten months. Many caterpillars which hibernate do so immediately after emerging from the egg and before having made the first moult. The great majority, however, hibernate after having passed one or more moults. With the approach of spring they renew their feeding upon the first reappearance of the foliage of their proper food-plant, or are transformed into chrysalids and presently emerge as perfect insects.

A few species live gregariously during the period of hibernation, constructing for themselves a shelter of leaves woven together with strands of silk.

Transformation.—The larval or caterpillar stage having been completed, and full development having been attained, the caterpillar is transformed into a pupa, or chrysalis. Of this, the third stage in the life of the insect, we now shall speak at length.

THE PUPA, OR CHRYSALIS

The caterpillars of many butterflies attach themselves by a button of silk to the under surface of a branch or stone, or other projecting surface, and are transformed into chrysalids,

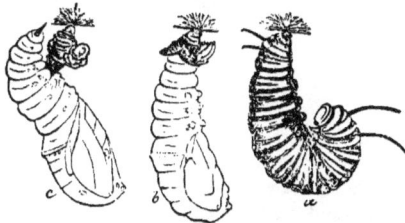

Fig. 23.—Caterpillar of *Anosia plexippus*, undergoing change into chrysalis: *a*, caterpillar just before rending of the skin; *b*, chrysalis just before the cremaster, or hook, at its end is withdrawn; *c*, chrysalis holding itself in place by the folds of the shed skin caught between the edges of the abdominal segments, while with the cremaster, armed with microscopic hooks, it searches for the button of silk from which it is to hang (Riley). (Compare Fig. 24, showing final form of the chrysalis.)

which are naked, and which hang perpendicularly from the surface to which they are attached. Other caterpillars attach themselves to surfaces by means of a button of silk which holds the anal extremity of the chrysalis, and have, in addition, a girdle of silk which passes around the middle of the chrysalis, holding it in place very much as a papoose is held on the back of an Indian squaw by a strap passed over her shoulders.

The Form of Chrysalids.—The forms assumed by the insect in this stage of its being vary very greatly, though there is a general resemblance among the different families and subfamilies, so that

11

it is easy for one who has studied the matter to tell approximately to what family the form belongs, even when it is not specifically known. Chrysalids are in most cases obscure in coloring, though a few are quite brilliant, and, as in the case of the common milkweed butterfly (*Anosia plexippus*), ornamented with golden-hued spots. The chrysalids of the *Nymphalidæ*, one of the largest

Fig. 24.—Chrysalis of *Anosia plexippus*, final form (Riley).

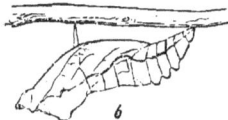

Fig. 25.—Chrysalis of *Papilio philenor: a*, front view; *b*, side view, showing manner in which it is held in place by the girdle of silk (Riley).

groups of butterflies, are all suspended. The chrysalids of the *Papilionidæ*, or swallowtail butterflies, are held in place by girdles, and generally are bifurcate or cleft at the upper end (Fig. 25), and are greenish or wood-brown in color.

A study of the structure of all chrysalids shows that within them there is contained the immature butterfly. The segments of the body are ensheathed in the corresponding segments of the chrysalis, and soldered over these segments are ensheathing plates of chitinous matter under which are the wings of the butterfly, as well as all the other organs necessary to its existence in the airy realm upon which it enters after emergence from the chrysalis. The practised eye of the observer is soon able to distinguish the location of the various parts of the butterfly in the chrysalis, and when the time for escape

Fig. 26.—*Pieris protodice: a*, caterpillar; *b*, chrysalis (Riley).

draws near, it is in many cases possible to discern through the thin, yet tough and hard, outer walls of the chrysalis the spots and colors on the wings of the insect.

Duration of Pupal Life.—Many butterflies remain in the chrysalis stage only for a few weeks; others hibernate in this state, and in temperate climates a great many butterflies pass the winter as chrysalids. Where, as is sometimes the case, there are two or three generations or broods of a species during the year, the life of one brood is generally longer than that of the others, because this brood is compelled to overwinter, or hibernate. There are a number of butterflies known in temperate North America which have three broods: a spring brood, emerging from chrysalids which have overwintered; an early summer brood; and a fall brood. The chrysalids in the latter two cases generally represent only a couple of weeks at most in the life of the insect. In tropical and semi-tropical countries many species remain in the chrysalis form during the dry season, and emerge at the beginning of the rains, when vegetation is refreshed and new and tender growths occur in the forests.

Fig. 27.—Chrysalis of *Pieris oleracea* (Riley).

The Transformation from the Chrysalis to the Imago.—The perfectly developed insect is known technically as the *imago.* When the time of maturity in the chrysalis state has been reached, the coverings part in such a way as to allow of the escape of the perfect insect, which, as it comes forth, generally carries with it some suggestion of its caterpillar state in the lengthened abdomen, which it with apparent difficulty trails after it until it secures a hold upon some object from which it may depend while a process of development (which lasts generally a few hours) takes place preparatory to flight. The imago, as it first emerges, is provided with small, flaccid wings, which, together with all the organs of sense, such as the antennæ, require for their complete development the injection into them of the vital fluids which, upon first emergence, are largely contained in the cavities of the thorax and abdomen. Hanging pendant on a projecting twig, or clinging to the side of a rock, the insect remains

Fig. 28.—Butterfly (*Papilio asterias*) just emerging from chrysalis.

13

fanning its wings, while by the strong process of circulation a rapid injection of the blood into the wings and other organs takes place, accompanied by their expansion to normal proportions, in which they gradually attain to more or less rigidity. Hardly anything in the range of insect life is more interesting than this rapid development of the butterfly after its first emergence from the chrysalis. The body is robbed of its liquid contents in a large degree; the abdomen is shortened up; the chitinous rings which compose its external skeleton become set and hardened; the wings are expanded, and then the moment arrives when, on airy pinions, the creature that has lived a worm-like life for weeks and months, or which has been apparently sleeping the sleep of death in its cerements, soars aloft in the air, the companion of the sunlight and the breezes.

ANATOMY OF BUTTERFLIES

The body of the butterfly consists of three parts—the head, the thorax, and the abdomen.

The Head.—The head is globular, its breadth generally exceeding its length. The top is called the *vertex;* the anterior portion,

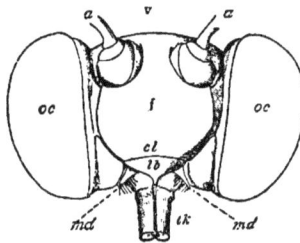

Fig. 29.—Head of milkweed butterfly, stripped of scales and greatly magnified (after Burgess): *v*, vertex; *f*, front; *cl*, clypeus; *lb*, labrum, or upper lip; *md*, mandibles; *a*, antennæ; *oc*, eyes; *tk*, spiral tongue, or proboscis.

corresponding in location to the human face, is called the *front.* Upon the sides of the head are situated the large *compound eyes*, between which are the *antennæ*, or "feelers," as they are sometimes called. Above the mouth is a smooth horny plate, the *clypeus.* The *labrum*, or upper lip, is quite small. On both sides of the mouth are rudimentary *mandibles*, which are microscopic objects. The true suctorial apparatus is formed by the *maxillæ*, which are produced in the form of semi-cylindrical tubes, which, being brought together and interlocking, form a com-

14

plete tube, which is known as the *proboscis*, and which, when not in use, is curled up spirally, looking like a watch-spring. At

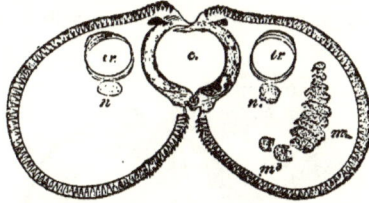

FIG. 30.—Cross-section of the sucking-tube of the milkweed butterfly, to show the way in which the halves unite to form a central canal (*c*): *tr*, tracheæ, or air-tubes; *n*, nerves; *m*, *m*3, muscles of one side. (Magnified 125 diameters.) (Burgess.)

the upper end of the proboscis, in the head, is a bulb-like en-largement, in the walls of which are inserted muscles which have

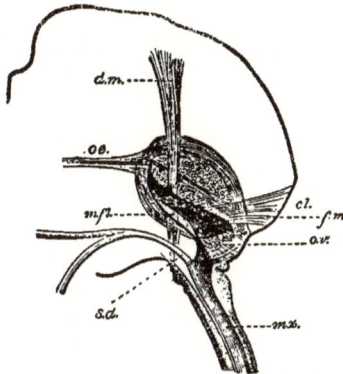

FIG. 31.—Longitudinal section of the head of the milkweed butterfly: *cl*, clypeus; *mx*, left maxilla, the right being removed; *mfl*, floor of mouth; *œ*, œsophagus, or gullet; *ov*, mouth-valve; *sd*, salivary duct; *dm* and *fm*, dorsal and frontal muscles, which open the sac. (Magnified 20 diameters). (Burgess.)

their origin on the inner wall of the head. When these muscles contract, the bulb-like cavity is enlarged, a vacuum is produced,

15

and the fluids in the cup of the flower flow up the proboscis and
into the bulb. The bulb is also surrounded by muscles, which,
when contracting, compress it. The external opening of the
tube has a flap, or valve, which, when the bulb is compressed,

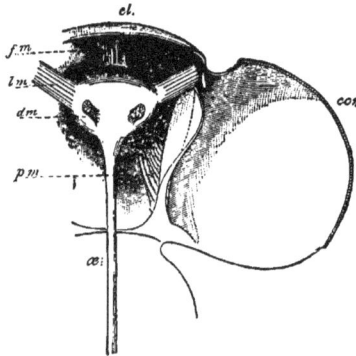

FIG. 32.—Interior view of head of milkweed butterfly: *cl*, clypeus; *cor*, cornea of the eye; *œ*, œsophagus, or
gullet; *fm*, frontal muscle; *dm*, dorsal muscles; *lm*, lateral muscles; *pm*, muscles moving the palpus (Burgess).

closes and causes the fluid in it to flow backward into the gullet
and the stomach. The arrangement is mechanically not unlike
that in a bulb-syringe used by physicians. The process of feeding
in the case of the butterfly is a process of pump-
ing honeyed water out of the flowers into the stomach.
The length of the proboscis varies; at its base and on
either side are placed what are known as the maxillary
palpi, which are very small. The lower lip, or *labium*, which is also almost obsolete in the butterflies,
has on either side two organs known as the *labial
palpi*, which consist of three joints. In the butterflies the labial palpi are generally well developed,
though in some genera they are quite small. The
antennæ of butterflies are always provided at the extremity with a club-shaped enlargement, and because
of this clubbed form of the antennæ the entire group are known
as the *Rhopalocera*, the word being compounded from the Greek

FIG. 33.—
Labial palpus
of *Colias*,
magnified 10
diameters.

16

word ῥώπαλον (*rhopalon*), which means a *club*, and the word κέρας (*keras*) which means a *horn*.

It will be observed from what has been said that the head in these creatures is to a large extent the seat of the organs of sense and alimentation. What the function of the antennæ may be is somewhat doubtful, the opinion of scientific men being divided. The latest researches would indicate that these organs, which have been regarded as the organs of smell and sometimes as the organs of hearing, have probably a compound function, possibly enabling the creature to hear, certainly to smell, but also, perhaps, being the seat of impressions which are not strictly like any which we receive through our senses.

Thorax.—The thorax is more or less oval in form, being somewhat flattened upon its upper surface. It is composed of three parts, or segments, closely united, which can only be distinguished from one another by a careful dissection. The anterior segment is known as the prothorax, the middle segment as the mesothorax, and the after segment as the metathorax. The legs are attached in pairs to these three subdivisions of the thorax, the anterior pair being therefore sometimes spoken of as the prothoracic legs, the second pair as the mesothoracic legs, and the latter pair as the metathoracic legs (Fig. 34). On either side of the mesothorax are attached the anterior pair of wings, over which, at

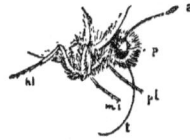

Fig. 34.—*Colias philodice:* a, antenna; p, extremity of palpus; *pl*, prothoracic leg; *ml*, mesothoracic leg; *hl*, metathoracic or hind leg; *t*, proboscis.

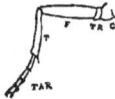

their insertion into the body, are the *tegulæ*, or lappets; on either side of the metathorax are the posterior pair of wings. It will be seen from what has been said that the thorax bears the organs of locomotion. The under side of the thorax is frequently spoken of by writers, in describing butterflies, as the *pectus*, or breast.

Fig. 35.—Leg of butterfly: *c*, coxa; *tr*, trochanter; *f*, femur; *t*, tibia; *tar*, tarsi.

The Abdomen.—The abdomen is formed normally of nine segments, and in most butterflies is shorter than the hind wings. On the last segment there are various appendages, which are mainly sexual in their nature.

The Legs.—Butterflies have six legs, arranged in three pairs, as we have already seen. Each leg consists of five parts, the

17

first of which, nearest the body, is called the *coxa*, with which articulates a ring-like piece known as the *trochanter*. To this is attached the *femur*, and united with the femur, forming an angle with it, is the *tibia*. To the tibia is attached the *tarsus*, or foot, the last segment of which bears the claws, which are often very minute and blunt in the butterflies, though in moths they are sometimes strongly hooked. The tibiæ are often armed with spines. In some groups of butterflies the anterior pair of legs is aborted, or dwarfed, either in one or both sexes, a fact which is useful in determining the location of species in their systematic order.

The Wings. — The wings of butterflies consist of a framework of horny tubes which are in reality double, the inner tube being

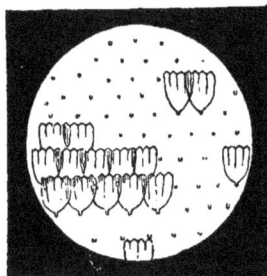

Fig. 36. — Magnified representation of arrangement of the scales on the wing of a butterfly.

Fig. 37. — Androconia from wings of male butterflies: *a*, *Neonympha eurytus*; *b*, *Argynnis aphrodite*; *c*, *Pieris oleracea*.

filled with air, the outer tube with blood, which circulates most freely during the time that the insect is undergoing the process of development after emergence from the chrysalis, as has been already described. After emergence the circulation of the blood in the outer portion of the tubes is largely, if not altogether, suspended. These horny tubes support a broad membrane, which is clothed in most species upon both sides with flattened scales which are attached to the membrane in such a way that they overlap one another like the shingles on a roof. These scales are very beautiful objects when examined under a microscope, and there is considerable diversity in their form as well as in their colors. The

18

Explanation of Plate III

Reproduced, with the kind permission of Dr. S. H. Scudder, from " The Butterflies of New England," vol. iii, Plate 74.

Caterpillars of Nymphalidæ

1. *Œneis semidea*. Penultimate stage.
2. *Œneis semidea*.
3. *Neonympha curytus*.
4. *Œneis semidea*.
5. *Anosia plexippus*.
6. *Neonympha eurytus*.
7. *Œneis semidea*. Just hatched.
8. *Neonympha phocion*.
9. *Satyrodes canthus*.
10. *Neonympha curytus*.
11. *Œneis jutta*. Just hatched.
12. *Neonympha phocion*.
13. *Neonympha eurytus*. Penultimate stage.
14. *Neonympha eurytus*. Plain and enlarged.
15. *Œneis semidea*.
16. *Debis portlandia*.
17. *Basilarchia astyanax*.
18. *Satyrus alope*.
19. *Basilarchia disippus*.
20. *Chlorippe clyton*.
21. *Basilarchia astyanax*.
22. *Basilarchia disippus*. Plain outline to show the attitude sometimes assumed.
23. *Grapta interrogationis*.
24. *Basilarchia disippus*.
25. *Basilarchia astyanax*. Plain.
26. *Basilarchia arthemis*.
27. *Grapta interrogationis*.
28. *Vanessa antiopa*.
29. *Junonia cœnia*.
30. *Junonia cœnia*.
31. *Grapta progne*.
32. *Grapta faunus*.
33. *Grapta satyrus*.
34. *Pyrameis huntera*.
35. *Pyrameis atalanta*.
36. *Vanessa milberti*.
37. *Pyrameis cardui*.
38. *Grapta comma*.

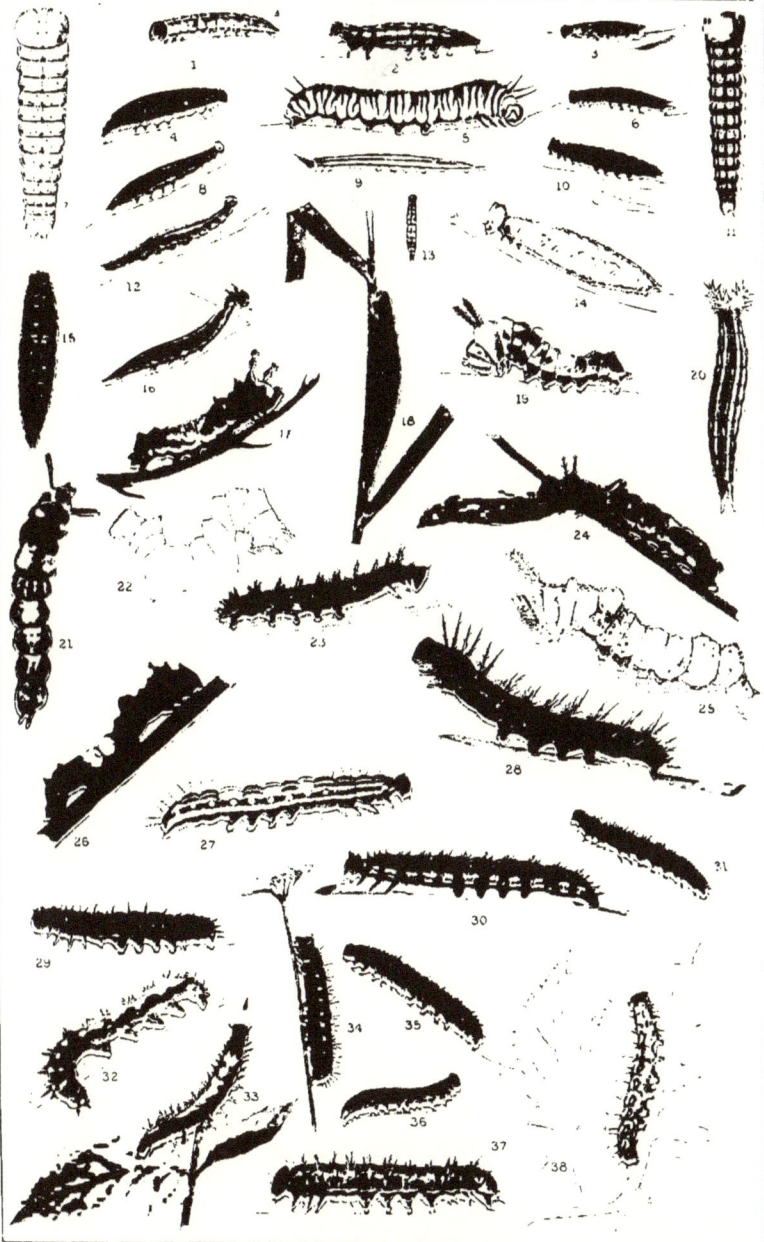

males of many species have peculiarly shaped scales arranged in tufts and folds, which are called androconia, and are useful in microscopically determining species (Fig. 37). The portion of the wings which is nearest to the thorax at the point where they are attached to the body is called the *base;* the middle third of the wing is known as the *median* or *discal area*, the outer third as the *limbal area*. The anterior margin of the wings is called the *costal margin;* the outer edge is known as the *external margin*, the inner edge as the *inner margin*. The shape of the wings varies very much. The tip of the front wing is called the apex, and this may be rounded, acute, falcate (somewhat sickle-shaped), or square. The angle formed by the outer margin of the front wing with the inner margin is commonly known as the *outer angle*. The corresponding angle on the hind wing is known as the *anal angle*, and the point which corresponds to the tip or apex of the front wing is known as the *external angle* (Fig. 38). A knowledge of these terms is necessary in order to understand the technical descriptions which are given by authors.

If a wing is examined with the naked eye, or even with a lens, a clear conception of the structure of the veins can rarely be formed. Therefore it is generally necessary to remove from the wings the scales which cover them, or else bleach them. The scales may be removed mechanically by rubbing them off. They may be made transparent by the use of chemical agents. In the case of specimens which are so valuable as to forbid a resort to these methods, a clear knowledge of the structure of the veins may be formed by simply moistening them with pure benzine or chloroform, which enables the structure of the veins to be seen for a few moments. The evaporation of these fluids is rapid, and they produce no ill effect upon the color and texture of the wings. In the case of common species, or in the case of such as are abundantly represented in the possession of the collector, and the practical destruction of one or two of which is a matter of no moment, it is easy to use the first method. The wing should be placed between two sheets of fine writing-paper which have been moistened by the breath at the points where the wing is laid, and then by lightly rubbing the finger-nail or a piece of ivory, bone, or other hard substance over the upper piece of paper, a good many of the scales may be removed. This process may be repeated until almost all of them have been taken off. This method is

19

efficient in the case of many of the small species when they are
still fresh; in the case of the larger species the scales may be removed by means of a camel's-hair pencil such as is used by paint-
ers. The chemical method of bleaching wings is simple and inexpensive. For this purpose the wing should be dipped in alcohol
and then placed in a vessel containing a bleaching solution of some
sort. The best agent is a solution of chloride of lime. After the
color has been removed from the wing by the action of the
chloride it should be washed in a weak solution of hydrochloric
acid. It may then be cleansed in pure water and mounted upon
a piece of glass, as microscopic slides are mounted, and thus preserved. When thus bleached the wing is capable of being minutely studied, and all points of its anatomy are brought clearly
into view.

The veins in both the fore and hind wings of butterflies
may be divided into simple and
compound veins. In the fore
wing the simple veins are the
costal, the radial, and the submedian; in the hind wing, the costal, the subcostal, the upper and
lower radial, the submedian, and
the internal are simple. The

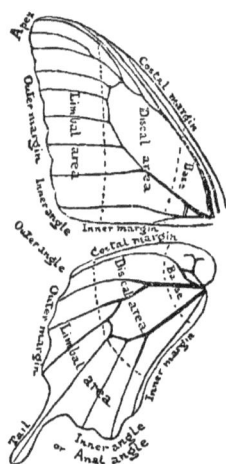

FIG. 38.—Outline of wing, giving names of parts.

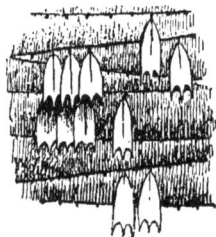

FIG. 39.—Arrangement of scales
on wing of butterfly.

costal vein in the hind wing is, however, generally provided near
the base with a short ascending branch which is known as the
precostal vein. In addition to these simple veins there are in the
fore wing two branching veins, one immediately following the

costal, known as the subcostal, and the other preceding the sub-median, known as the median vein. The branches of these compound veins are known as nervules. The median vein always has three nervules. The nervules of the subcostal veins branch upwardly and outwardly toward the costal margin and the apex of the fore wing. There are always from four to five subcostal nervules. In the hind wing the subcostal is simple. The median vein in the hind wing has three nervules as in the fore wing. Between the subcostal and the median veins, toward the base in both wings, is inclosed the cell, which may be wholly or partially open at its outer extremity, or closed. The veinlets which close the cell at its outward extremity are known as the discocellular veins, of which there are normally three. From the point of union of these discocellular veins go forth the radial veins known respectively as the upper and lower radials, though the upper radial in many genera is emitted from the lower margin of the subcostal.

An understanding of these terms is, however, more readily derived from a study of the figure in which the names of these parts are indicated (Fig. 40).

Butterflies generally hold their wings erect when they are at rest, with their two upper surfaces in proximity, the under surfaces alone displaying their colors to the eye. Only in a few genera of the larger butterflies, and these tropical species, with which this book does not deal, is there an exception to this rule, save in the case of the *Hesperiidæ*, or "skippers," in which very frequently, while the anterior wings are folded together, the posterior wings lie in a horizontal position.

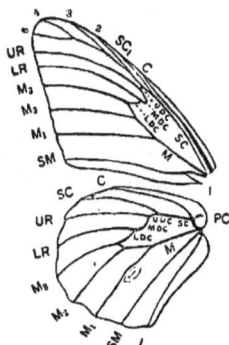

Fig. 40.—Wing of *Anosia plexippus*, showing the names of the veins and nervules; *C, C,* costal veins; *SC,* subcostal vein; *SC₁,*etc., subcostal nervules; *UR,* upper radial; *LR,* lower radial; *M,* median veins; *M₁, M₂, M₃,* median nervules; *SM,* submedian veins; *I,* internal veins; *PC,* precostal nervule; *UDC, MDC, LDC,* upper, middle, and lower discocellulars.

Internal Organs.—Thus far we have considered only the external organs of the butterfly. The internal organs have been made the subject of close study and research by many writers, and a volume might be prepared upon this subject. It will, however, suffice for us to call the attention of the student to the principal facts.

The muscular system finds its principal development in the thorax, which bears the organs of locomotion. The digestive system consists of the proboscis, which has already been described, the gullet, or œsophagus, and the stomach, over which is a large, bladder-like vessel called the food-reservoir, a sort of crop preceding the true stomach, which is a cylindrical tube; the intestine is a slender tube, varying in shape in different genera, divided into the small intestine, the colon, and the rectum. Butterflies breathe through spiracles, little oval openings on the sides of the segments of the body, branching from which inwardly are the tracheæ, or bronchial tubes. The heart, which is located in the same relative position as the spine in vertebrate animals, is a tubular structure. The nervous system lies on the lower or ventral side of the body, its position being exactly the reverse of that which is found in the higher animals. It consists of nervous cords and ganglia, or nerve-knots, in the different segments. Those in the head are more largely developed than elsewhere, forming a rudimentary brain, the larger portion of which consists of two enormous optic nerves. The student who is desirous of informing

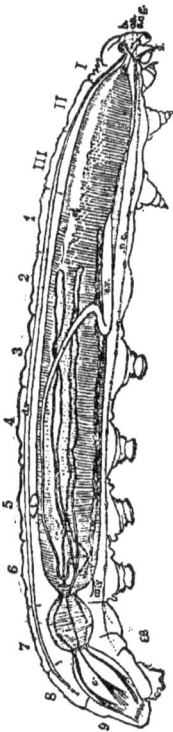

FIG. 41.—Longitudinal section through the larva of *Anosia plexippus*, *δ*, to show the internal anatomy (the Roman numerals indicate the thoracic, the Arabic the abdominal segments): *b*, brain; *sog*, subœsophageal ganglion; *nc*, nervous cord; *œ*, œsophagus; *st*, stomach; *i*, intestine; *c*, colon; *se*, spinning-vessel of one side; *s*, spinneret; *mv*, Malpighian vessel, of which only the portions lying on the stomach are shown, and not the multitudinous convolutions on the intestine; *t*, testis; *dv*, dorsal vessel; the salivary glands are not shown. (Magnified 3 diameters.) (Burgess.)

himself more thoroughly and accurately as to the internal anatomy of these insects may consult with profit some of the treatises which are mentioned in the list of works dealing with the subject which is given elsewhere in this book.

Polymorphism and Dimorphism.—Species of butterflies often show great differences in the different broods which appear. The brood which emerges in the springtime from the chrysalis, which has passed the winter under the snows, may differ very strikingly from the insect which appears in the second or summer brood; and the insects of the third or fall brood may differ again from either the spring or the summer brood. The careful student notes these differences. Such species are called polymorphic, that is, appearing under different forms. Some species reveal a singular difference between the sexes, and there may be two forms of the same sex in the same species. This is most common in the case of the female butterfly, and where there are two forms of the female or the male such a species is said to have dimorphic females or males. This phenomenon is revealed in the case of the well-known Turnus Butterfly; in the colder regions of the continent the females are yellow banded with black, like the males, but in more southern portions of the continent black females are quite common, and these dark females were once

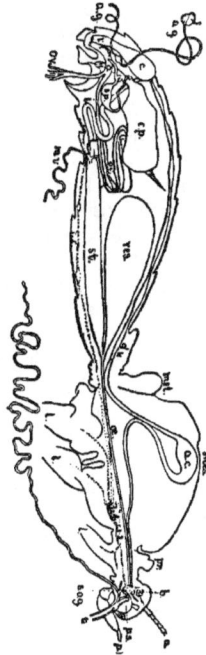

Fig. 42.—Longitudinal section through the imago of *Anosia plexippus*, ♀, to show the internal anatomy: *t*, tongue; *p*, palpus; *a*, antenna; *pr*, prothorax; *mes*, mesothorax; *met*, metathorax; *ps*, pharyngeal sac; *b*, brain; *sog*, subœsophageal ganglion; 1–2, blended first and second ganglia of the larva; 3–4, blended third and fourth ganglia of the larva; *l*, *l*, *l*, the three legs; *œc*, aortal chamber; *dv*, dorsal vessel; *œ*, œsophagus; *rvs*, reservoir for air or food; *st*, stomach; *int*, Malpighian vessels; *i*, intestine; *c*, colon; *r*, rectum; *cp*, copulatory pouch; *o*, oviduct; *ag*, accessory glands; *sp*, spermatheca; *ov*, ovaries (not fully developed); *nc*, nervous cord. (Magnified 3 diameters.) (Burgess.)

23

thought, before the truth was known, to constitute a separate species.

Albinism and Melanism.—Albinos, white or light-colored forms, are quite common among butterflies, principally among the females. On the other hand, melanism, or a tendency to the production of dark or even black forms, reveals itself. Melanism is rather more common in the case of the male sex than in the female sex. The collector and student will always endeavor, if possible, to preserve these curious *aberrations*, as they are called. We do not yet entirely understand what are the causes which are at work to produce these changes in the color, and all such aberrant specimens have interest for the scientific man.

Monstrosities.—Curious malformations, producing monstrosities, sometimes occur among insects, as in other animals, and such malformed specimens should likewise be preserved when found. One form of malformation which is not altogether uncommon consists in an apparent confusion of sexes in specimens, the wings of a male insect being attached to the body of a female, or half of an insect being male and half female.

Mimicry.—One of the most singular and interesting facts in the animal kingdom is what has been styled mimicry. Certain colors and forms are possessed by animals which adapt them to their surroundings in such wise that they are in a greater or less degree secured from observation and attack. Or they possess forms and colors which cause them to approximate in appearance other creatures, which for some reason are feared or disliked by animals which might prey upon them, and in consequence of this resemblance enjoy partial or entire immunity. Some butterflies, for instance, resemble dried leaves, and as they are seated upon the twigs of trees they wholly elude the eye. This illustrates the first form of mimicry. Other butterflies so closely approximate in form and color species which birds and other insects will not attack, because of the disagreeable juices which their bodies contain, that they are shunned by their natural enemies, in spite of the fact that they belong to groups of insects which are ordinarily greedily devoured by birds and other animals. A good illustration of this fact is found in the case of the Disippus Butterfly, which belongs to a group which is not specially protected, but is often the prey of insect-eating creatures. This butterfly has assumed almost the exact color and markings of the milkweed butterfly, *Anosia plexippus*, which is distasteful to birds, and hence enjoys peculiar freedom from the attacks of enemies. Because this adaptation of one form to another evidently serves the purpose of defense this phenomenon has been called " protective mimicry." The reader who is curious to know more about the subject will do well to consult the writings of Mr. Alfred Russel Wallace and Mr. Darwin, who have written at length upon mimicry among butterflies. There is here a field of most interesting inquiry for the student.

The Distribution of Butterflies.—Butterflies are found everywhere that plant life suited to the nourishment of the caterpillars is found. There are some species which are arctic and are found in the brief summer of the cold North and upon the lofty summits of high mountains which have an arctic climate. Most of them are, however, children of the sun, and chiefly abound in the temperate and tropical regions of the earth. While the number of species which are found in the tropics vastly exceeds the number of species found in the temperate zone, it is apparently true that the number of specimens of certain species is far more numerous in temperate regions than in the tropics. Very rarely in tropical countries are great assemblages of butterflies to be seen, such as may be found in the summer months in the United States, swarming around damp places, or hovering over the fields of blooming clover or weeds. In the whole vast region extending from the Rio Grande of Texas to the arctic circle it is doubtful whether more than seven hundred species of butterflies are found. On the continent of Europe there are only about four hundred and fifty species. The number of species of butterflies and the number of species of birds in the United States are very nearly the same.

CHAPTER II

THE CAPTURE, PREPARATION, AND PRESERVATION OF SPECIMENS

> "What hand would crush the silken-wingèd fly,
>   The youngest of inconstant April's minions,
>   Because it cannot climb the purest sky,
>   Where the swan sings, amid the sun's dominions?
>   Not thine."   SHELLEY.

> "Do not mash your specimens!"—THE PROFESSOR.

COLLECTING APPARATUS

Nets.—In the capture of insects of all orders, and especially of butterflies and moths, one of the most important instruments is the net. German naturalists make use of what are known as shears (*Scheren*), which are made like gigantic scissors, having at the end two large oval rings upon which wire gauze or fine netting is stretched. With this implement, which looks like an old-fashioned candle-snuffer of colossal size, they succeed in collecting specimens without doing much injury. Shears are, however, not much in vogue among the naturalists of other countries. The favorite instrument for the ordinary collector is the net. Nets may be made in various ways and of various materials. There are a multitude of devices which have been invented for enabling the net to be folded up so as to occupy but little space when not in use. The simplest form of the net, which can be made almost anywhere, is constructed as follows: A rod—preferably of bamboo, or some other light, stiff material—is used as the handle, not more than five feet in length. Attached to this at its upper end, a loop or ring made of metal, or some moderately stiff

26

yet flexible material, should be tied securely. Upon this there should be sewed a bag of fine netting, preferably tarletan. The

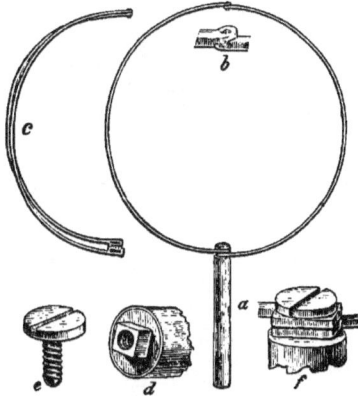

FIG. 43.—Plan for folding net-ring: c, halves of ring detached; b, upper joint of the halves; a, ring set; d, cap of ferrule; f, cap of ferrule, showing screw in place; e, screw (Riley).

bag should be quite long, not less than eighteen inches deep; the ring should be not less than a foot in diameter. Such a net can be made at a cost of but a few cents, and will be, in most cases, as efficient as any of the more expensive nets which are more carefully constructed. A good, cheap ring for a net may be made by using the brass ferrule of a fishing-rod. The ferrule should be

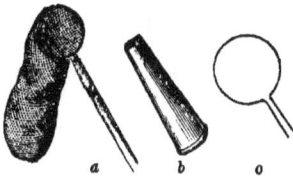

FIG. 44.—a, net; b, ferrule to receive handle; c, wire hoop to be fastened in the upper end of the ferrule (Riley).

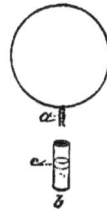

FIG. 45.—a, ring of metal tied with wire at a; b, ferrule; c, plug put in before pouring in solder(Riley).

at least three quarters of an inch in diameter. Into this insert the ends of a metal ring made by bending brass, aluminium, or iron wire into the proper form. When the ends have been inserted into the ferrule, melted solder or lead may be poured into it, and the ends of the wire forming the ring will be thus firmly secured in the ferrule. The ferrule can then be inserted into its mate placed at the end of a bamboo rod. I have commonly obtained for this purpose the last joint or butt of a fishing-rod as the handle of a net. Such a handle can often be purchased for a small sum from a dealer in fishing-rods. It can be made very cheaply. Any kind of a stick, if not too heavy, will do. It is sometimes convenient to have it in your power to lengthen the handle of your net so as to reach objects that are at some elevation above the head, and for this purpose I have had nets made with handles capable of being lengthened by jointed extensions. In collecting in tropical countries, among tall shrubbery and undergrowth, nets thus made, capable of having their handles greatly lengthened, have often proved serviceable. One of the most successful collectors I have ever had in my employment made his net by simply bending a piece of bamboo into the form of the frame of an Indian snow-shoe, to which he attached a handle about a foot and a half in length, and to this he affixed a bag of netting. He was, however, a Japanese, and possessed a singular dexterity in the capture of specimens with this simple apparatus to which I myself never attained. When tarletan cannot be had, ordinary mosquito-netting will do as the material for the bag. It is, however, too coarse in the mesh for many delicate and minute species. Very fine netting for the manufacture of the bags is made in Switzerland, and can be obtained from reputable dealers.

In order to protect and preserve the net, it is well to bind it with some thin muslin at the point where it is joined to the ring. Nets are sometimes made with a strip of muslin, about two inches wide, attached to the entire circumference of the ring, and to this strip of muslin the bag is sewed. For my part, I prefer gray or green as the color for a net. White should be avoided, as experience shows that a white net will often alarm an insect when a net of darker material will not cause it to fly before the collector is ready to bring the net down over the spot where it is settled.

Collecting-Jars.—In killing insects various methods have been used. In practice the most approved method is to employ a jar charged with cyanide of potash or with carbonate of ammonia.

For large moths and butterflies cyanide of potash and carbonate of ammonia serve very well, but it must be remembered that carbonate of ammonia bleaches insects which are green in color. It is well, in my judgment, to use a drop or two of chloroform in the jar charged with carbonate of ammonia, for the collection of diurnal lepidoptera. By putting a few drops of chloroform into the jar, the insect is anesthetized, and its struggles are made quickly to cease. The principal objection to chloroform is the fact that it induces rigidity of the thoracic muscles, which subsequently sometimes interferes with handsome setting.

In the preparation of the poisoning-jar it is well to use a jar which has a ground-glass stopper, and the mouth of which is about three inches in diameter. This will be large enough for most specimens. The one-pound hydrate of chloral jars, provided with glass stoppers and sold by Schering, make the neatest collecting-jars that are known to the writer. I have found it well to have such jars partly covered with leather after the fashion of a drinking-flask. An opening in the leather is left on either side, permitting an inspection of the contents of the jar. The leather protects from

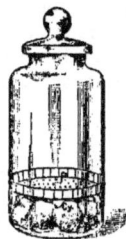

Fig. 46.—Cyanide-jar prepared for use: *P*, perforated cardboard; *Cy*, lumps of cyanide of potash.

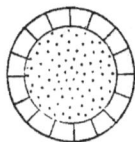

breakage. At the bottom of such a jar a few lumps of cyanide of potash, about the size of a filbert, should be placed. Over this may be laid a little cotton, to prevent the lumps from rattling about loosely at the bottom of the jar. Over the cotton there is pasted a sheet of strong white paper, perforated with a multitude of holes. In securing the white paper over the cyanide, the writer has resorted to a simple method which is explained in the annexed diagram. A piece of paper is placed under the jar, and a circle the size of the inside of the jar is traced upon it. Then a disk is cut out about three quarters of an inch greater in diameter than the original circle (Fig. 47). The paper is punctured over the entire surface included within the inner line, and then, with a scissors, little gashes are made from the outer circumference inward, so as to permit of the folding up of the edge of the disk. A little gum tragacanth is

Fig. 47.—Piece of paper punctured and slit for pasting over the cyanide in the collecting-jar.

29

then applied to these upturned edges; and it is inserted into the
jar and pasted securely over the cyanide by the upturned flaps.
A jar thus charged will last for a long time, if kept properly closed
when not in use. Cyanide of potash has a tendency to deliquesce,
or melt down in the presence of moisture, and in very humid climates or damp places, if the jar is not kept well stoppered, the
cyanide will quickly become semi-fluid, the paper will become
moist, and specimens placed in the jar will be injured or completely ruined. It is well, however, to bear in mind the fact that
the fumes of hydrocyanic acid (prussic acid), which are active in
producing the death of the insect, will not be given off in sufficient volume unless there is some small amount of moisture pres-

Fig. 48. — Method of disabling a butterfly by pinch-
ing it when in the net.

ent in the jar; and in a very dry climate the writer has found it
sometimes necessary to add a drop or two of water from time to
time to the cyanide. The same method which has been described
for charging a jar with cyanide of potash can be employed in
charging it with carbonate of ammonia.

Field-Boxes. — In collecting butterflies it is often possible to
kill, or half kill, the specimens contained in the net by a smart
pinch administered to the insect by the thumb and the first finger,
the pressure being applied from without the net (Fig. 48). This
mode of procedure, however, unless the operator is careful, is apt
to somewhat damage the specimens. The writer prefers to hold
the insect firmly between the thumb and the first finger, and apply
a drop or two of chloroform from a vial which should be carried in

Explanation of Plate IV

Reproduced, with the kind permission of Dr. S. H. Scudder, from " The Butterflies of New England," vol. iii, Plate 83.

Chrysalids in Color and in Outline — Nymphalidæ

1. *Anosia plexippus.* Side view.
2. *Anosia plexippus.* In outline.
3. *Anosia plexippus.* Dorsal view.
4. *Œneis semidea.*
5. *Œneis semidea.* Dorsal view.
6. *Debis portlandia.*
7. *Satyrus nephele.*
8. *Satyrus nephele.* Dorsal view.
9. *Satyrodes canthus.* Side view.
10. *Neonympha phocion.* Side view.
11. *Neonympha phocion.* Side view.
12. *Basilarchia astyanax.* Side view.
13. *Basilarchia astyanax.* Side view.
14. *Basilarchia arthemis.* Side view.
15. *Chlorippe clyton.* Side view.
16. *Chlorippe clyton.* Side view.
17. *Chlorippe clyton.* Dorsal view.
18. *Basilarchia disippus.* Ventral view.
19. *Basilarchia disippus.* Side view.
20. *Basilarchia disippus.* Side view.
21. *Grapta interrogationis.* Dorsal view.
22. *Grapta interrogationis.* Side view.
23. *Basilarchia arthemis.* Dorsal view.
24. *Grapta interrogationis.* Outline of mesothoracic tubercle from the side.
25. *Grapta interrogationis.*
26. *Grapta interrogationis.* Outline of head from in front.
27. *Grapta comma.* Outline of head from in front ; enlarged.
28. *Neonympha eurytus.* Side view.
29. *Grapta comma.* Outline of mesothoracic tubercle from the side.
30. *Grapta comma.* The same from another specimen.
31. *Grapta faunus.* Outline of head from in front.
32. *Grapta progne.* Outline of head from in front.
33. *Grapta faunus.* Side view.
34. *Grapta faunus.* Side view in outline.
35. *Grapta faunus.* Ventral view in outline.
36. *Vanessa j-album.* Outline of mesothoracic tubercle from the side.
37. *Grapta progne.* Side view.
38. *Grapta progne.* Side view.
39. *Grapta comma.* Side view.
40. *Grapta interrogationis.* Side view.
41. *Grapta satyrus.* Side view.
42. *Grapta satyrus.* Ventral view.
43. *Vanessa milberti.* Side view.
44. *Vanessa j-album.* Side view.
45. *Vanessa j-album.* Ventral view.
46. *Grapta comma.* Side view.
47. *Grapta comma.* Side view
48. *Grapta comma.* Dorsal view.
49. *Vanessa milberti.* Side view
50. *Vanessa milberti.* Dorsal view.
51. *Vanessa antiopa.* Side view.
52. *Pyrameis atalanta.* Side view.
53. *Pyrameis atalanta.* Dorsal view.
54. *Pyrameis huntera.* Side view.
55. *Pyrameis atalanta.* Side view.
56. *Junonia cœnia.* Side view.
57. *Junonia cœnia.* Dorsal view.
58. *Vanessa antiopa.* Side view.
59. *Vanessa antiopa.* Dorsal view.
60. *Pyrameis cardui.* Side view.
61. *Pyrameis cardui.* Side view.
62. *Pyrameis cardui.* Dorsal view.
63. *Pyrameis huntera.* Dorsal view.
64. *Pyrameis huntera.* Side view, with nest woven before pupation.
65. *Junonia cœnia.* Side view.
66. *Junonia cœnia.* Side view.
67. *Junonia cœnia.* Side view.

Plate IV.

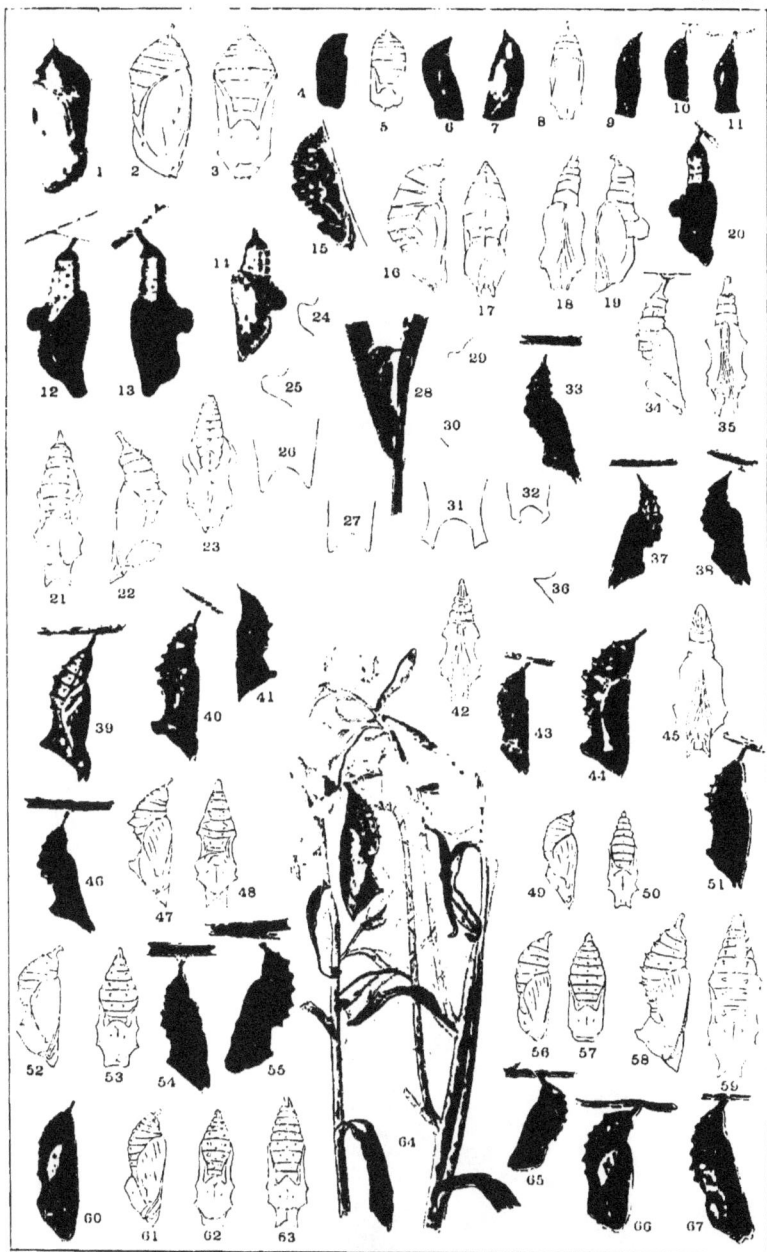

the upper left-hand vest-pocket. The application of the chloroform will cause the insect to cease its struggles immediately, and
it may then be placed in the poisoning-jar, or it may be pinned
into the field-box. The field-box, which should be worn at the
side, securely held in its place by a strap going over the shoulder
and by another strap around the waist, may be provided with the
poisoning apparatus or may be without it. In the former case the
box should be of tin, and should have securely fastened in one corner some lumps of cyanide, tied in gauze. The box should be very
tight, so that when it is closed the fumes of the cyanide may be
retained. The bottom should be covered with cork, upon which
the specimens, as they are withdrawn from the poisoning-jar,
should be pinned. It is well to bear strictly in mind that it is a
mistake to continue to put one specimen after another into the
poisoning-jar until it is half filled or quite filled with specimens.
In walking about the field, if there are several insects in the jar at
a time, they are likely to become rubbed and their beauty partially
destroyed by being tossed about as the collector moves from place
to place; and a large insect placed in a jar in which there are one
or two smaller insects will in its death-struggles possibly injure the
latter. So, as fast as the insects are partially asphyxiated, or deprived of the power of motion, they should be removed from the
poisoning-jar to the poisoning-box, where they are pinned in place
and prevented from rubbing one against the other. Some collectors prefer simply to stun the insects, and then pin them into the
field-box, where they are left, in whole or in part, to recover their
vitality, to be subsequently put to death upon the return of the
collector from the field. This mode of procedure, while undoubtedly it yields in the hands of a skilful operator the most beautiful specimens, appears to the writer to be somewhat cruel, and he
does not therefore approve of it.

The Use of the Net.—In the use of the net the old saying is
true that " practice makes perfect." The bag of the net should be
sufficiently long to allow of its being completely closed when
hanging from the ring on either side. It is possible to sweep
into the net an insect which is fluttering through the air, and then
by a turn of the hand to close the bag and to capture the specimen. When the insect has alighted upon the ground it is best
to clap the net over it and then to raise the net with one hand.
Very many species have the habit of flying upward. This is par-

ticularly true of the skippers, a group of very vigorous and swift-flying butterflies. The writer prefers, if possible, to clap the net over the specimens and then to allow them to rise, and, by inserting the wide-mouthed collecting-jar below, to capture them without touching them at all with the fingers. So far as possible the fingers should not be allowed to come in contact with specimens, whether in or out of the net, though some persons acquire an extremely delicate yet firm touch which enables them to handle the wings of frail species without removing any of the scales. Nothing is more unsightly in a collection than specimens that have been caught and rubbed by the fingers.

Baits. — Moths are frequently taken by the method of collecting known as "sugaring." But it may also be employed for butterflies. For this purpose a mixture of beer and cheap brown sugar may be used. If the beer be stale drippings, so much the better. In fact, it is well, if the collector intends to remain in one locality for some time, to make a mixture of beer and sugar some hours or a day in advance of its application. In semi-tropical countries a mixture of beer and sugar is hardly as good as a mixture of molasses and water into which a few tablespoonfuls of Jamaica rum have been put. A mixture thus prepared seems to attract more effectually than the first prescription. Having provided a pail with a quart or two of the mixture, the collector resorts to the point where he proposes to carry on his work. With an ordinary whitewash brush the mixture is applied to the trunks of trees, stumps, fence-rails, and other objects. It is well to apply the mixture to a series of trees and posts located on the side of a bit of woodland, or along a path through forests, if comparatively open and not too dense. The writer has rarely had success in sugaring in the depths of forests. His greatest success has always been on paths and at the edge of woods. Many beetles and other insects come to the tempting sweets, and separate jars for capturing these should be carried in the pocket. The collector never should attempt to kill beetles in the same jar into which he is putting butterflies. The hard, horny bodies and spiny legs of beetles will make sad havoc with the delicate wings of butterflies.

Many other baits besides this may be employed to attract insects. Some writers recommend a bait prepared by boiling dried apples and mashing them into a pulp, adding a little rum to the mixture, and applying this to the bark of trees. In tropical countries bananas, especially rotten bananas, seem to have a charm for insects. The cane-trash at sugar-mills is very attractive. If possible, it is well to obtain a quantity of this trash and scatter it along forest paths. Some insects have very peculiar appetites and are attracted by things loathsome. The ordure of carnivorous animals seems to have a special charm for some of the most magnificently colored and the rarest of tropical butterflies. A friend of mine in Africa, who collected for me for a number of years, used to keep civet-cats, the ordure of which was collected and placed at appropriate points in the forest paths; and he was richly rewarded by obtaining many insects which were not obtained in any other way. Putrid fish have a charm for other species, and dead snakes, when rankly high, will attract still others. It may be observed that after the trees have been treated for a succession of days or nights with the sweetening mixture spoken of above, they become very productive. When collecting in Japan I made it a rule to return in the morning to the spots that I had sugared for moths the evening before, and I was always amply repaid by finding multitudes of butterflies and even a good many day-flying moths seated upon the mossy bark, feasting upon the remnants of the banquet I had provided the evening before. There is no sport—I do not except that of the angler—which is more fascinating than the sport derived by an enthusiastic entomologist from the practice of "sugaring." It is well, however, to know always where your path leads, and not to lay it out in the dusk, as the writer once did when staying at a well-known summer resort in Virginia. The path which he had chosen as the scene of operations was unfortunately laid, all unknown to himself, just in the rear of the poultry-house of a man who sold chickens to the hotel; and when he saw the dark lantern mysteriously moving about, he concluded that some one with designs upon his hens was hidden in the woods, and opened fire with a seven-shooter, thus coming very near to terminating abruptly the career of an ardent entomologist.

Beating.—There are many species which are apparently not attracted by baits such as we have spoken of in the preceding paragraph. The collector, passing through the grove, searches diligently with his eye and captures what he can see, but does not fail also with the end of his net-handle to tap the trunks of trees and to shake the bushes, and as the insects fly out, to note the point where they settle, and then make them his prey. It is well in this work, as in all collecting, to proceed somewhat leisurely, and to keep perfectly cool. The caricature sometimes found in newspapers of the ardent lepidopterist running like a " quarter-back " across a ten-acre lot in quest of some flying insect does not represent the truly skilful collector, whose movements are more or less stealthy and cautious.

THE BREEDING OF SPECIMENS

By breeding it is possible to obtain specimens in the most perfect condition. Bred specimens which have not had an opportunity to fly are always preferred on account of their freshness of color and perfection of form. A great many species which apparently are exceedingly rare may often be obtained in considerable numbers by the process of breeding, the caterpillar being more readily found than the perfect insect. Although the process of breeding involves a good deal of labor and care, it affords a most delightful field for observation, and the returns are frequently of the very greatest value.

How to Get the Eggs of Butterflies.—The process of breeding may begin with the egg. The skilful eye of the student will detect the eggs of butterflies upon the leaves upon which they have been deposited. The twig may be cut and placed in a vase, in water, and kept fresh until the minute caterpillar emerges, and then from time to time it may be transferred to fresh leaves of the same species of plant, and it will continue to make its moults until at last it is transformed into a chrysalis, and in due season the butterfly emerges. Eggs may frequently be obtained in considerable numbers by confining the female under gauze, with the appropriate food-plant. A knowledge of the food-plant may often be obtained by watching the female and observing upon what plants she deposits her eggs. The exceedingly beautiful researches of Mr. W. H. Edwards were largely promoted by his skill in inducing females to oviposit upon their food-plants. He did this generally by confining the female with the food-plant in a barrel or nail-keg, the bottom of which had been knocked out, and over the top of which he tied mosquito-netting. The plant was placed under the keg. The insects thus confined may be fed with a mixture of honey and water placed upon the leaves.

In collecting caterpillars it is well to have on hand a number of small boxes in which to place them, and also a botany-box in which to bring from the field a supply of their appropriate food.

The process of breeding may begin with the caterpillar. The collector, having discovered the caterpillar feeding upon the branch of a certain plant, provides the creature with a constant supply of the fresh foliage of the same plant, until it finally pupates.

Breeding-Cages. — Various devices for breeding caterpillars and rearing moths and butterflies are known. One of the most important of these devices is the breeding-cage, which is sometimes called a vivarium. The simplest form of the vivarium is often the best. In breeding some species the best method is simply to pot a plant of the species upon which the larva is known to feed, and to place the potted plant in a box over which some mosquito-netting is tied. The writer frequently employs for this purpose cylinders of glass over the top of which per-

Fig. 49. — Cheap form of breeding-cage: *G*, lid covered with mosquito-netting; *E*, pan of earth; *B*, bottle for food-plant.

forated cardboard is placed. This method, however, can be resorted to only with the more minute forms and with plants that do not attain great height. Another form of vivarium is represented in the adjoining woodcut (Fig. 50). The writer has successfully employed, for breeding insects upon a large scale, ordinary store boxes provided with a lid made by fastening together four pieces of wood, making a frame large enough to cover the top of the box, and covering it with gauze. The food-plant is kept fresh in bottles or jars which are set into the boxes. Be careful, however, after you have put the branches upon which the caterpillars are feeding into the jars, to stuff something into the neck of the jar so as to prevent the caterpillar from accidentally getting into the water and drowning himself—a mishap which otherwise might occur. When breeding is undertaken on a still larger scale, it may be well to set apart for this purpose a room, preferably in

35

an outbuilding, all the openings leading from which should be carefully closed so as to prevent the escape of the caterpillars.

How to Find Caterpillars.—Many species of caterpillars are not hard to discover; they are more or less conspicuous objects, and strike the eye. Some species conceal themselves by weaving together the leaves of the plant on which they feed, or by bending

FIG. 50.—Breeding-cage: *a*, base, battened at *g* to prevent warping; *b*, removable body of cage, inclosing zinc pan, *f*, *f*, containing jar for plant, *d*, and filled with five inches of soil, *e*; *c*, removable top, covered with wire gauze. The doors and sides are of glass (Riley).

a single leaf into a curved receptacle in which they lie hidden. Others conceal themselves during the daytime about the roots of trees or under bark or stones, only emerging in the night-time to feed upon the foliage. The collector will carefully search for these. The presence of caterpillars is generally indicated by the ravages which they have committed upon the foliage. By carefully scanning a branch the collector will observe that the leaves

36

have been more or less devoured. Generally underneath the tree will be found the frass, or ejectamenta, of the caterpillar. The presence of the ejectamenta and the evidence of the ravages committed by the larvæ upon the foliage will give the collector a clue to the whereabouts of the caterpillar. ·The writer has found it generally advantageous to search for caterpillars that feed upon trees along the wide, sandy margins of brooks and rivers. The frass is easily discovered upon the sand, and by casting the eye upward into the foliage it is often easy to detect the insect. The pavements in towns and cities which are bordered by trees may also very well be scanned for evidence of the presence of caterpillars. A favorite collecting-ground of the writer is one of the large cemeteries of the city in which he lives, in which there are numerous trees and a great quantity of shrubbery. Wood-boring species, as a rule, are more difficult to obtain and rear than those that feed upon the foliage.

Hibernating Caterpillars.—While some difficulty attends the preservation of chrysalids in the case of those species which pupate in the fall and pass the winter in the chrysalis state under the ground, far more difficulty attends the preservation of species which hibernate in the caterpillar state. As a rule, it is found best to expose the boxes containing these species in an ice-house or other cold place, keeping them there until there is available an abundant supply of the tender shoots of the plant upon which they are in the habit of feeding. They may then be brought forth , from cold storage and placed in proximity to the food-plant, upon which they will proceed to feed.

THE PRESERVATION OF SPECIMENS

Papering Specimens.—When time and opportunities do not suffice for the proper preparation of butterflies for display in the permanent collection, the collector may, in the case of the larger species, conveniently place them in envelopes, with their wings folded (Fig. 51), and they may then be stored in a box until such time as he is able to relax the specimens and properly mount them. Thousands of insects are thus annually collected. The small drug envelopes, or the larger pay-roll envelopes, which may be bought in boxes by the thousand of any stationer for a comparatively small sum, are preferable because of their convenience. Many collectors, however, paper their specimens in envelopes which they make of oblong bits of paper adapted to the size of the insect. The process of making the envelope and of papering the insect is accurately depicted in the accompanying cut

FIG. 51.—Butterfly in envelope.

(Fig. 52). The writer finds it good in the case of small butterflies to place them in boxes between layers of cheap plush or velvet. A small box, a few inches long, may be provided, and at its bottom a layer of velvet is placed; upon this a number of small butterflies are laid. Over them is placed a layer of velvet, with its soft pile facing the same side of the velvet at the bottom. On top of this another piece of velvet is laid, with its pile upward, and other specimens are again deposited, and over this another piece of velvet is laid, and so on. If the box is not filled full at once, it is well to have enough pieces of velvet cut to fill it, or else place cotton on top, so as to keep the layers of velvet from moving or shaking about. A yard or two of plush or velvet will suffice for the packing of a thousand specimens of small butterflies.

Mounting Butterflies.—When the collector has time enough at his disposal he should at once mount his specimens as they are intended to be displayed in the collection. We shall now proceed to explain the manner in which this is most advantageously accomplished.

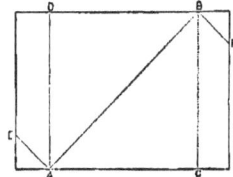

FIG. 52.—Method of folding paper for envelopes: first fold on line *AB*; then on *AD* and *CB*; then on *BF* and *EA*.

The insect should first of all be pinned. The pin should be thrust perpendicularly through the thorax, midway between the wings, and at a considerable elevation upon the pin. It should then be placed upon the setting-board or setting-block. Setting-boards or setting-blocks are pieces of wood having a groove on the upper surface of sufficient depth to accommodate the body of the insect and to permit the wings to be brought to the level of the upper surface of the board (Fig. 53). They should also be provided either with a cleft or a hole which will permit the pin to be thrust down below the body of the insect for a considerable

38

distance. As a rule, the wings of all specimens should be mounted at a uniform elevation of about seven eighths of an inch above the point of the pin. This is known as the " continental method " of mounting, and is infinitely prefer-able to the old-fashioned " English method," in which the insect was pinned low down upon the pin, so that its wings touched the surface of the box.

Setting-blocks are most advanta-geously employed in setting small species, especially the *Hesperiidæ*, the wings of which are refractory. When the insect has been pinned upon the setting-board or setting-block, the next step is to set the wings in the position which they are to maintain when the specimen is thoroughly dry. This is accom-

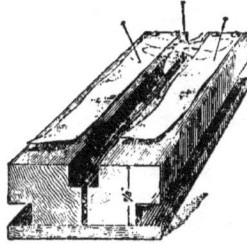

Fig. 53.—Setting-board designed by the author. The wings of the insect are held in place by strips of tracing-muslin, such as is used by engineers. The grooves at the side serve to hold the board in place in the drying-box. (See Fig. 59.)

plished by means of what are known as "setting-needles" (Fig. 56). Setting-needles may be easily made by simply stick-ing ordinary needles into wooden matches from which the tips have been removed. In drawing the wings into position, care should be taken to plant the setting-needle behind the strong nervure on the costal margin of the wing; otherwise the wings are lia-ble to be torn and disfigured. The rule in setting lep-idoptera is to draw the anterior wing forward in such a manner that the

Fig. 54.—Setting-block: *A*, holes to enable the pin to reach to the cork; *C*, cork, filling groove on the bottom of the block; *B*, slit to hold thread.

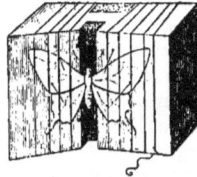

Fig. 55.—Setting-block with butterfly expanded upon it.

posterior margin of this wing is at right angles to the axis of the body, the axis of the body being a line drawn through the head to the extremity of the abdomen. The hind wing should then be moved forward, its anterior margin lying under the op-posing margin of the front wing. When the wings have thus

39

been adjusted into the position which they are to occupy, slips of tracing-muslin or of paper should be drawn down over them and securely pinned, the setting-needles being removed.

In pinning down the strips which are to hold the wings in place, be careful to pin around the wing, but never, if possible, through it. When the wings have been adjusted in the position in which they are to remain, the antennæ, or feelers, should be attended to and drawn forward on the same plane as the wings and secured in place. This may ordinarily be done by setting pins in such a position as to hold them where they are to stay. Then the body, if it has a tendency to sag down at the end of the abdomen, should be raised. This may also be accomplished by means of pins thrust beneath on either side. The figure on the next page shows more clearly what is intended. When the insect has been set, the board should be put aside in a place where it will not be molested or attacked by pests, and the specimens upon it allowed to dry. A box with shelves in it is often used for this purpose. This box should have a door at the front covered with wire gauze, and the back should also be open, covered with gauze, so as to allow a free circulation of air. A few balls of naphthaline placed in it will tend to keep away mites and other pests. The time during which the specimen should remain on the board until it is dried varies with its size and the condition of the atmosphere. Most butterflies and moths in dry weather will be sufficiently dried to permit of their removal from the setting-boards in a week; but large, stout-bodied moths may require as much as two weeks, or even more time, before they are dry enough to be taken off the boards. The process of drying may be hastened by placing the boards in an oven, but the temperature of the oven must be quite low. If too much heat is applied, great injury is sure to result. Only a careful and expert operator should resort to the use of the oven, a temperature above 120° F. being sure to work mischief.

Fig. 56.—Setting-needle.

Fig. 57.—Setting-board with moth expanded upon it (Riley).

Relaxing Specimens.—When butterflies or moths have been put up in papers or mounted on pins without having their wings expanded and set it becomes necessary, before setting them, to relax them. This may be accomplished in several ways. If the specimens have been pinned it is best to place them on pieces of sheet-cork on a tray of sand which has been thoroughly moistened and treated with a good dose of carbolic acid. Over all a bell-glass is put. A tight tin box will serve the same purpose, but a broad sheet of bibulous paper should always be put

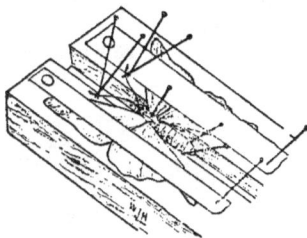

Fig. 58.—Butterfly pinned on board, showing method of holding up body and pinning down antennæ.

over the box, under the lid, before closing it, and in such a way as to leave the edges of the paper projecting around the edges of the lid. This is done to absorb the moisture which might settle by condensation upon the lid and drop upon the specimens. In a bell-glass the

Fig. 50.—Drying-box: *a*, setting-board partly pulled out; *b*, T-shaped strip working in groove on setting-board; *c*, front door, sliding down by tongue, *d*, working in a groove at side in front.

moisture generally trickles down the sides. Earthenware crocks with closely fitting lids are even better than tin boxes, but they must have paper put over them, before closing, in the same way as is done when tin boxes are used. When specimens have been preserved in papers or envelopes these should be opened a little and laid upon damp, carbolized sand under a bell-glass or in a closed receptacle of some kind. Papered specimens may also be placed in their envelopes between clean towels, which have been moistened in water to which a little carbolic acid has been added. The towels should be wrung out quite dry before using them.

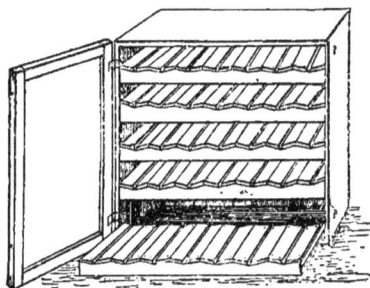

FIG. 60.—Drying-box (Riley).

The method of placing between towels should never be used in the case of very small and delicate species and those which are blue or green in color. Great care must be exercised not to allow the insects to become soaked or unduly wet. This ruins them. They should, however, be damp enough to allow the wings and other organs to be freely moved. When the insects have been relaxed they may be pinned and expanded on setting-boards like freshly caught specimens. It is well in setting the wings of re-laxed specimens, after having thrust the pin through the body, to take a small forceps and, seizing the wings just where they join the body, gently move them so as to open them and make their movement easy before pinning them upon the setting-board. The skilful manipulator in this way quickly ascertains whether they have been sufficiently relaxed to admit of their being readily set. If discovered to be too stiff and liable to break they must be still further relaxed. Dried specimens which have been relaxed and then mounted generally require only a short time to dry again, and need rarely be kept more than twenty-four hours upon the setting-boards.

The process of setting insects upon setting-blocks is exactly the same as when setting-boards are used, with the simple difference that, instead of pinning strips of paper or tracing-muslin over the wings, the wings are held in place by threads or very narrow tapes, which are wound around the block. When the wings are not covered with a very deep and velvety covering of scales the threads or tapes may be used alone; but when the wings are thus clothed it becomes necessary to put bits of paper or cardboard over the wings before wrapping with the threads. Unless this is done the marks of the threads will be left upon the wings. Some little skill, which is easily acquired by practice, is necessary in order to employ setting-blocks to advantage, but in the case of small species and species which have refractory wings they are much to be preferred to the boards.

The Preparation and Preservation of Eggs.—The eggs of butterflies may be preserved by simply putting them into tubes containing alcohol, or they may be placed in vials containing dilute glycerine or a solution of common salt. The vials should be kept tightly corked and should be marked by a label written with a lead-pencil and placed within the bottle, upon which the name of the species and the date of collection should be noted, or a reference made to the collector's note-book. Unless the eggs of insects are preserved in fluid they are apt in many cases to dry up and become distorted, because, on account of their small size, it is impossible to void them of their contents. The larvæ escaping from eggs often void the shell very neatly, leaving, however, a large orifice. Such remnants of shells may be preserved, as they often are useful in showing some of the details of marking; but great vigilance in securing them should be exercised, for almost all the larvæ of butterflies have the curious habit of whetting their appetites for future repasts by turning around and either wholly or partially devouring the shell of the egg which they have quitted. Eggs are most neatly mounted in the form of microscopic slides in glycerine jelly contained in cells of appropriate depth and diameter. It is best, if possible, to mount several specimens upon the same slide, showing the side of the egg as well as the end. A cabinet filled with the eggs of butterflies thus mounted is valuable and curious.

The Preservation of Chrysalids.—Chrysalids may be deprived of their vitality by simply immersing them in alcohol, or they may be killed by means of chloroform, and they may then be fastened upon pins like the imago, and arranged appropriately in the collection with the species. Some chrysalids, however, lose their color when killed in this way, and it is occasionally well to void them of their contents by making an opening and carefully removing the parts that are contained within, replacing with some material which will prevent the chrysalis from shrinking and shriveling. This method of preserving need, however, be resorted to only in exceptional cases. When a butterfly has escaped from its chrysalis it frequently leaves the entire shell behind, with the parts somewhat sundered, yet, nevertheless, furnishing a clear idea of the structure of the chrysalis. If no other specimen of the chrysalis can be obtained than these voided shells they should be preserved.

The Preservation of Caterpillars.—The caterpillars of butterflies when they first emerge from the egg, and before they make the first moult, are, for the most part, extremely small, and are best preserved as microscopic objects in cells filled with glycerine. After each successive moult the larva increases rapidly in size. These various stages in the development of the caterpillar should all be noted and preserved, and it is customary to put up these collections in vials filled with alcohol or a solution of formaline (which latter, by the by, is preferable to alcohol), or to inflate them. The method of inflation secures the best specimens.

In inflating larvæ the first step is carefully to remove the contents of the larval skin. This may be done by making an incision with a stout pin or a needle at the anal extremity, and then, between the folds of a soft towel or cloth, pressing out the contents of the abdominal cavity. The pressure should be first applied near the point where the pellicle has been punctured, and should then be carried forward until the region of the head is reached. Care must be exercised to apply only enough pressure to expel the contents of the skin without disturbing the tissues which lie nearest to the epidermis, in which the pigments are located, and not to remove the hairs which are attached to the body. Pressure sufficient to bruise the skin should never be applied. A little practice soon imparts the required dexterity. The contents of the larval skin having been removed, the next step is to inflate and dry the empty skin. A compact statement of the method of performing this operation is contained in Hornaday's "Taxidermy

44

EXPLANATION OF PLATE V

Reproduced, with the kind permission of Dr. S. H. Scudder, from "The Butterflies of New England," vol. iii, Plate 84.

CHRYSALIDS IN COLOR AND IN OUTLINE—NYMPHALIDÆ, LYCÆNIDÆ, PIERINÆ

1. *Argynnis cybele.* Side view.
2. *Argynnis cybele.* Dorsal view.
3. *Argynnis cybele.* Side view.
4. *Argynnis idalia.* Side view.
5. *Argynnis aphrodite.* Side view.
6. *Argynnis atlantis.* Side view.
7. *Melitæa phaëton.* Side view.
8. *Euptoieta claudia.* Side view.
9. *Euptoieta claudia.* Side view.
10. *Brenthis bellona.* Side view.
11. *Brenthis bellona.* Side view.
12. *Brenthis myrina.* Side view.
13. *Brenthis myrina.* Side view.
14. *Brenthis myrina.* Dorsal view.
15. *Melitæa phaëton.* Side view.
16. *Melitæa phaëton.* Dorsal view.
17. *Melitæa harrisi.* Side view.
18. *Melitæa harrisi.* Dorsal view.
19. *Phyciodes nycteis.* Side view.
20. *Phyciodes tharos.* Dorsal view.
21. *Phyciodes tharos.* Side view.
22. *Phyciodes tharos.* Side view.
23. *Libythea bachmani.* Side view.
24. *Libythea bachmani.* Side view.
25. *Thecla calanus.* Side view.
26. *Thecla irus.* Side view, enlarged.
27. *Thecla calanus.* Side view.
28. *Thecla liparops.* Side view.
29. *Thecla edwardsi.* Side view.
30. *Thecla damon.* Side view.
31. *Thecla damon.* Side view, enlarged.
32. *Thecla irus.* Dorsal view.
33. *Thecla irus.* Side view.
34. *Thecla irus.* Side view.
35. *Thecla acadica.* Side view.
36. *Lycæna pseudargiolus.* Side view.
37. *Thecla titus.* Side view.
38. *Thecla niphon.* Side view.
39. *Thecla melinus.* Side view. Copied from Abbot's drawing in the British Museum.

40. *Thecla niphon.* Side view. Copied from Abbot's drawing in Dr. Boisduval's library.
41. *Lycæna scudderi.* Side view, enlarged.
42. *Lycæna comyntas.* Side view. Copied from Abbot's drawing in Dr. Boisduval's library.
43. *Lycæna pseudargiolus.* Side view, enlarged. Copied from Abbot's drawing in Dr. Boisduval's library.
44. *Lycæna pseudargiolus.* Side view.
45. *Feniseca tarquinius.* Side view.
46. *Feniseca tarquinius.* Side view. Copied from Abbot's drawing in the British Museum.
47. *Lycæna comyntas.* Side view, enlarged.
48. *Lycæna comyntas.* Side view.
49. *Chrysophanus hypophlæas.* Side view.
50. *Chrysophanus thoë.* Side view.
51. *Terias nicippe.* Side view.
52. *Terias nicippe.* Dorsal view.
53. *Colias eurytheme.* Side view.
54. *Colias philodice.* Dorsal view.
55. *Colias philodice.* Side view.
56. *Terias lisa.* Side view.
57. *Pieris napi,* var. *oleracea.* Side view.
58. *Pieris rapæ.* Side view.
59. *Euchloë genutia.* Side view.
60. *Callidryas eubule.* Side view.
61. *Callidryas eubule.* Side view.
62. *Callidryas eubule.* Dorsal view.
63. *Pieris napi,* var. *oleracea.* Side view.
64. *Pieris napi,* var. *oleracea.* Dorsal view.
65. *Pieris rapæ.* Dorsal view.
66. *Pieris protodice.* Dorsal view.
67. *Pieris protodice.* Side view.

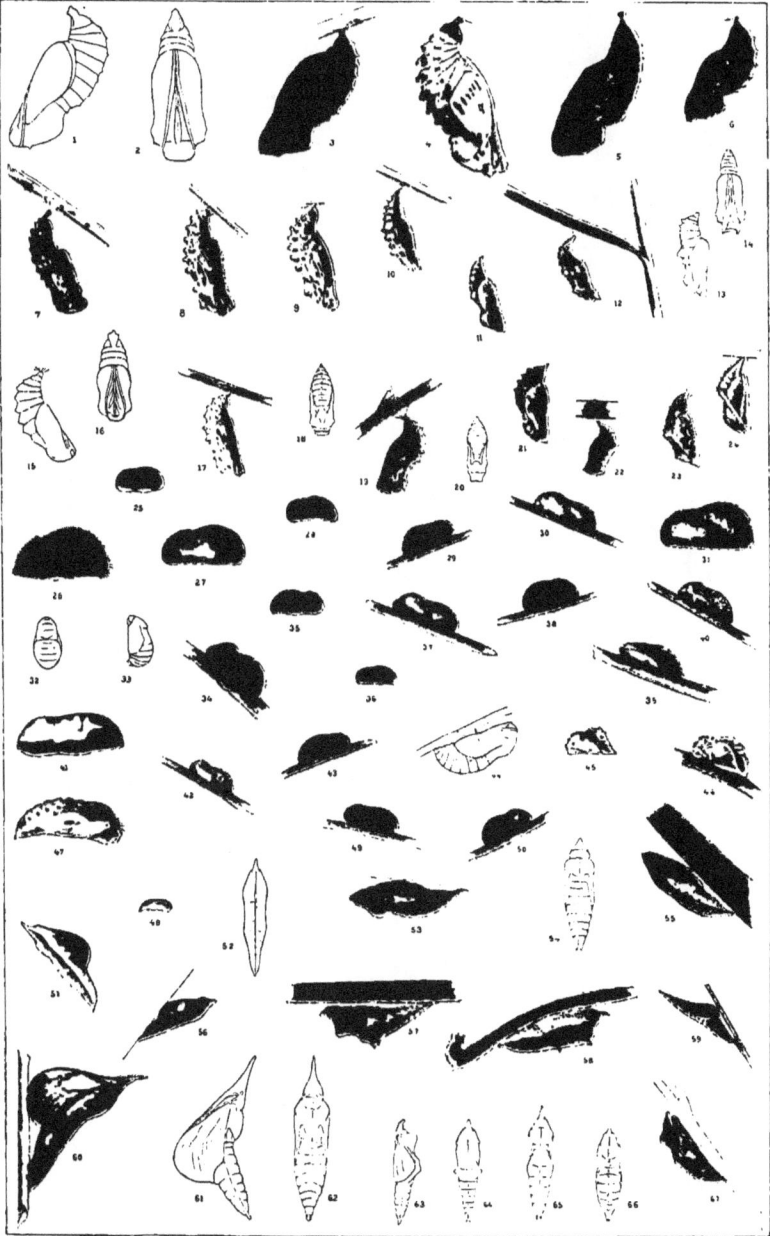

and Zoölogical Collecting," from the pen of the writer, and I herewith reproduce it:

"The simplest method of inflating the skins of larvæ after the contents have been withdrawn is to insert a straw or grass stem of appropriate thickness into the opening through which the contents have been removed, and then by the breath to inflate the specimen, while holding over the chimney of an Argand lamp,

Fig. 61.—Apparatus for inflating larvæ: B, foot-bellows; K, rubber tube; C, flask; D, anhydrous sulphuric acid; E, overflow-flask; F, rubber tube from flask; G, standard with cock to regulate flow of air; H, glass tube with larva upon it; I, copper drying-plate; J, spirit-lamp.

the flame of which must be regulated so as not to scorch or singe it. Care must be taken in the act of inflating not to unduly distend the larval skin, thus producing a distortion, and also to dry it thoroughly. Unless the latter precaution is observed a subsequent shrinking and disfigurement will take place. The process of inflating in the manner just described is somewhat laborious, and while some of the finest specimens which the writer has ever seen were prepared in this primitive manner, various expedients for lessening the labor involved have been devised, some of which are to be highly commended.

" A comparatively inexpensive arrangement for inflating larvæ is a modification of that described in the ' Entomologische Nachrichten' (1879, vol. v, p. 7), devised by Mr. Fritz A. Wachtel (Fig. 61). It consists of a foot-bellows such as is used by chemists in the laboratory, or, better still, of a small cylinder such as is used for holding gas in operating the oxyhydrogen lamp of a sciopticon. In the latter case the compressed air should not have a pressure exceeding twenty pounds to the square inch, and the cock regulating the flow from the cylinder should be capable of very fine adjustment. By means of a rubber tube the air is conveyed from the cylinder to a couple of flasks, one of which contains concentrated sulphuric acid, and the other is intended for the reception of any overflow of the hydrated sulphuric acid which may occur. The object of passing the air through sulphuric acid is to rob it, so far as possible, of its moisture. It is then conveyed into a flask, which is heated upon a sand-bath, and thence by a piece of flexible tubing to a tip mounted on a joint allowing vertical and horizontal motion and secured by a standard to the working-table. The flow of air through the tip is regulated by a cock. Upon the tip is fastened a small rubber tube, into the free extremity of which is inserted a fine-pointed glass tube. This is provided with an armature consisting of two steel springs fastened upon opposite sides, and their ends bent at right angles in such a way as to hold the larval skin firmly to the extremity of the tube. The skin having been adjusted upon the fine point of the tube, the bellows is put into operation, and the skin is inflated. A drying apparatus is provided in several ways. A copper plate mounted upon four legs, and heated by an alcohol-lamp placed below, has been advocated by some. A better arrangement, used by the writer, consists of a small oven heated by the flame of an alcohol-lamp or by jets of natural gas, and pro-

Fig. 62.—Tip of inflating-tube, with armature for holding larval skin.

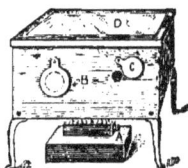

Fig. 63.—Drying-oven: A, lamp; B, pin to hold door open; C, door open; D, glass cover.

vided with circular openings of various sizes, into which the larval skin is introduced (Fig. 63).

" A less commendable method of preserving larvæ is to place them in alcohol. The larvæ should be tied up in sacks of light gauze netting, and a label of tough paper, with the date and locality of capture, and the name, if known, written with a lead-pencil, should be attached to each such little sack. Do not use ink on labels to be immersed, but a hard lead-pencil. Alcoholic specimens are liable to become shriveled and discolored, and are not nearly as valuable as well-inflated and dried skins.

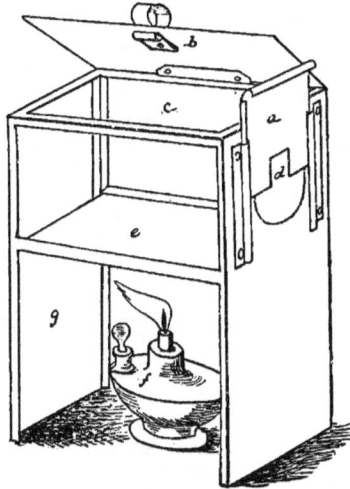

Fig. 64.—Drying-oven: *a*, sliding door; *b*, lid; *c*, body of oven with glass sides; *d*, opening for inserting inflating-tube; *e*, copper bottom; *f*, spirit-lamp; *g*, base (Riley).

" When the skins have been inflated they may be mounted readily by being placed upon wires wrapped with green silk, or upon annealed aluminium wire. The wires are bent and twisted together for a short distance and then made to diverge. The diverging ends are pressed together, a little shellac is placed upon their tips, and they are then inserted into the opening at the anal

47

extremity of the larval skin. Upon the release of pressure they spread apart, and after the shellac has dried the skin is firmly held by them. They may then be attached to pins by simply twisting the free end of the wire about the pin, or they may be placed upon artificial imitations of the leaves and twigs of their appropriate food-plants."

THE PRESERVATION AND ARRANGEMENT OF COLLECTIONS

The secret of preserving collections of lepidoptera in beautiful condition is to exclude light, moisture, and insect pests. Light ultimately bleaches many species, moisture leads to mould and mildew, and insect pests devour the specimens. The main thing is therefore to have the receptacles in which the specimens are placed dark and as nearly as possible hermetically sealed and kept in a dry place. In order to accomplish this, various devices have been resorted to.

Boxes. — Boxes for the preservation of specimens are made with a tongue on the edges of the bottom fitting into a groove upon the lid, or they may be made with inside pieces fastened around the inner edge of the bottom and projecting so as to catch the lid. The accompanying outlines show the method of joining different forms of boxes (Figs. 65 – 67). The bottom of the box should be lined with some substance which will enable the specimens to

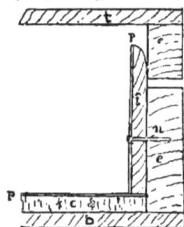

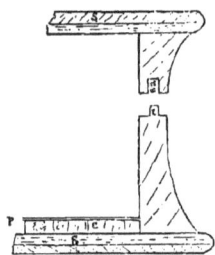

Fig. 65. — Detail drawing of front of box, made to resemble a book: *s, s,* sides, made of two pieces of wood glued together across the grain; *l,* tongue; *g,* groove; *c,* cork; *p,* paper covering the cork.

Fig. 66. — Detail drawing of front of box: *t,* top; *b,* bottom; *a,* side; *f,* strip, nailed around inside as at *n*; *c,* cork; *p,* paper lining.

be pinned into it securely. For this purpose sheet-cork about a quarter of an inch thick is to be preferred to all other substances. Ground cork pressed into layers and covered with white paper is manufactured for the purpose of lining boxes. Turf compressed into sheets about half an inch thick and covered with paper is used by many European collectors. Sheets of aloe-pith or of the wood of the yucca, half an inch thick, are used, and the pith of corn-stalks (Indian corn or maize) may also be employed, laid into the box and glued neatly to the bottom. The corn-pith should be cut into pieces about half an inch square and joined together neatly, covering it with thin white paper after the surface has been made quite even and true. Cork is, however, the best material, for, though more expensive than the other things named, it has greater power to hold the pins, and unless these are securely fixed and held in place great damage is sure to result. A loose specimen in a box will work incalculable damage. Boxes should be made of light, thoroughly seasoned wood, and should be very tight. They are sometimes made so that specimens may be pinned both upon the top and the bottom, but this is not to be commended. The depth of the box should be sufficient to admit of the use of the longest insect-pin in use, and a depth between top and bottom of two and a quarter inches is therefore sufficient. Boxes are sometimes made with backs in imitation of books, and

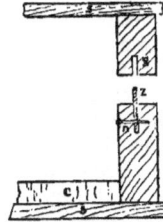

Fig. 67.—Detail drawing of box, in which the tongue, *t*, is made of strips of zinc let into a groove and fastened as at *n*; *g*, groove to catch tongue; *s, s,* top and bottom; *c,* cork.

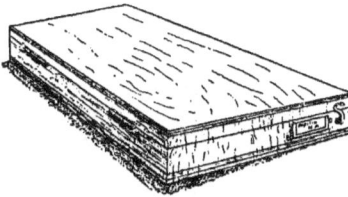

Fig. 68.—Insect-box for preservation of collections.

a collection arranged in such boxes presents an attractive external appearance. A very good box is made for the United States Department of Agriculture and for the Carnegie Museum in Pittsburgh (Fig. 68). This box is thirteen inches long, nine inches wide, and three inches thick (external measurement). The depth between the bottom and the lid on the inside is two and one eighth inches. The ends and sides are dovetailed; the top and bottom are each made of two pieces of light stuff, about one

49

eighth of an inch thick, glued together in such a way that the grain of the two pieces crosses at right angles, and all cracking and warping are thus prevented. The lids are secured to the bottoms by brass hooks fitting into eyelets. Such boxes provided with cork do not cost more than fifty-five cents apiece when bought in quantities. Boxes may be made of stout pasteboard about one eighth or three sixteenths of an inch thick, with a rabbet-tongue on the inside. Such boxes are much used in France and England, and when well and substantially made are most excellent. They may be obtained for about thirty-five cents apiece lined with compressed cork.

Cabinets and Drawers.—Large collections which are intended to be frequently consulted are best preserved in cabinets fitted with glass-covered drawers. A great deal of variety exists in the plans which are adopted for the display of specimens in cabinets. Much depends upon the taste and the financial ability of the collector. Large sums of money may be expended upon cabinets, but the main thing is to secure the specimens from dust, mould, and insect pests. The point to be observed most carefully is so to arrange the drawers that they are, like the boxes, practically air-tight. The writer employs as the standard size for the drawers in his own collection and in the Carnegie Museum a drawer which is twenty-two inches long, sixteen inches wide, and two inches deep (inside measurement). The outside dimensions are: length, twenty-three inches exclusive of face; breadth, seventeen inches; height, two and three eighths inches. The covers are glazed with double-strength glass. They are held upon the bottoms by a rabbet placed inside of the bottom and nearly reaching the lower surface of the glass on the cover when closed. The drawers are lined upon the bottom with cork five sixteenths of an inch thick, and are papered on the bottom and sides with good linen paper, which does not easily become discolored. Each drawer is faced with cherry and has a knob. These drawers are arranged in cabinets built in sections for convenience in handling. The two lower sections each contain thirty drawers, the upper section nine. The drawers are arranged in three perpendicular series and are made interchangeable, so that any drawer will fit into any place in any one of the cabinets. This is very necessary, as it admits of the easy rearrangement of collections. On the sides of each drawer a pocket is cut on the inner surface, which communicates through an opening in the rabbet with the interior. The paper lining the inside is perforated over this opening with a number of small holes. The pocket is kept filled with naphthaline crystals, the fumes of which pass into the interior and tend to keep away pests. The accompanying figure gives the details of construction (Fig. 69). Such drawers can be made at a cost of about $3.50 apiece, and the cost of a cabinet finished and supplied with them is about $325, made of cherry, finished in imitation of mahogany.

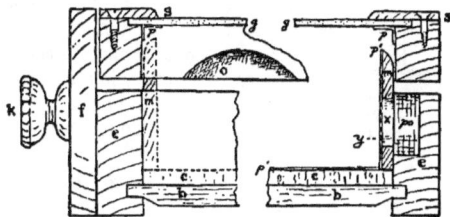

FIG. 69.—Detail drawing of drawer for cabinet: *e, e,* ends; *b,* bottom; *c,* cork; *p, p,* paper strips in corners of lid to exclude dust; *g, g,* glass of cover, held in place by top strips, *s, s; m, m,* side pieces serving as rabbets on inside; *po,* pocket in ends and sides, sawn out of the wood; *x,* opening through the rabbet into this pocket; *y,* holes through the paper lining, *p',* allowing fumes of naphthaline to enter interior of drawer; *f,* front; *k,* knob; *o,* lunette cut in edge of the top piece to enable the lid to be raised by inserting the fingers.

Some persons prefer to have the bottoms as well as the tops of the drawers in their cabinets made of glass. In such cases the specimens are pinned upon narrow strips of wood covered with cork, securely fastened across the inside of the drawers. This arrangement enables the under side of specimens to be examined and compared with as much freedom as the upper side, and without removing them from the drawers; but the strips are liable at times to become loosened, and when this happens great havoc is wrought among the specimens if the drawer is moved carelessly. Besides, there is more danger of breakage.

Another way of providing a cheap and very sightly lining for the bottom of an insect-box is illustrated in Fig. 70. A frame of wood like a slate-frame is provided, and on both sides paper is stretched. To stretch the paper it ought to be soaked in water before pasting to the frame; then when it dries it is as tight and smooth as a drum-head.

The beginner who has not a long purse will do well to preserve his collections in boxes such as have been described. They can

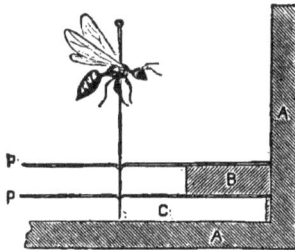

be obtained quite cheaply and are most excellent. Cabinets are more or less of a luxury for the amateur, and are only a necessity in the case of great collections which are constantly being consulted. The boxes may be arranged upon shelves. Some of the largest and best collections in the world are preserved in boxes, notably those of the United States National Museum.

Fig. 70.—A, A, side and bottom of box; B, frame fitting into box; C, space which must be left between frame and bottom of box; P, P, paper stretched on frame.

Labeling.—Each specimen should have on the pin below the specimen a small label giving the date of capture, if known, and the locality. Below this should be a label of larger size, giving its scientific name, if ascertained, and the sex. Labels should be neat and uniform in size. A good size for labels for large species is about one inch long and five eighths of an inch wide. The labels should be written in a fine but legible hand. Smaller labels may be used for smaller species. A crowquill pen and India ink are to be preferred in writing labels.

Arrangement of Specimens.—Specimens are best arranged in rows. The males should be pinned in first in the series, after them the females. Varieties should follow the species. After these should be placed any aberrations or monstrosities which the collector may possess. The name

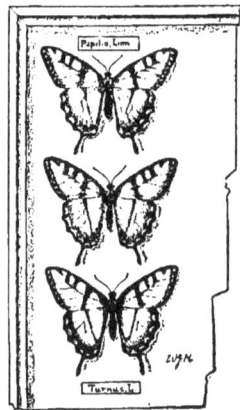

Fig. 71.—Manner of arranging specimens in cabinet.

of the genus should precede all the species contained in the collection, and after each species the specific name should be placed. Fig. 71 shows the manner of arrangement.

Insect Pests.—In order to preserve collections, great care must be taken to exclude the various forms of insect pests, which are likely, unless destroyed and kept from attacking the specimens, to ruin them utterly in comparatively a short time. The pests which are most to be feared are beetles belonging to the genera *Dermestes* and *Anthrenus*. In addition to these beetles, which commit their ravages in the larval stage, moths and mites prey upon collections. Moths are very infrequently, however, found in collections of insects, and in a long experience the writer has known only one or two instances in which any damage was inflicted upon specimens by the larvæ of moths. Mites are much more to be dreaded.

In order to prevent the ravages of insects, all specimens, before putting them away into the boxes or drawers of the cabinet in which they are to be preserved, should be placed in a tight box in which chloroform, or, better, carbon bisulphide, in a small pan is put, and they should be left here for at least twenty-four hours, until it is certain that all life is extinct. Then they should be transferred to the tight boxes or drawers in which they are to be kept. The presence of insect pests in a collection is generally first indicated by fine dust under the specimen, this dust being the excrement of the larva which is committing depredations upon the specimen. In case the presence of the larva is detected, a liberal dose of chloroform should at once be administered to the box or tray in which the specimen is contained. The specimen itself ought to be removed, and may be dipped into benzine.

Naphthaline crystals or camphor is generally employed to keep out insect pests from boxes. They are very useful to deter the entrance of pests, but when they have once been introduced into a collection neither naphthaline nor camphor will kill them. Naphthaline is prepared in the form of cones attached to a pin, and these cones may be placed in one corner of the box. They are made by Blake & Co. of Philadelphia, and are in vogue among entomologists. However, a good substitute for the cones may very easily be made by taking the ordinary moth-balls which are sold everywhere. By heating a pin red-hot in the flame of an alcohol-lamp it may be thrust into the moth-ball; as it enters it melts the naphthaline, which immediately afterward cools and

FIG. 72.—Naphthaline cone.

holds the pin securely fixed in the moth-ball. In attaching these pins to moth-balls, hold the pin securely in a forceps while heating it in the flame of the lamp, and thrust the red-hot pin into the center of the ball. Naphthaline crystals and camphor may be secured in the corner of the box by tying up a quantity of them in a small piece of netting and pinning the little bag thus made in the corner of the tray. By following these directions insect pests may be kept out of collections. It is proper to observe that while carbon bisulphide is more useful even than chloroform in killing pests, and is also cheaper, it should be used with great care, because when mixed with atmospheric air it is highly explosive, and its use should never take place where there are lamps burning or where there is fire. Besides, its odor is extremely unpleasant, unless it has been washed in mercury.

Greasy Specimens.—Specimens occasionally become greasy. When this happens they may be cleansed by pinning them down on a piece of cork secured to the bottom of a closed vessel, and gently filling it with benzine, refined gasoline, or ether. After leaving them long enough to remove all the grease they may be taken out of the bath and allowed to dry in a place where there is no dust. This operation should not take place near a lighted lamp or a fire.

Mould.—When specimens have become mouldy or mildewed it is best to burn them up if they can be spared. If not, after they have been thoroughly dried remove the mould with a sable or camel's-hair pencil which has been rubbed in carbolic acid (crystals liquefied by heat). Mildew in a cabinet is hard to eradicate, and heat, even to burning, is about the only cure, except the mild use of carbolic acid in the way suggested.

Repairing Specimens.—Torn and ragged specimens are to be preferred to none at all. "The half of a loaf is better than no bread." Until the torn specimen can be replaced by a better, it is always well to retain it in a collection. But it is sometimes possible to repair torn specimens in such a way as to make them more presentable. If an antenna, for instance, has been broken off, it may be replaced neatly, so that only a microscopic examination will disclose the fact that it was once away from the place where it belonged. If a wing has been slit, the rent may be mended so neatly that only a very careful observer can detect the fact. If a piece has been torn out of a wing, it may be replaced

54

by the corresponding portion of the wing of another specimen of the same sex of the same species in such a way as almost to defy detection. The prime requisites for this work are patience, a steady hand, a good eye, a great deal of "gumption," a few setting-needles, a jeweler's forceps, and a little shellac dissolved in alcohol. The shellac used in replacing a missing antenna should be of a thickish consistency; in repairing wings it should be well thinned down with alcohol. In handling broken antennæ it is best to use a fine sable pencil, which may be moistened very lightly by applying it to the tip of the tongue. With this it is possible to pick up a loose antenna and place it wherever it is desired. Apply the shellac to the torn edges of a broken wing with great delicacy of touch and in very small quantity. Avoid putting on the adhesive material in "gobs and slathers." Repairing is a fine art, which is only learned after some patient experimentation, and is only to be practised when absolutely necessary. The habit of some dealers of patching up broken specimens with parts taken from other species is highly to be reprobated. Such specimens are more or less caricatures of the real thing, and no truly scientific man will admit such scarecrows into his collection, except under dire compulsion.

Packing and Forwarding Specimens. —It often becomes necessary to forward specimens from one place to another. If it is intended to ship specimens which have been mounted upon pins they should be securely pinned in a box lined with cork. A great many expanded specimens may be pinned in a box by resorting to the method known as "shingling," which is illustrated in Fig. 73. By causing the wings of specimens to overlap, as is shown in the figure, a great many can be accommodated in a small space. When the specimens have been packed the box should be securely closed, its edges shut with paper, after some drops of chloroform have been poured into the box, and then this box should be placed in an outer box containing excelsior, hay, cotton, or loose shavings in sufficient abundance to prevent the jarring of the inner box and consequent breakage. Where specimens are forwarded in envelopes, having been collected in the

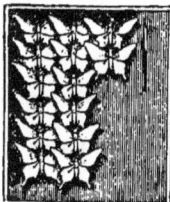

FIG. 73.—Butterflies pinned into a box overlapping one another, or "shingled."

55

field, and are not pinned, the precaution of surrounding them with packing such as has been described is not necessary, but the box in which they are shipped should always be strong enough to resist breakage. Things forwarded by mail or by express always receive rough treatment, and the writer has lost many fine specimens which have been forwarded to him because the shipper was careless in packing.

Pins. — In the preceding pages frequent reference has been made to insect-pins. These are pins which are made longer and thinner than is the case with ordinary pins, and are therefore adaptable to the special use to which they are put. There are a number of makers whose pins have come into vogue. What are known as Karlsbader and Kläger pins, made in Germany, are the most widely used. They are made of ordinary pin-metal in various sizes. The Karlsbader pins have very fine points, but, owing to the fineness of the points and the softness of the metal, they are very apt to buckle, or turn up at the points. The Kläger pins are not exposed to the same objection, as the points are not quite so fine. The best pins, however, which are now made are those which have recently been introduced by Messrs. Kirby, Beard, & Co. of England. They are made of soft steel, lacquered, possessing very great

Fig. 74. — Butterfly-forceps, half-size.

strength and considerable flexibility. The finest-sized pin of this make has as much strength as the largest pin of the other makes that have been mentioned, and the writer has never known them to buckle at the tip, even when pinned through the hardest insect tissues. While these pins are a little more expensive than others, the writer does not fail to give them an unqualified preference.

The Forceps. — An instrument which is almost indispensable to the student of entomology is the forceps. There are many forms of forceps, and it is not necessary to speak at length in reference to the various shapes; but for the use of the student of butterflies the forceps made by the firm of Blake & Co. of Philadelphia is to be preferred to all others. The head of this firm is himself a famous entomologist, and he has given us in the forceps which is illustrated in Fig. 74 an instrument which comes as near perfection as the art of the maker of instruments can produce. The small forceps represented in Fig. 75 is very useful in pinning small specimens. In handling mounted specimens it

FIG. 75.—Insect-forceps.

is well always to take hold of the pin below the specimen with the forceps, and insert it into the cork by the pressure of the forceps. If the attempt is made to pin down a specimen with the naked fingers holding the pin by the head, the finger is apt to slip and the specimen to be ruined.

IMMORTALITY

A butterfly basked on a baby's grave,
　Where a lily had chanced to grow:
"Why art thou here with thy gaudy dye,
When she of the blue and sparkling eye
　Must sleep in the churchyard low?"

Then it lightly soared thro' the sunny air,
　And spoke from its shining track:
"I was a worm till I won my wings,
And she, whom thou mourn'st, like a seraph sings;
　Would'st thou call the blest one back?"
　　　　　　　　　　SIGOURNEY.

57

CHAPTER III

" Winged flowers, or flying gems."
MOORE.

At the base of all truly scientific knowledge lies the principle of order. There have been some who have gone so far as to say that science is merely the orderly arrangement of facts. While such a definition is defective, it is nevertheless true that no real knowledge of any branch of science is attained until its relationship to other branches of human knowledge is learned, and until a classification of the facts of which it treats has been made. When a science treats of things, it is necessary that these things should become the subject of investigation, until at last their relation to one another, and the whole class of things to which they belong, has been discovered. Men who devote themselves to the discovery of the relation of things and to their orderly classification are known as systematists.

The great leader in this work was the immortal Linnæus, the "Father of Natural History," as he has been called. Upon the foundation laid by him in his work entitled " Systema Naturæ," or " The System of Nature," all who have followed after him have labored, and the result has been the rise of the great modern sciences of botany and zoölogy, which treat respectively of the vegetable and animal kingdoms.

The Place of Butterflies in the Animal Kingdom. —The animal kingdom, for purposes of classification, has been subdivided into various groups known as subkingdoms. One of these subkingdoms contains those animals which, being without vertebræ, or an internal skeleton, have an external skeleton, composed of a series of horny rings, attached to which are various organs. This

EXPLANATION OF PLATE VI

Reproduced with the kind permission of Dr. S. H. Scudder, from "The Butterflies of New England," vol. iii, Plate 85.

CHRYSALIDS IN COLOR AND IN OUTLINE—PAPILIONINÆ AND HESPERIIDÆ

1. *Papilio turnus.*
2. *Papilio turnus.* Dorsal view.
3. *Papilio turnus.*
4. *Papilio turnus.*
5. *Papilio troilus.* Dorsal view.
6. *Papilio troilus.*
7. *Papilio troilus.*
8. *Papilio cresphontes.*
9. *Papilio cresphontes.* Dorsal view.
10. *Papilio cresphontes.*
11. *Papilio ajax.*
12. *Papilio ajax.* Dorsal view.
13. *Papilio asterias.*
14. *Papilio philenor.* Dorsal view.
15. *Papilio philenor.* Dorsal view.
16. *Papilio philenor.*
17. *Papilio philenor.*
18. *Papilio asterias.* Dorsal view.
19. *Papilio asterias.*
20. *Papilio philenor.*
21. *Achalarus lycidas.*
22. *Epargyreus tityrus.*
23. *Eudamus proteus.* From the original by Abbot in the British Museum.
24. *Thorybes bathyllus.* From the original by Abbot in the British Museum.
25. *Epargyreus tityrus.*
26. *Epargyreus tityrus.*
27. *Thanaos icelus.*
28. *Thorybes pylades.*
29. *Pholisora catullus.* From the original by Abbot in the British Museum.
30. *Thanaos lucilius.*
31. *Thanaos lucilius.* Dorsal view.
32. *Thanaos lucilius.*
33. *Thanaos juvenalis.*
34. *Thanaos persius.*
35. *Hesperia montivaga.* From the original by Abbot in the British Museum.
36. *Pholisora catullus.*
37. *Thanaos martialis.* From the original by Abbot in the British Museum.
38. *Thanaos brizo.* From the original by Abbot in Dr. Boisduval's library.
39. *Hylephila phylæus.* From the original by Abbot in Dr. Boisduval's library.
40. *Amblyscirtes vialis.*
41. *Pholisora catullus.*
42. *Thymelicus ætna.* From the original by Abbot in Dr. Boisduval's library.
43. *Atalopedes huron.*
44. *Limochores taumas.*
45. *Amblyscirtes samoset.* After the original by Abbot in the British Museum.
46. *Lerema accius.* After the original by Abbot in Boston Society of Natural History.
47. *Atalopedes huron.*
48. *Calpodes ethlius.*

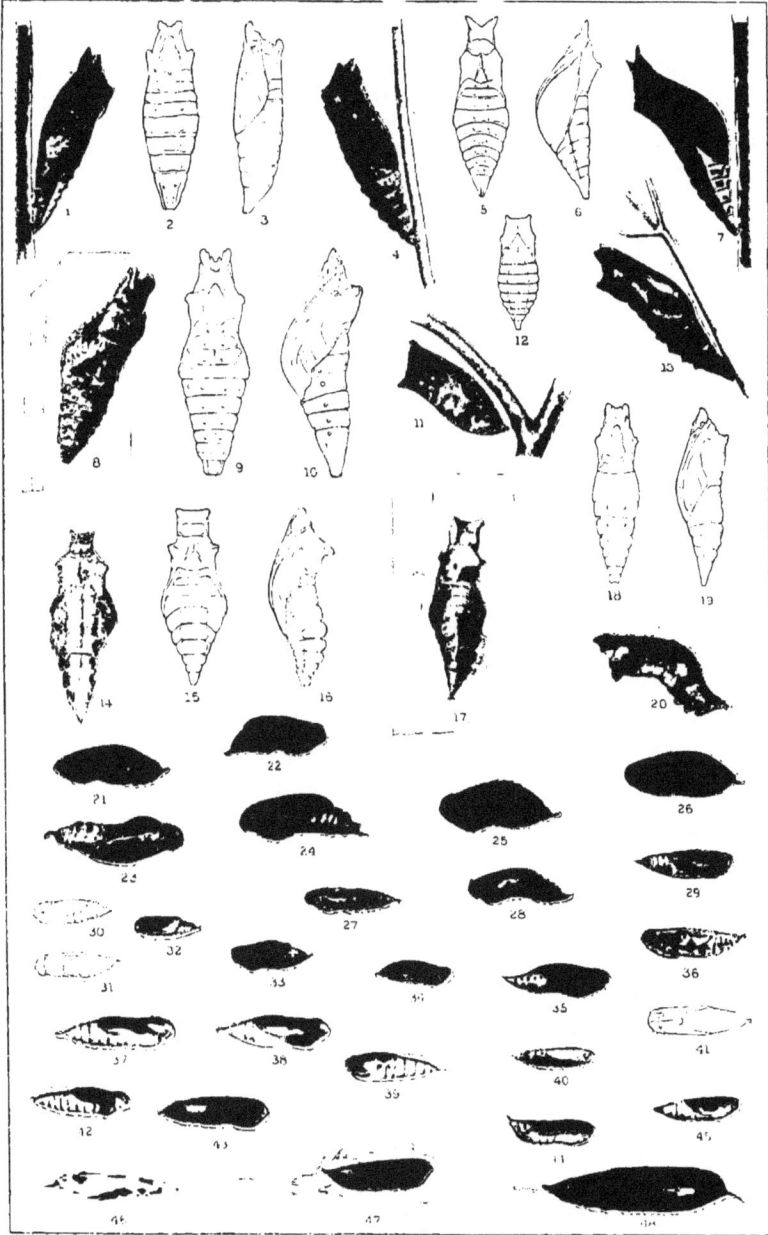

subkingdom is known by naturalists under the name of the *Arthropoda*. The word *Arthropoda* is derived from the Greek language, and is compounded of two words, ἄρθρον (*arthron*), meaning a *joint*, and ποῦς (*pous*), meaning a *foot*. The *Arthropoda* seem at first sight to be made up of jointed rings and feet; hence the name.

The subkingdom of the *Arthropoda* is again subdivided into six classes. These are the following:

Class I. The *Crustacea* (Shrimps, Crabs, Water-fleas, etc.).

Class II. The *Podostomata* (King-crabs, Trilobites [fossil], etc.).

Class III. The *Malacopoda* (*Peripatus*, a curious genus of worm-like creatures, found in the tropics, and allied to the Myriapods in some important respects).

Class IV. The *Myriapoda* (Centipedes, etc.).

Class V. The *Arachnida* (Spiders, Mites, etc.).

Class VI. The *Insecta* (Insects).

That branch of zoölogy which treats of insects is known as entomology.

The *Insecta* have been variously subdivided by different scientific writers, but the following subdivision has much in it to commend it, and will suffice as an outline for the guidance of the advanced student.

CLASS VI. INSECTA (INSECTS PROPER)

HETEROMETABOLA

For the most part undergoing only a partial metamorphosis in the development from the egg to the imago.

ORDERS

1. *Thysanura*.
   Suborders:
   Collembola (Podura, Springtails).
   Symphyla (Scolopendrella).
   Cinura (Bristletails, etc.).
2. *Dermatoptera* (Earwigs).
3. *Pseudoneuroptera*.
   Suborders:
   Mallophaga (Bird-lice).
   Platyptera (Stone-flies, Termites, etc.).
   Odonata (Dragon-flies, etc.).
   Ephemerina (May-flies, etc.).
4. *Neuroptera* (Corydalis, Ant-lion, Caddis-flies, etc.).
5. *Orthoptera* (Cockroach, Mantis, Mole-cricket, Grasshopper, Katydid, etc.).

6. *Hemiptera.*
   Suborders:
   Parasita (Lice).
   Sternorhyncha (Aphids, Mealy Bugs, etc.).
   Homoptera (Cicada, Tree-hoppers, etc.).
   Heteroptera (Ranatra, Belostoma, Water-spiders, Squash-bugs, Bedbugs, etc.)

7. *Coleoptera.*
   Suborders:
   Cryptotetramera (Lady-birds, etc.).
   Cryptopentamera (Leaf-beetles, Longhorns, Weevils, etc.).
   Heteromera (Blister-beetles, Meal-beetles, etc.).
   Pentamera (Fire-flies, Skipjacks, June-bugs, Dung-beetles, Stag-beetles, Rove-beetles, Tiger-beetles, etc.).

METABOLA

Undergoing for the most part a complete metamorphosis from egg, through larva and pupa, to imago.

ORDERS

8. *Aphaniptera* (Fleas).

9. *Diptera.*
   Suborders:
   Orthorhapha (Hessian Flies, Buffalo-gnats, Mosquitos, Crane-flies, Horse-flies).
   Cyclorhapha (Syrphus, Bot-flies, Tsetse, House-flies, etc.).

10. *Lepidoptera.*
    Suborders:
    Rhopalocera (Butterflies).
    Heterocera (Moths).

11. *Hymenoptera.*
    Suborders:
    Terebrantia (Saw-flies, Gall-wasps, Ichneumon-flies, etc.).
    Aculeata (Ants, Cuckoo-flies, Digger-wasps, True Wasps, Bees).

It will be seen by glancing at the foregoing table that the butterflies and moths are included as suborders in the tenth group of the list, to which is applied the name *Lepidoptera*. This word, like most other scientific words, is derived from the Greek, and is compounded of the noun λεπίς (*lepis*), which signifies a *scale*, and the noun πτερόν (*pteron*), which signifies a *wing*. The butterflies and moths together constitute the order of scale-winged insects. The appropriateness of this name will no doubt be at once recognized by every reader, who, having perhaps unintentionally rubbed off some of the minute scales which clothe the wings of a butterfly, has taken the trouble to examine them under a microscope, or who has attentively read what has been said upon this subject in the first chapter of this book. By referring again to the classification which has been given, it will be noted that the last four orders in the list agree in that the creatures included within them undergo for the most part what is known as a complete metamorphosis; that is to say, they pass through four successive stages of development, existing first as eggs, then as worm-like larvæ, or caterpillars, then as pupæ, and finally as perfect, fully developed insects, gifted for the most part with the power of flight, and capable of reproducing their kind. All of this has been to some extent already elucidated in the first chapter of the present volume, but it may be well to remind the reader of these facts at this point.

A question which is frequently asked by those who are not familiar with the subject relates to the manner in which it is possible to distinguish between moths and butterflies. A partial answer can be made in the light of the habits of the two classes of lepidoptera. Butterflies are diurnal in their habits, flying between sunrise and dusk, and very rarely taking the wing at night. This habit is so universal that these insects are frequently called by entomologists "the diurnal lepidoptera," or are simply spoken of as "diurnals." It is, however, true that many species of moths are also diurnal in their habits, though the great majority of them are nocturnal, or crepuscular, that is, flying at the dusk of the evening, or in the twilight of the early morning. Upon the basis of mere habit, then, we are able only to obtain a partial clue to the distinction between the two suborders. A more definite distinction is based upon structure, and specifically upon the structure of the antennæ. Butterflies have long, thread-like antennæ, provided with a swelling at the extremity, giving them a somewhat club-shaped appearance (Fig. 76). This form of an-

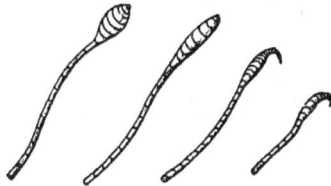

Fig. 76.—Antennæ of butterflies.

tennæ is very unusual among the moths, and only occurs in a few rare genera, found in tropical countries, which seem to represent connecting-links between the butterflies and the moths. All the true moths which are found within the limits of the United

61

States and Canada have antennæ which are not club-shaped, but are of various other forms. Some moths have thread-like antennæ tapering to a fine point; others have feather-shaped antennæ; others still have antennæ which are prismatic in form, and provided with a little hook, or spur, at the end; and there are many modifications and variations of these forms. The club-shaped form of the antennæ of butterflies has led naturalists to call them *Rhopalocera*, as has been already explained in speaking of this subject on page 17. Moths are called *Heterocera*. The word *Heterocera* is compounded of the Greek word ἕτερον (*heteron*), meaning *other*, and the Greek word κέρας (*keras*), meaning a *horn*. They are lepidoptera which have antennæ which are *other than club-shaped*. Besides the distinctions which exist in the matter of the form of the antennæ, there are distinctions in the veins of the wings, and in the manner of carrying them when at rest or in flight, which are quite characteristic of the two groups; but all of these things the attentive student will quickly learn for himself by observation.

Fig. 77.—Antennæ of moths.

Scientific Arrangement.—Having thus cast a passing glance at the differences which exist between moths and butterflies, we take up the question of the subdivision of the butterflies into natural groups. Various systems of arranging butterflies have been suggested from time to time by learned writers, and for a knowledge of these systems the student may consult works which treat of them at length. It is sufficient for beginners, for whom this book is principally written, to observe that in modern science, for purposes of convenience, as well as from regard for essential truth, all individuals are looked upon as belonging to a *species*. A species includes all those individuals, which have a common ancestry, and are so related in form and structure as to be manifestly separable from all other similarly constituted assemblages of individuals. For instance, all the large cats having a tawny skin, and in the male a shaggy mane, constitute a species, which we call the lion; the eagles in the eastern United States, which in adult plumage have a snow-white head and neck and a white tail, constitute a species, which we know as the "white-headed" or "bald-headed" eagle. Species may then be grouped together, and those which are manifestly closely related to one another are regarded as forming a natural assemblage of species, to which we give the name of a *genus*. For example, all the large cats, such as the lion, the tiger, the puma, and the jaguar, are grouped together by naturalists, and form a genus to which is given the Latin name *Felis*, meaning *cat*. The name of the genus always comes before that of the species. Thus the tiger is spoken of scientifically as *Felis tigris*. The genera which are closely related to one another may again be assembled as *sub-families;* and the subfamilies may be united to form *families*. For instance, all the various genera of cats form a family, which is known as the *Felidæ*, or the Cat Family. A group of families constitutes a *suborder* or an *order*. The cats belong to the *Carnivora*, or order of flesh-eating animals.

In zoölogy family names are formed with the termination -*idæ*, and subfamily names with the termination -*inæ*.

Everything just said in regard to the classification of the higher animals applies likewise to butterflies. Let us take as an illustration the common milkweed butterfly. Linnæus for a fanciful reason gave this insect the name *Plexippus*. This is its specific name, by which it is distinguished from all other butter-flies. It belongs to the genus *Anosia*. The genus *Anosia* is one of the genera which make up the subfamily of the *Euplœinæ*. The *Euplœinæ* belong to the great family of the *Nymphalidæ*. The *Nymphalidæ* are a part of the suborder of the *Rhopalocera*, or true butterflies, one of the two great subdivisions of the order *Lepidoptera*, belonging to the great class *Insecta*, the highest class in the subkingdom of the *Arthropoda*. The matter may be represented in a tabular form, in the reverse order from that which has been given:

Subkingdom, *Arthropoda*.
Class, *Insecta*.
Order, *Lepidoptera*.
Suborder, *Rhopalocera*.
Family, *Nymphalidæ*.
Subfamily, *Euplœinæ*.
Genus, *Anosia*.
Species, *Plexippus* (Milkweed Butterfly).

63

Varieties.—A still further subdivision is in some cases recognized as necessary. A species which has a wide range over an extensive territory may vary in different parts of the territory within which it is found. The butterflies of certain common European species are found also in Japan and Corea, but, as a rule, they are much larger in the latter countries than they are in Europe, and in some cases more brightly colored. Naturalists have therefore distinguished the Asiatic from the European form by giving the former what is known as a varietal name. Similar differences occur among butterflies on the continent of North America. The great yellow and black-barred swallowtail butterfly known as *Papilio turnus* occurs from Florida to Alaska. But the specimens from Alaska are always much smaller than those from other regions, and have a very dwarfed appearance. This dwarfed form constitutes what is known as a local race, or variety, of the species. The members of a species which occur upon an island frequently differ in marked respects from specimens which occur upon the adjacent mainland. By insulation and the process of through-breeding the creature has come to acquire characteristics which separate it in a marked degree from the closely allied continental form, and yet not sufficiently to justify us in treating it as a distinct species. It represents what is known as an insular race, or variety, and we give it therefore a varietal name. Naturalists also distinguish between seasonal, dimorphic, melanic, and albino forms. Names descriptive or designatory of these forms are frequently applied to them. All of this will become plainer in the course of the study of the succeeding pages, and in the effort to classify specimens which the student will make.

Sex.—The designation of the sex is important in the case of all well-ordered collections of zoölogical specimens. As a measure of convenience, the male is usually indicated by the sign of Mars, ♂, while the female is indicated by the sign of Venus, ♀. The inscription, "*Argynnis Diana,* ♂," therefore means that the specimen is a male of *Argynnis Diana*, and the inscription, "*Argynnis Diana,* ♀," means that the specimen is a female of the same species. These signs are invariably employed by naturalists to mark the sexes.

The Division of Butterflies into Families.—Without attempting to go deeply into questions of classification at the present point,

64

it will be well for us to note the subdivisions which have been made into the larger groups, known as families, and to show how butterflies belonging to one or the other of these may be distinguished from one another. There are five of these families represented within the territory of which this book takes notice. These five families are the following:

1. The NYMPHALIDÆ, or "Brush-footed Butterflies."
2. The LEMONIIDÆ, or "Metal-marks."
3. The LYCÆNIDÆ, or "Blues," "Coppers," and "Hair-streaks."
4. The PAPILIONIDÆ, or the "Swallowtails" and their allies.
5. The HESPERIIDÆ, or the "Skippers."

The NYMPHALIDÆ, the "Brush-footed Butterflies."

The butterflies of this family may be distinguished as a great class from all other butterflies by the fact that *in both sexes the first, or prothoracic, pair of legs is greatly dwarfed, useless for walking, and therefore carried folded up against the breast.* From this peculiarity they have also been called the "Four-footed Butterflies." This is the largest of all the families of the butterflies, and has been subdivided into many subfamilies. Some of the genera are composed of small species, but most of the genera are made up of medium-sized or large species. The family is geologically very ancient, and most of the fossil butterflies which have been discovered belong to it. The caterpillars are in most of the subfamilies provided with horny or fleshy projections. The chrysalids always hang suspended by the tail.

The LEMONIIDÆ, the "Metal-marks."

This family is distinguished from others by the fact that *the males have four ambulatory or walking feet, while the females have six such feet. The antennæ are relatively longer than in the Lycænidæ.* The butterflies belonging to this great group are mostly confined to the tropics of the New World, and only a few genera and species are included in the region covered by this volume. They are usually quite small, but are colored in a bright and odd manner, spots and checkered markings being very common. Many are extremely brilliant in their colors. *The caterpillars are small and contracted. Some are said to have chrysalids which are suspended; others have chrysalids girdled and attached at the anal extremity, like the Lycænidæ. The butterflies in many genera have the habit of alighting on the under side of leaves, with their wings expanded.*

65

The LYCÆNIDÆ, the " Gossamer-winged Butterflies."

This great family comprises the butterflies which are familiarly known as the "hair-streaks," the "blues," and the "coppers." *The males have four and the females six walking feet. The caterpillars are small, short, and slug-shaped. The chrysalids are provided with a girdle, are attached at the end of the abdomen, and lie closely appressed to the surface upon which they have undergone transformation.* Blue is a very common color in this family, which includes some of the gayest of the small forms which are found in the butterfly world. *In alighting they always carry their wings folded together and upright.*

The PAPILIONIDÆ, the "Swallowtails" and their allies.

These butterflies *have six walking feet in both sexes. The caterpillars are elongate, and in some genera provided with osmateria, or protrusive organs secreting a powerful and disagreeable odor. The chrysalids are elongate, attached at the anal extremity, and held in place by a girdle of silk, but not closely appressed to the surface upon which they have undergone transformation.*

The HESPERIIDÆ, or the "Skippers."

They are generally *small in size, with stout bodies, very quick and powerful in flight. They have six walking feet in both sexes. The tibiæ of the hind feet, with few exceptions, have spurs. The caterpillars are cylindrical, smooth, tapering forward and backward from the middle, and generally having large globular heads. For the most part they undergo transformation into chrysalids which have a girdle and an anal hook, or cremaster, in a loose cocoon, composed of a few threads of silk,* and thus approximate the moths in their habits. The genus *Megathymus* has the curious habit of burrowing in its larval stage in the underground stems of the yucca.

To one or the other of these five families all the butterflies, numbering about six hundred and fifty species, which are found from the Rio Grande of Texas to the arctic circle, can be referred.

Scientific Names. —From what has been said it is plain to the reader that the student of this delightful branch of science is certain to be called upon to use some rather long and, at first sight, uncouth words in the pursuit of the subject. But experience, that best of teachers, will soon enable him to master any little difficulties which may arise from this source, and he will come finally to recognize how useful these terms are in designating distinctions which exist, but which are often wholly overlooked by the uneducated and unobservant. It is not, however, necessary that the student should at the outset attempt to tax his memory with all of the long scientific names which he encounters in this and similar books. The late Dr. Horn of Philadelphia, who was justly regarded, during the latter years of his life, as the most eminent student of the *Coleoptera*, or beetles, of North America, once said to the writer that he made it a religious duty not to try to remember all the long scientific names belonging to the thousands of species in his collection, but was content to have them attached to the pins holding the specimens in his cabinets, where he could easily refer to them. The student who is engaged in collecting and studying butterflies will very soon come, almost without effort, to know their names, but it is not a sin to forget them.

In writing about butterflies it is quite customary to abbreviate the generic name by giving merely its initial. Thus in writing about the milkweed butterfly, *Anosia plexippus*, the naturalist will designate it as "*A. plexippus.*" To the specific name he will also attach the name of the man who gave this specific name to the insect. As Linnæus was the first to name this insect, it is proper to add his name, when writing of it, or to add an abbreviation of his name, as follows: "*A. plexippus*, Linnæus," or "Linn." In speaking about butterflies it is quite common to omit the generic name altogether and to use only the specific name. Thus after returning in the evening from a collecting-trip, I might say, "I was quite successful to-day. I took twenty *Aphrodites*, four *Myrinas*, and two specimens of *Atlantis.*" In this case there could be no misunderstanding of my meaning. I took specimens of three species of the genus *Argynnis—A. aphrodite, A. myrina,* and *A. atlantis;* but it is quite enough to designate them by the specific names, without reference to their generic classification.

Synonyms.—It is a law among scientific men that the name first given to an animal or plant shall be its name and shall have priority over all other names. Now, it has happened not infrequently that an author, not knowing that a species has been described already, has redescribed it under another name. Such a name applied a second time to a species already described is called a *synonym*, and may be published after the true name. Sometimes species have had a dozen or more different names

67

applied to them by different writers, but all such names rank as synonyms according to the law of priority.

Popular Names.—Common English names for butterflies are much in vogue in England and Scotland, and there is no reason why English names should not be given to butterflies, as well as to birds and to plants. In the following pages this has been done to a great extent. I have used the names coined by Dr. S. H. Scudder and by others, so far as possible, and have in other cases been forced myself to coin names which seemed to be appropriate, in the hope that they may come ultimately to be widely used. The trouble is that ordinary people do not take pains to observe and note the distinctions which exist among the lower animals. The vocabulary of the common farmer, or even of the ordinary professional man, is bare of terms to point out correctly the different things which come under the eye. All insects are "bugs" to the vulgar, and even the airy butterfly, creature of grace and light, is put into the same category with roaches and fleas. Apropos of the tendency to classify as "bugs" all things which creep and are small, it may be worth while to recall the story, which Frank Buckland tells in his "Log-book of a Fisherman and Naturalist," of an adventure which he had, when a school-boy, at the booking-office of the London, Chatham, and Dover Railway Company in Dover. He had been for a short trip to Paris, and had bought a monkey and a tortoise. Upon his return from sunny France, as he was getting his ticket up to London, Jocko stuck his head out of the bag in which his owner was carrying him. The ticket-agent looked down and said, "You will pay half-fare for him." "How is that?" exclaimed young Buckland. "Well, we charge half-fare for dogs." "But this is not a dog," replied the indignant lad; "this is a monkey." "Makes no difference," was the answer; "you must pay half-fare for him." Reluctantly the silver was laid upon the counter. Then, thrusting his hands into the pocket of his greatcoat, Buckland drew forth the tortoise, and, laying it down, asked, "How much do you charge for this?" The ancient receiver of fares furbished his spectacles, adjusted them to his nose, took a long look, and replied, "We don't charge nothin' for them; them 's insects." It is to be hoped that the reader of this book will in the end have a clearer view of facts as to the classification of animals than was possessed by the ticket-agent at Dover.

CHAPTER IV

Early Writers.—The earliest descriptions of North American butterflies are found in writings which are now almost unknown, except to the close student of science. Linnæus described and named a number of the commoner North American species, and some of them were figured by Charles Clerck, his pupil, whose work entitled "Icones" was published at Stockholm in the year 1759. Clerck's work is exceedingly rare, and the writer believes that he has in his possession the only copy in North America. Johann Christian Fabricius, a pupil of Linnæus, who was for some time a professor in Kiel, and attached to the court of the King of Denmark, published between the year 1775 and the year 1798 a number of works upon the general subject of entomology, in which he gave descriptions, very brief and unsatisfactory, of a number of North American species. His descriptions were written, as were those of Linnæus, in the Latin language. About the same time that Fabricius was publishing his works, Peter Cramer, a Dutchman, was engaged in giving to the world the four large quartos in which he endeavored to figure and describe the butterflies and moths of Asia, Africa, and America. Cramer's work was entitled "Papillons Exotiques," and contained recognizable illustrations of quite a number of the North American forms. The book, however, is rare and expensive to-day, but few copies of it being accessible to American students.

Jacob Hübner, who was born at Augsburg in the year 1761, undertook the publication, in the early part of the present century, of an elaborate work upon the European butterflies and moths, parallel with which he undertook a publication upon the butterflies and moths of foreign lands. The title of his work is "Samm-

69

lung Exotischer Schmetterlinge." To this work was added, as an appendix, partly by Hübner and partly by his successor and co-laborer, Karl Geyer, another, entitled "Zuträge zur Sammlung Exotischer Schmetterlinge." The two works together are illustrated by six hundred and sixty-four colored plates. This great publication contains some scattered figures of North American species. A good copy sells for from three hundred and fifty to four hundred dollars, or even more.

The first work which was devoted exclusively to an account of the lepidoptera of North America was published in England by Sir James Edward Smith, who was a botanist, and who gave to the world in two volumes some of the plates which had been drawn by John Abbot, an Englishman who lived for a number of years in Georgia. The work appeared in two folio volumes, bearing the date 1797. It is entitled "The Natural History of the Rarer Lepidopterous Insects of Georgia." It contains one hundred and four plates, in which the insects are represented in their various stages upon their appropriate food-plants. Smith and Abbot's work contains original descriptions of only about half a dozen of the North American butterflies, and figures a number of species which had been already described by earlier authors. It is mainly devoted to the moths. This work is now rare and commands a very high price.

The next important work upon the subject was published by Dr. J. A. Boisduval of Paris, a celebrated entomologist, who was assisted by Major John E. Leconte. The work appeared in the year 1833, and is entitled "Histoire Générale et Monographie des Lepidoptères et des Chenilles de l'Amérique Septentrionale." It contains seventy-eight colored plates, each representing butterflies of North America, in many cases giving figures of the larva and the chrysalis as well as of the perfect insect. The plates were based very largely upon drawings made by John Abbot, and represent ninety-three species, while in the text there are only eighty-five species mentioned, some of which are not figured. What has been said of all the preceding works is also true of this: it is very rarely offered for sale, can only be found upon occasion, and commands a high price.

In the year 1841 Dr. Thaddeus William Harris published "A Report on the Insects of Massachusetts which are Injurious to Vegetation." This work, which was originally brought out in pursuance of an order of the legislature of Massachusetts, by the Commissioners of the Zoölogical and Botanical Survey of the State, was republished in 1842, and was followed by a third edition in 1852. The last edition, revised and improved by Charles L. Flint, Secretary of the Massachusetts State Board of Agriculture, appeared in 1862. This work contains a number of figures and descriptions of the butterflies of New England, and, while now somewhat obsolete, still contains a great deal of valuable information, and is well worth being rescued by the student from the shelves of the second-hand book-stalls in which it is now and then to be found. For the New England student of entomology it remains to a greater or less extent a classic.

In 1860 the Smithsonian Institution published a "Catalogue of the Described Lepidoptera of North America," a compilation prepared by the Rev. John G. Morris. This work, though very far from complete, contains in a compact form much valuable information, largely extracted from the writings of previous authors. It is not illustrated.

With the book prepared by Dr. Morris the first period in the development of a literature relating to our subject may be said to close, and the reader will observe that until the end of the sixth decade of this century very little had been attempted in the way of systematically naming, describing, and illustrating the riches of the insect fauna of this continent. Almost all the work, with the exception of that done by Harris, Leconte, and Morris, had been done by European authors.

Later Writers. — At the close of the Civil War this country witnessed a great intellectual awakening, and every department of science began to find its zealous students. In the annals of entomology the year 1868 is memorable because of the issue of the first part of the great work by William H. Edwards, entitled "The Butterflies of North America." This work has been within the last year (1897) brought to completion with the publication of the third volume, and stands as a superb monument to the scientific attainments and the inextinguishable industry of its learned author. The three volumes are most superbly illustrated, and contain a wealth of original drawings, representing all the stages in the life-history of numerous species, which has never been surpassed. Unfortunately, while including a large number of the species known to inhabit North America, the book is nevertheless

71

not what its title would seem to imply, and is far from complete, several hundreds of species not being represented in any way, either in the text or in the illustrations. In spite of this fact it will remain to the American student a classic, holding a place in the domain of entomology analogous to that which is held in the science of ornithology by the "Birds of America," by Audubon.

A work even more elaborate in its design and execution, contained in three volumes, is "The Butterflies of New England," by Dr. Samuel Hubbard Scudder, published in the year 1886. No more superbly illustrated and exhaustive monograph on any scientific subject has ever been published than this, and it must remain a lasting memorial of the colossal industry and vast learning of the author, one of the most eminent scientific men whom America has produced.

While the two great works which have been mentioned have illustrated to the highest degree not only the learning of their authors, but the vast advances which have been made in the art of illustration within the last thirty years, they do not stand alone as representing the activity of students in this field. A number of smaller, but useful, works have appeared from time to time. Among these must be mentioned "The Butterflies of the Eastern United States," by Professor G. H. French. This book, which contains four hundred and two pages and ninety-three figures in the text, was published in Philadelphia in 1886. It is an admirable little work, with the help of which the student may learn much in relation to the subject; but it greatly lacks in illustration, without which all such publications are not attractive or thoroughly useful to the student. In the same year appeared "The Butterflies of New England," by C. J. Maynard, a quarto containing seventy-two pages of text and eight colored plates, the latter very poor. In 1878 Herman Strecker of Reading, Pennsylvania, published a book entitled "Butterflies and Moths of North America," which is further entitled "A Complete Synonymical Catalogue." It gives only the synonymy of some four hundred and seventy species of butterflies, and has never been continued by the author, as was apparently his intention. It makes no mention of the moths, except upon the title-page. For the scientific student it has much value, but is of no value to a beginner. The same author published in parts a work illus-

trated by fifteen colored plates, entitled "Lepidoptera—Rhopalo-
ceres and Heteroceres—Indigenous and Exotic," which came out
from 1872 to 1879, and contains recognizable figures of many
North American species.

In 1891 there appeared in Boston, from the pen of C. J. Maynard, a work entitled "A Manual of North American Butterflies."
This is illustrated by ten very poorly executed plates and a number of equally poorly executed cuts in the text. The work is
unfortunately characterized by a number of serious defects which
make its use difficult and unsatisfactory for the correct determination of species and their classification.

In 1893 Dr. Scudder published two books, both of them useful, though brief, one of them entitled "The Life of a Butterfly,"
the other, "A Brief Guide to the Commoner Butterflies of the
Northern United States and Canada." Both of these books were
published in New York by Messrs. Henry Holt & Co., and contain valuable information in relation to the subject, being to a
certain extent an advance upon another work published in 1881
by the same author and firm, entitled "Butterflies."

Periodical Literature.—The reader must not suppose that the
only literature relating to the subject that we are considering is to
be found in the volumes that have been mentioned. The original
descriptions and the life-histories of a large number of the species
of the butterflies of North America have originally appeared in the
pages of scientific periodicals and in the journals and proceedings
of different learned societies. Among the more important publications which are rich in information in regard to our theme
may be mentioned the publications relating to entomology issued
by the United States National Museum, the United States Department of Agriculture, and by the various American commonwealths,
chief among the latter being Riley's "Missouri Reports." Exceedingly valuable are many of the papers contained in the
" Transactions of the American Entomological Society," " Psyche,"
the "Bulletin of the Brooklyn Entomological Society" (1872–
85), "Papilio" (1881–84), "Entomologica Americana" (1885–90),
the "Journal of the New York Entomological Society," the
"Canadian Entomologist," and "Entomological News." All of
these journals are mines of original information, and the student
who proposes to master the subject thoroughly will do well to
obtain, if possible, complete sets of these periodicals, as well as

of a number of others which might be mentioned, and to subscribe for such of them as are still being published.

There are a number of works upon general entomology, containing chapters upon the diurnal lepidoptera, which may be consulted with profit. Among the best of these are the following: " A Guide to the Study of Insects," by A. S. Packard, Jr., M. D. (Henry Holt & Co., New York, 1883, pp. 715, 8vo); " A Textbook of Entomology," by Alpheus S. Packard, M. D., etc. (The Macmillan Company, New York, 1898, pp. 729, 8vo); " A Manual for the Study of Insects," by John Henry Comstock (Comstock Publishing Company, Ithaca, New York, 1895, pp. 701, 8vo).

HUGO'S "FLOWER TO BUTTERFLY"

" Sweet, live with me, and let my love
    Be an enduring tether;
  Oh, wanton not from spot to spot,
    But let us dwell together.

" You 've come each morn to sip the sweets
    With which you found me dripping,
  Yet never knew it was not dew,
    But tears, that you were sipping.

" You gambol over honey meads
    Where siren bees are humming;
  But mine the fate to watch and wait
    For my beloved's coming.

" The sunshine that delights you now
    Shall fade to darkness gloomy;
  You should not fear if, biding here,
    You nestled closer to me.

" So rest you, love, and be my love,
    That my enraptured blooming
  May fill your sight with tender light,
    Your wings with sweet perfuming.

" Or, if you will not bide with me
    Upon this quiet heather,
  Oh, give me wing, thou beauteous thing,
    That we may soar together."

EUGENE FIELD.

74

THE BUTTERFLIES

OF

NORTH AMERICA NORTH OF MEXICO

"Lo, the bright train their radiant wings unfold!
With silver fringed, and freckled o'er with gold:
On the gay bosom of some fragrant flower
They, idly fluttering, live their little hour;
Their life all pleasure, and their task all play,
All spring their age, and sunshine all their day."

MRS. BARBAULD.

ORDER LEPIDOPTERA

SUBORDER RHOPALOCERA (BUTTERFLIES)

FAMILY I

NYMPHALIDÆ (THE BRUSH-FOOTED BUTTERFLIES)

THE family of the Nymphalidæ is composed of butterflies which are of medium and large size, though a few of the genera are made up of species which are quite small. They may be distinguished from all other butterflies by the fact that the first pair of legs in both sexes is atrophied or greatly reduced in size, so that they cannot be used in walking, but are carried folded up upon the breast. The fore feet, except in the case of the female of the snout-butterflies (Libytheinæ), are without tarsal claws, and hence the name " Brush-footed Butterflies " has been applied to them. As the anterior pair of legs is apparently useless, they have been called "The Four-footed Butterflies," which is scientifically a misnomer.

Egg.—The eggs of the Nymphalidæ, for the most part, are dome-shaped or globular, and are marked with raised longitudinal lines extending from the summit toward the base over the entire surface or over the upper portion of the egg. Between these elevations are often found finer and less elevated cross-lines. In a few genera the surface of the eggs is covered with reticulations arranged in geometrical patterns (see Fig. 1).

Caterpillar.—The caterpillars of the Nymphalidæ, as they emerge from the egg, have heads the diameter of which is larger than that of the body, and covered with a number of wart-like

77

elevations from which hairs arise. The body of the immature larva generally tapers from before backward (see Plate III, Figs. 7 and 11). The mature larva is cylindrical in form, sometimes, as in the Satyrinæ, thicker in the middle. Often one or more of the segments are greatly swollen in whole or in part. The larvæ are generally ornamented with fleshy projections or branching spines.

Chrysalids.—The chrysalids are for the most part angular, and often have strongly marked projections. As a rule, they hang with the head downward, having the cremaster, or anal hook, attached to a button of silk woven to the under surface of a limb of a tree, a stone, or some other projecting surface. A few boreal species construct loose coverings of threads of silk at the roots of grasses, and here undergo their transformations. The chrysalids are frequently ornamented with golden or silvery spots.

This is the largest of all the families of butterflies, and it is also the most widely distributed. It is represented by species which have their abode in the cold regions of the far North and upon the lofty summits of mountains, where summer reigns for but a few weeks during the year; and it is enormously developed in equatorial lands, including here some of the most gloriously colored species in the butterfly world. But although these insects appear to have attained their most superb development in the tropics, they are more numerous in the temperate regions than other butterflies, and a certain fearlessness, and fondness for the haunts of men, which seems to characterize some of them, has brought them more under the eyes of observers. The literature of poetry and prose which takes account of the life of the butterfly has mainly dealt with forms belonging to this great assemblage of species.

In the classification of the brush-footed butterflies various subdivisions have been suggested by learned authors, but the species found in the United States and the countries lying northward upon the continent may be all included in the following six groups, or subfamilies:

1. The *Euplœinæ*, the Euplœids.
2. The *Ithomiinæ*, the Ithomiids.
3. The *Heliconiinæ*, the Heliconians.
4. The *Nymphalinæ*, the Nymphs.
5. The *Satyrinæ*, the Satyrs.
6. The *Libytheinæ*, the Snout-butterflies.

78

The insects belonging to these different subfamilies may be distinguished by the help of the following analytical table, which is based upon that of Professor Comstock, given in his "Manual for the Study of Insects" (p. 396), which in turn is based upon that of Dr. Scudder, in "The Butterflies of New England" (vol. i, p. 115).

KEY TO THE SUBFAMILIES OF THE NYMPHALIDÆ OF THE UNITED STATES AND CANADA

I. With the veins of the fore wings not greatly swollen at the base.
    A. Antennæ naked.
        (*a*) Fore wings less than twice as long as broad—*Euplœinæ*.
        (*b*) Fore wings twice as long as broad and often translucent, the abdomen extending far beyond the inner margin of the hind wings —*Ithomiinæ*.
    B. Antennæ clothed with scales, at least above.
        (*a*) Fore wings at least twice as long as broad—*Heliconiinæ*.
        (*b*) Fore wings less than twice as long as broad.
          1. Palpi not as long as the thorax—*Nymphalinæ*.
          2. Palpi much longer than the thorax—*Libytheinæ*.

II. With some of the veins of the fore wings greatly swollen at the base—*Satyrinæ*.

We now proceed to present the various genera and species of this family which occur within the territorial limits of which this book treats. The reader will do well to accompany the study of the descriptions, which are at most mere sketches, by a careful examination of the figures in the plates. In this way a very clear idea of the different species can in most instances be obtained. But with the study of the book should always go, if possible, the study of the living things themselves. Knowledge of nature founded upon books is at best second-hand. To the fields and the woods, then, net in hand! Splendid as may be the sight of a great collection of butterflies from all parts of the world, their wings

    "Gleaming with purple and gold,"

no vision is so exquisite and so inspiring as that which greets the true aurelian as in shady dell or upon sun-lit upland, with the blue sky above him and the flowers all around him, he pursues his pleasant, self-imposed tasks, drinking in health at every step.

SUBFAMILY EUPLŒINÆ (THE MILKWEED BUTTERFLIES)

" Lazily flying
Over the flower-decked prairies, West;
Basking in sunshine till daylight is dying,
And resting all night on Asclepias' breast;
Joyously dancing,
Merrily prancing,
Chasing his lady-love high in the air,
Fluttering gaily,
Frolicking daily,
Free from anxiety, sorrow, and care! "
C. V. RILEY.

Butterfly. — Large butterflies; head large; the antennæ inserted on the summit, stout, naked, that is to say, not covered with scales, the club long and not broad; palpi stout; the thorax somewhat compressed, with the top arched. The abdomen is moderately stout, bearing on the eighth segment, on either side, in the case of the male, clasps which are quite conspicuous. The fore wings are greatly produced at the apex and more or less excavated about the middle of the outer border; the hind wings are rounded and generally much smaller than the fore wings; the outer margin is regular, without tails, and the inner margin is sometimes channeled so as to enfold the abdomen. The fore legs are greatly atrophied in the male, less so in the female; these atrophied legs are not provided with claws, but on the other legs the claws are well developed.

Egg. — The eggs are ovate conical, broadly flattened at the base and slightly truncated at the top, with many longitudinal ribs and transverse cross-ridges (see Fig. 4).

Caterpillar. — On emerging from the chrysalis the head is not larger than the body; the body has a few scattered hairs on each

80

EXPLANATION OF PLATE VII

ANOSIA AND BASILARCHIA

1. *Anosia plexippus*, Linnæus, ♂.   3. *Anosia berenice*, var. *strigosa*, Bates, ♀.
2. *Anosia berenice*, Cramer, ♂.   4. *Basilarchia disippus*, Godart, ♂.
   5. *Basilarchia hulsti*, Edwards, ♂.

PLATE VII.

segment. On reaching maturity the head is small, the body large, cylindrical, without hair, and conspicuously banded with dark stripes upon a lighter ground, and on some of the segments there are generally erect fleshy processes of considerable length (see Fig. 16). The caterpillars feed upon different species of the milkweed (*Asclepias*).

Chrysalis.—The chrysalis is relatively short and thick, rounded, with very few projections, tapers very rapidly over the posterior part of the abdomen, and is suspended by a long cremaster from a button of silk (see Fig. 24). The chrysalis is frequently ornamented with golden or silver spots.

This subfamily reaches its largest development in the tropical regions of Asia. Only one genus is represented in our fauna, the genus *Anosia*.

Genus ANOSIA, Hübner

Butterfly.—Large-sized butterflies; fore wings long, greatly produced at the apex, having a triangular outline, the outer margin approximately as long as the inner margin; the costal border is regularly bowed; the outer border is slightly excavated, the outer angle rounded; the hind wings are well rounded, the costal border projecting just at the base, the inner margin likewise projecting at the base and depressed so as to form a channel clasping the abdomen. On the edge of the first median nervule of the male, about its middle, there is a scent-pouch covered with scales.

Egg.—The egg is ovate conical, ribbed perpendicularly with many raised cross-lines between the ridges. The eggs are pale green in color.

Fig. 78.—Neuration of the genus *Anosia*.

Caterpillar.—The caterpillar is cylindrical, fleshy, transversely wrinkled, and has on the second thoracic and eighth abdominal segment pairs of very long and slender fleshy filaments; the body is ornamented by dark bands upon a greenish-yellow ground-color; the filaments are black.

81

Chrysalis.—The chrysalis is stout, cylindrical, rapidly taper-ing on the abdomen, and is suspended from a button of silk by a long cremaster. The color of the chrysalis is pale green, orna-mented with golden spots.

The larvæ of the genus *Anosia* feed for the most part upon the varieties of milkweed (*Asclepias*), and they are therefore called "milkweed butterflies." There are two species of the genus found in our fauna, one, *Anosia plexippus*, Linnæus, which is distributed over the entire continent as far north as southern Canada, and the other, *Anosia berenice*, Cramer, which is con-fined to the extreme southwestern portions of the United States, being found in Texas and Arizona.

(1) **Anosia plexippus**, Linnæus, Plate VII, Fig. 1, ♂ (The Monarch).

Butterfly.—The upper surface of the wings of this butterfly is bright reddish, with the borders and veins broadly black, with two rows of white spots on the outer borders and two rows of pale spots of moderately large size across the apex of the fore wings. The males have the wings less broadly bordered with black than the females, and on the first median nervule of the hind wings there is a black scent-pouch.

Egg.—The egg is ovate conical, and is well represented in Fig. 4 in the introductory chapter of this book.

Caterpillar.—The caterpillar is bright yellow or greenish-yel-low, banded with shining black, and furnished with black fleshy thread-like appendages before and behind. It likewise is well delineated in Fig. 16, as well as in Plate III, Fig. 5.

Chrysalis.—The chrysalis is about an inch in length, pale green, spotted with gold (see Fig. 24, and Plate IV, Figs. 1-3).

The butterfly is believed to be polygoneutic, that is to say, many broods are produced annually; and it is believed by writers that with the advent of cold weather these butterflies migrate to the South, the chrysalids and caterpillars which may be un-developed at the time of the frosts are destroyed, and that when these insects reappear, as they do every summer, they represent a wave of migration coming northward from the warmer regions of the Gulf States. It is not believed that any of them hibernate in any stage of their existence. This insect sometimes appears in great swarms on the eastern and southern coasts of New Jersey in late autumn. The swarms pressing southward are arrested by the ocean. The writer has seen stunted trees on the New Jersey coast in the middle of October, when the foliage has already fallen, so completely covered with clinging masses of these butterflies as to present the appearance of trees in full leaf (Fig. 79).

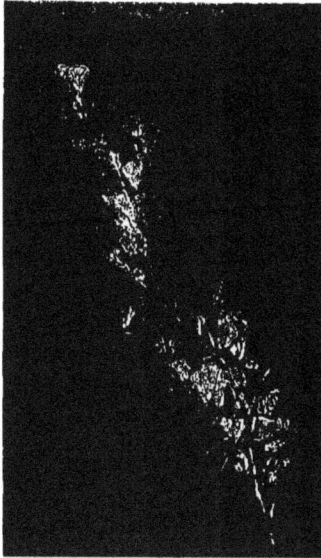

FIG. 79.—Swarms of milkweed butterflies resting on a tree. Photographed at night by Professor C. F. Nachtrieb. (From "Insect Life," vol. v, p. 206, by special permission of the United States Department of Agriculture.)

This butterfly is a great migrant, and within quite recent years, with Yankee instinct, has crossed the Pacific, probably on merchant vessels, the chrysalids being possibly concealed in bales of hay, and has found lodgment in Australia, where it has greatly multiplied in the warmer parts of the Island Continent, and has thence spread northward and westward, until in its migrations it has reached Java and Sumatra, and long ago took possession of the Philippines. Moving eastward on the lines of travel, it has established a more or less precarious foothold for itself in southern England, as many as two or three dozen of these butterflies having been taken in a single year in the United Kingdom. It is well established at the Cape Verde Islands, and in a short time we may expect to hear of it as having taken possession of the continent of Africa, in which the family of plants upon which the caterpillars feed is well represented.

(2) **Anosia berenice**, Cramer, Plate VII, Fig. 2, ♂ (The Queen).

This butterfly is smaller than the Monarch, and the ground-color of the wings is a livid brown. The markings are somewhat similar to those in *A. plexippus*, but the black borders of the hind wings are relatively wider, and the light spots on the apex of the fore wings are whiter and differently located, as may be learned from the figures given in Plate VII.

There is a variety of this species, which has been called **Anosia strigosa** by H. W. Bates (Plate VII, Fig. 3, ♂), which differs only in that on the upper surface of the hind wings the veins as far as the black outer margin are narrowly edged with grayish-white, giving them a streaked appearance. This insect is found in Texas, Arizona, and southern New Mexico.

All of the Euplœinæ are "protected" insects, being by nature provided with secretions which are distasteful to birds and predaceous insects. These acrid secretions are probably due to the character of the plants upon which the caterpillars feed, for many of them eat plants which are more or less rank, and some of them even poisonous to the higher orders of animals. Enjoying on this account immunity from attack, they have all, in the process of time, been mimicked by species in other genera which have not the same immunity. This protective resemblance is well illustrated in Plate VII. The three upper figures in the plate represent, as we have seen, species of the genus *Anosia;* the two lower figures represent two species of the genus *Basilarchia*. Fig. 4 is the male of *B. disippus*, a very common species in the northern United States, which mimicks the Monarch. Fig. 5 represents the same sex of *B. hulstii*, a species which is found in Arizona, and there flies in company with the Queen, and its variety, *A. strigosa*, which latter it more nearly resembles.

SUBFAMILY ITHOMIINÆ (THE LONG-WINGS)

"There be Infects with little hornes proaking out before their eyes, but weak and tender they be, and good for nothing; as the Butterflies."—PLINY, PHILEMON HOLLAND'S Translation.

Butterfly.—This subfamily is composed for the most part of species of moderate size, though a few are quite large. The fore wings are invariably greatly lengthened and are generally at least twice as long as broad. The hind wings are relatively small, rounded, and without tails. The wings in many of the genera are transparent. The extremity of the abdomen in both sexes extends far beyond the margin of the hind wings, but in the female not so much as in the male. The antennæ are not clothed with scales, and are very long and slender, with the club also long and slender, gradually thickening to the tip, which is often drooping. The fore legs are greatly atrophied in the males, the tibia and tarsi in this sex being reduced to a minute knob-like appendage, but being more strongly developed in the females.

The life-history of none of the species reputed to be found in our fauna has been carefully worked out. The larvæ are smooth, covered in most genera with longitudinal rows of conical prominences.

The chrysalids are said to show a likeness to those of the Euplœinæ, being short, thick, and marked with golden spots. Some authors are inclined to view this subfamily as merely constituting a section of the Euplœinæ. The insects are, however, so widely unlike the true Euplœinæ that it seems well to keep them separate in our system of classification. In appearance they approach the Heliconians more nearly than the Euplœids. Ithomiid butterflies swarm in the tropics of the New World, and several hundreds of species are known to inhabit the hot lands of Central and South America. But one genus is found in the Old World, *Hamadryas*, confined to the Australian region. They are

85

protected like the Euplœids and the Heliconians. In flight they are said to somewhat resemble the dragon-flies of the genus *Agrion*, their narrow wings, greatly elongated bodies, and slow, flitting motion recalling these insects, which are known by schoolboys as "darning-needles."

Three genera are said to be represented in the extreme southwestern portion of the United States. I myself have never received specimens of any of them which indisputably came from localities within our limits, and no such specimens are found in the great collection of Mr. W. H. Edwards, which is now in my possession. A paratype of Reakirt's species, *Mechanitis californica*, is contained in the collection of Theodore L. Mead, which I also possess. Mr. Mead obtained it from Herman Strecker of Reading, Pennsylvania. Reakirt gives Los Angeles as the locality from which his type came; but whether he was right in this is open to question, inasmuch, so far as is known, the species has not been found in that neighborhood since described by Reakirt.

Genus MECHANITIS, Fabricius

Butterfly.—Butterflies of moderate size, with the fore wings greatly produced, the inner margin bowed out just beyond the base, and deeply excavated between this projection and the inner angle. The lower discocellular vein in the hind wings is apparently continuous with the median vein, and the lower radial vein being parallel with the median nervules, the median vein has in consequence the appearance of being four-branched. The submedian vein of the fore wings is forked at the base. The costal margin of the hind wings is clothed with tufted erect hairs in the male sex. The fore legs of the male are greatly atrophied, the tarsi and the tibia being fused and reduced to a small knob-like appendage. The fore legs of the female are also greatly reduced, but the tarsi and tibia are still recognizable as slender, thread-like organs.

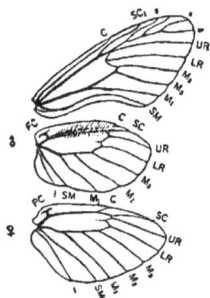

Fig. 80.—Neuration of the genus *Mechanitis*. The letters refer to the names of the veins. (See Fig. 40.)

86

The caterpillars are smooth, cylindrical, ornamented with rows of short fleshy projections.

The chrysalids are short and stout, suspended, and marked with golden spots.

There are numerous species belonging to this genus, all natives of tropical America. The only species said to be found within the limits of the United States occurs, if at all, in southern California. It is, however, probably only found in the lower peninsula of California, which is Mexican territory. No examples from Upper California are known to the writer.

(1) **Mechanitis californica**, Reakirt, Plate VIII, Fig. 2, ♂ (The Californian Long-wing).

The original description given by Reakirt in the " Proceedings of the Entomological Society of Philadelphia," vol. v, p. 223, is as follows:

" Expanse, 2.45–2.56 inches. Fore wing above, brownish-black; a basal streak over the median nervure, and two rounded spots near the inner angle, orange-tawny; of these the outer is the largest, sometimes the inner is yellow, and sometimes both are nearly obsolete; a spot across the cell near its termination, much narrower than in *M. isthmia*, and in one example reduced to a mere dot on the median nervure; a more or less interrupted belt across the wing from the costa to near the middle of the outer margin, and an oblong subapical spot, yellow; in the specimen just mentioned there is an additional yellow spot below the medio-central veinlet.

" Beneath the same, suffused with orange-tawny at the base and the inner angle, with a row of eight or nine submarginal white spots along the outer margin.

" Hind wing above, orange-tawny, with a broad mesial band, entire, and a narrow outer border, from the middle of the costa to the anal angle, brownish-black.

" Beneath the same, a yellow spot on the root of the wing; a band runs along the subcostal nervure from the base to the margin, where it is somewhat dilated; immediately below its termination, a mark in the form of an irregular figure 2, usually with the upper part inordinately enlarged; between this and the base, on the central line of the band above, three small subtriangular spots; all these markings blackish-brown; a submarginal row of seven white spots on the outer margin.

87

"Body brownish; wing-lappets and thorax spotted with tawny-orange; antennæ yellowish, with the base dusky.

"*Hab.*—Los Angeles, California."

The species is probably only a local race of the insect known to naturalists as *M. polymnia*, Linnæus, as Reakirt himself admits. The figure in the plate is from one of Reakirt's paratypes.

Genus CERATINIA, Fabricius

Butterfly.—Butterflies of medium size, very closely related in structure to the butterflies of the genus *Mechanitis*. The peculiarity of this genus, by which it may be distinguished from others belonging to this subfamily, is the fact that the *lower* discocellular vein in the hind wing of the male sex is strongly in-angled, while in the genus *Mechanitis* it is the *middle* discocellular vein of the hind wing which is bent inwardly.

Early Stages.—Unknown for the most part.

There are at least fifty species belonging to this genus found in the tropical regions of America; only one is said to occur occasionally within the limits of the region covered by this volume.

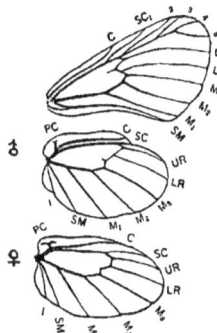

FIG. 81.—Neuration of the genus *Ceratinia*. (For explanation of lettering, see Fig. 40.)

(1) **Ceratinia lycaste,** Fabricius, Plate VIII, Fig. 3, ♂ (Lycaste).

Butterfly.—The butterfly is rather small, wings semi-transparent, especially at the apex of the fore wings. The ground-color is pale reddish-orange, with the border black. There are a few irregular black spots on the discal area of the fore wings, and a row of minute white spots on the outer border. There is a black band on the middle of the hind wings, curved to correspond somewhat with the outline of the outer border. The markings on the under side are paler. The variety *negreta*, which is represented in the plate, has a small black spot at the end of the cell of the hind wings, replacing the black band in the form common upon the Isthmus of Panama.

EXPLANATION OF PLATE VIII

1. *Dircenna klugi*. Hübner, ♂.
2. *Mechanitis californica*, Reakirt, ♂.
3. *Ceratinia lycaste*, Fabricius, ♂.
4. *Colænis delila*, Fabricius, ♂.
5. *Heliconius charitonius*, Linnæus, ♂.
6. *Colænis julia*, Fabricius, ♂.
7. *Dione vanillæ*, Linnæus, ♂.
8. *Euptoieta hegesia*, Cramer, ♂.
9. *Euptoieta claudia*, Cramer, ♂.

PLATE VII

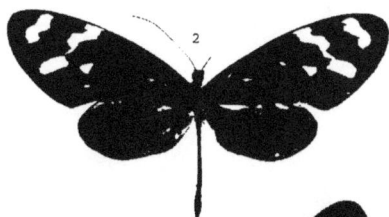

Early Stages.—Unknown.

Reakirt says that this butterfly occurs about Los Angeles, in California, and the statement has been repeated by numerous authors, who have apparently based their assertions upon Reakirt's report. I have no personal knowledge of the occurrence of the species within our borders. It is very abundant, however, in the warmer parts of Mexico and Central America, and it may possibly occur as a straggler within the United States.

Genus DIRCENNA, Doubleday

Butterfly.—Medium-sized butterflies, for the most part with quite transparent wings. The most characteristic features of this genus, separating it from its near allies, are the thread-like front feet of the females, furnished with four-jointed tarsi (Fig.83), the very hairy palpi, and the wide cell of the hind wing, abruptly terminating about the middle of the wing. Furthermore, in the male sex the hind wing is strongly bowed out about the middle of the costal margin, and the costal vein tends to coalesce with the subcostal about the middle.

Fig. 83.—Fore leg of *Dircenna klugii*, ♀, greatly magnified.

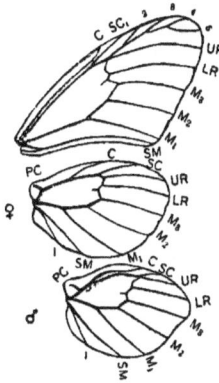

Fig. 82.—Neuration of the genus *Dircenna*.

Early Stages.—Very little is as yet known about the early stages of these insects, and what has been said of the characteristics of the caterpillars and chrysalids of the subfamily of the Ithomiinæ must suffice us here.

This genus numbers a large array of species which are found in the hottest parts of the tropics of the New World. They fairly swarm in wooded paths amid the jungle of the Amazonian region, and no collection, however small, is ever received from those parts without containing specimens belonging to the group.

(1) **Dircenna klugii**, Hübner, Plate VIII, Fig. 1, ♂ (Klug's Dircenna).

Butterfly.—Fore wings transparent gray, broken by clear, transparent, colorless spots at the apex, on the outer borders, and on the middle of the wing. The inner margin of the fore wing is black. The hind wings are transparent yellowish, with a narrow black outer border marked with small whitish spots. The body is black, with the thorax spotted with white. Expanse, 2.75 inches.

The specimen figured in the plate is from Mexico. Whether the insect has ever been taken within the limits of the United States is uncertain. It is another of the species attributed to our fauna by Reakirt, but which since his day has not been caught in the nets of any of the numerous butterfly-hunters who have searched the region in which he said it occurs. It may, however, be found upon the borders of Mexico, in the hotter parts of which country it is not at all uncommon. The " gentle reader " will kindly look for it when visiting Brownsville, Texas, and southern California, and, when finding it, herald the fact to the entomological world.

SUPERSTITIONS

"If a butterfly alights upon your head, it foretells good news from a distance. This superstition obtains in Pennsylvania and Maryland.

"The first butterfly seen in the summer brings good luck to him who catches it. This notion prevails in New York.

"In western Pennsylvania it is believed that if the chrysalids of butterflies be found suspended mostly on the under sides of rails, limbs, etc., as it were to protect them from rain, there will soon be much rain, or, as it is termed, a 'rainy spell'; but, on the contrary, if they are found on twigs and slender branches, that the weather will be dry and clear."—FRANK COWAN, *Curious History of Insects*, p. 229.

SUBFAMILY HELICONIINÆ (THE HELICONIANS)

MEDIUM or large-sized butterflies, with the fore wings twice as long as they are broad; the hind wings relatively small and rounded upon the outer margin; without tails. The palpi are produced. The antennæ, which are nearly as long as the body, are provided at the tip with a gradually tapering club, thicker and stouter than in the Ithomiinæ, and are clothed with scales on the upper surface. The fore legs are very feebly developed in both sexes. The eggs are cylindrical, twice as high as wide, tapering rather abruptly toward the apex, which is truncated; they are ribbed longitudinally, with strongly developed cross-ridges, giving the egg a somewhat pitted appearance. The caterpillar, when emerging from the egg, has the head somewhat larger than the body; each segment is clothed with hairs, which upon the first moult are replaced by branching spines. The caterpillar, when it reaches maturity, is provided with six branching spines on each segment. The chrysalis is very peculiar in shape, and is strongly angulated and covered with curious projections, which cause it to somewhat resemble a shriveled leaf.

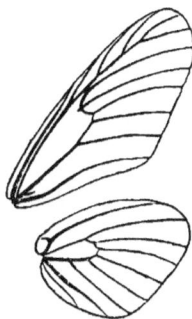

FIG. 84.—Neuration of the genus *Heliconius*.

These butterflies are extremely numerous in the tropics of the New World, and are there represented by a number of genera which are rich in species. Most of them are very gaily colored, the prevalent tints being black banded with yellow or crimson, sometimes marked with a brilliant blue luster. They are evidently very strongly protected. Belt, in his " Naturalist in Nicaragua," tells

us that birds and other animals observed by him invariably re-
fused to eat these butterflies, although they swarm in the forests;
and he vainly endeavored to induce a monkey which was very
fond of insects to eat them, the creature revealing by his grimaces
that they were extremely distasteful to him. Mr. Wallace believes
their immunity from attack is owing to a "strong, pungent, semi-
aromatic, or medicinal odor, which seems to pervade all the juices
of their system."

Genus HELICONIUS, Latreille

The description of the subfamily applies to the genus sufficiently
well to obviate the necessity of a more particular description, as
there is but a single species in our fauna.

(1) **Heliconius charitonius**, Linnæus, Plate VIII, Fig. 5, 6
(The Yellow-barred Heliconian; The Zebra).

This insect is a deep black, the fore wings crossed by three
bands of yellow: one near the apex; another running from the
middle of the costa to the middle of the outer margin; a third
running along the lower edge of the cell, and bending at an
obtuse angle from the point where the first median nervule
branches toward the outer angle, at its outer extremity followed
by a small yellow dot. The hind wings are crossed by a somewhat broad band of yellow running from the inner margin near
the base toward the outer angle, which it does not reach, and by
a submarginal curved band of paler yellow spots, gradually
diminishing in size from the inner margin toward the outer angle.
There are also a number of small twinned whitish spots on the
margin of the hind wing near the anal angle. The body is black,
marked with yellow spots and lines; on the under side both
wings are touched with crimson at their base, and the hind wings
have some pale pinkish markings near the outer angle.

The caterpillar feeds upon the passion-flower. The chrysalis,
which is dark brown, has the power when disturbed of emitting
a creaking sound as it wriggles about, a property which is reported to be characteristic of all the insects in the genus. This
butterfly is found in the hotter portions of the Gulf States, and is
rather abundant in Florida, in the region of the Indian River and
on the head waters of the St. Johns. It ranges southward all
over the lowlands of Mexico, Central America, and the Antilles.

SUBFAMILY NYMPHALINÆ (THE NYMPHS)

" Entomology extends the limits of being in new directions, so that I walk in nature with a sense of greater space and freedom. It suggests, besides, that the universe is not rough-hewn, but perfect in its details. Nature will bear the closest inspection; she invites us to lay our eye level with the smallest leaf and take an insect view of its plane."—THOREAU.

" My butterfly-net and pocket magnifying-glass are rare companions for a walk in the country."—WILLIAM HAMILTON GIBSON, *Sharp Eyes*, p. 117.

Butterfly.—The butterflies of this subfamily are mainly of moderate or large size, though some of the genera contain quite small species. The antennæ are always more or less heavily clothed with scales, and are usually as long as the abdomen, and in a few cases even longer. The club is always well developed; it is usually long, but in some genera is short and stout. The palpi are short and stout, densely clothed with scales and hairs. The thorax is relatively stout, in some genera exceedingly so. The fore wings are relatively broad, the length being to the breadth in most cases in the ratio of 5 to 3, or 3 to 2, though in a few mimetic forms these wings are greatly produced, and narrow, patterning after the outline of the Heliconians and Ithomiids, which they mimic. The fore wings are in most genera produced at the apex, and more or less strongly excavated on the outer margin below the apex. The discoidal cell is usually less than half the length of the wing from base to tip. It is occasionally open, but is more generally closed at its outer extremity by discocellular veins diminishing in thickness from the upper to the lower outer angle of the cell. The costal nervure usually terminates midway between the end of the cell and the tip. The two inner subcostal nervules usually arise before the end of the cell; the outer subcostal nervules invariably arise beyond the end of the cell.

The hind wings are rounded or angulated, with the outer

93

border scalloped or tailed; the inner border always affords a channel for the reception of the abdomen. The costal nervule invariably terminates at the external angle of this wing. The discoidal cell is frequently open, or simply closed by a slender veinlet, which it is not always easy to detect; the anal vein is never lacking.

The fore legs are greatly reduced in the male, less so in the female.

Egg.— The egg is either somewhat globular, or else barrel-shaped, with the sides marked with net-like elevations, or vertically ribbed (see Figs. 1, 8, 10).

Caterpillar.—When first emerging from the egg the caterpillar is generally furnished with long hairs rising singly from wart-like elevations which are arranged either in longitudinal rows or in geometric patterns (Fig. 85).

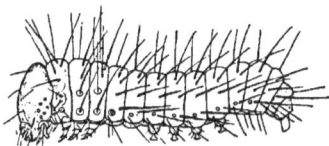

Fig. 85.— Caterpillar of *Vanessa antiopa*, just hatched. (Greatly magnified.) (After Scudder.)

As the caterpillars pass their successive moults the hairs are transformed into branching spines or tubercles (see Plate III, Figs. 28–38).

Chrysalis.—The chrysalis invariably hangs suspended from a button of silk, and is frequently furnished, especially on the dorsal or upper surface, with a number of prominences; the head is usually bifurcate, or cleft (see Plate IV, Figs. 21, 39, etc.).

This is the largest of all the subfamilies of the butterflies, and is widely distributed, including many of the most beautifully colored and most vigorous species which are known. There are twenty-six genera represented in our fauna, containing about one hundred and seventy species.

Genus COLÆNIS, Doubleday

Butterfly.— Butterflies of moderately large size, the fore wings greatly produced and relatively narrow; the hind wings evenly rounded and relatively small, of bright reddish-brown color, with darker markings. The species are mimics, and in the elongation of their wings reveal the influence of the Heliconians, protected species, which abound in the regions in which the genus attains its greatest development. The median vein in the upper wing is characterized by the presence at the base of a minute, thorn-like, external projection; the second subcostal nervule is emitted beyond the cell; the cell of the hind wing is open.

The life-history of the two species found within our fauna has not as yet been carefully worked out, and aside from a knowledge of the fact that the caterpillars closely resemble in many respects the caterpillars of the two succeeding genera, being provided with branching spines on their bodies, we do not know as yet enough to give any complete account of the early stages of these insects.

(1) **Colænis julia,** Fabricius, Plate VIII, Fig. 6, ♂ (Julia).

FIG. 86.—Neuration of the genus Co-lænis, slightly less than natural size.

The upper side is dark reddish-orange, the borders are black, a black band extends from the costa at the end of the cell to the outer margin on the line of the third median nervule; the costal area on the hind wings is silver-gray; the wings on the under side are pale rusty-red, mottled with a few darker spots, principally on the costa, at the end of the cell, and at the apex of the primaries. There are a few crimson marks at the base of the hind wings, and two light-colored lunules near the inner angle of the hind wings. Expanse of wing, 3.50 inches.

This butterfly, which mimics the genus *Heliconius* in the outline of the wings, is very common in the tropics of America, and only appears as an occasional visitant in southern Texas.

(2) **Colænis delila,** Fabricius, Plate VIII, Fig. 4, ♂ (Delila).

The Delila Butterfly very closely resembles Julia, and principally differs in being paler in color and without the black band extending from the costa to the outer margin of the primaries. This species has nearly the same form and the same size as the preceding, and, like it, is occasionally found in southern Texas. It is very common in Central America and the West Indies. One of the earliest memories of my childhood relates to a collection of Jamaican butterflies in which were a number of specimens of this butterfly, which I have always much admired.

Genus DIONE, Hübner
(Agraulis, *Boisd.-Lec.*)

Butterfly.—Head large, the antennæ moderately long, with the club flattened; the tip of the abdomen does not extend beyond the inner margin of the hind wings; the cell of the hind wings is open; the primaries are elongated, nearly twice as long as broad, with the exterior margin excavated; the secondaries at the outer margin denticulate. The prevalent color of the upper side of the wings is fulvous, adorned with black spots and lines, the under side of the wings paler brown, in some of the species laved with pink and brilliantly adorned with large silvery spots, as in the genus *Argynnis*.

Egg.—Conoidal, truncated on top, with fourteen ribs running from the apex to the base, between which are rows of elevated striæ, causing the surface to appear to be covered with quadrangular pits.

Larva.—The caterpillar is cylindrical in its mature stage, tapering a little from the middle toward the head, which is somewhat smaller than the body. The head and each segment of the body are adorned with branching spines.

FIG. 87.—Neuration of the genus *Dione*.

Chrysalis.—The chrysalis is suspended, and has on the dorsal surface of the abdomen a number of small projections. At the point where the abdominal and thoracic segments unite on the dorsal side there is a deep depression, succeeded on the middle of the thorax by a rounded elevation composed of the wing-cases. At the vertex of the chrysalis there is a conical projection; on the ventral side the chrysalis is bowed outwardly.

This genus is confined to the New World, and contains five species. It is closely related to the genus *Colænis* on the one hand and to the genus *Argynnis* on the other. It is distinguished from *Colænis* by the more robust structure of the palpi, which closely approximate in form the palpi of the genus *Argynnis*. It is distinguished from the species of the genus *Argynnis* by the form of the wings and by the open cell of the secondaries. The larva feeds upon the different species of the genus *Passiflora.*

96

I cannot at all agree with those who have recently classed this butterfly with the Heliconians. In spite of certain resemblances in the early stages between the insect we are considering and the early stages of some of the Heliconians, and in spite of the shape of the wings, which are remarkably elongated, there are structural peculiarities enough to compel us to keep this insect in the ranks of the Nymphalinæ, where it has been placed for sixty years by very competent and critical observers. In a popular work like this it manifestly is out of place to enter into a lengthy discussion of a question of this character, but it seems proper to call attention to the fact that in the judgment of the writer the location of this genus in the preceding subfamily does violence to obvious anatomical facts.

(1) **Dione vanillæ**, Linnæus, Plate VIII, Fig. 7, ♂ (The Gulf Fritillary).

Butterfly.—The upper side is bright fulvous; the veins on the fore wings are black, very heavy near the tip; there are four black spots on the outer border, and three discal spots of the same color; there are three irregular black spots toward the end of the cell, pupiled with white; the hind wings have a black border inclosing rounded spots of the ground-color; between the base and the outer margin there are three or four black spots; the under side of the fore wings is light orange, the markings of the upper side showing through upon the under side; the apex of the front wing is brown, inclosing light silvery spots; the secondaries are brown, with numerous elongated bright silver spots and patches. The female does not differ from the male, except that she is darker and the markings are heavier. Expanse, 2.50 –3.25 inches.

Caterpillar.—The caterpillar is cylindrical, with the head somewhat smaller than the body, pale yellowish-brown in color, marked with longitudinal dark-brown bands, of which the two upon the side are deeper in color than the one upon the back, which latter is sometimes almost entirely effaced; the base is slaty-black. There are orange spots about the spiracles. There are six rows of black branching spines upon the body, and two similar spines upon the head, these latter somewhat recurved. The feet and legs are black. The caterpillar feeds upon the various species of passion-flower which are found in the Southern States.

Chrysalis.— The chrysalis is dark brown, marked with a few small pale spots.

This species ranges from the latitude of southern Virginia southward to Arizona and California. It is abundant also in the Antilles and Mexico.

Genus EUPTOIETA, Doubleday

Butterfly.— Butterflies of medium size, having wings of a yellowish-brown color, marked with black, the under side of the wings devoid of silvery spots such as are found in the genera *Dione* and *Argynnis.* The palpi have the second joint strongly developed, increasing in thickness from behind forward, and thickly covered with long hair; the third joint is very small and pointed; the antennæ are terminated by a conspicuous pear-shaped club. The cell of the fore wing is closed by a very feeble lower discocellular vein, which unites with the median vein at the origin of the second median nervule; the cell of the hind wing is open, though occasionally there are traces of a feebly developed lower discocellular vein on this wing.

FIG. 88.—Neuration of the genus *Euptoieta.*

The outer margin of the fore wing is slightly excavated below the apex; the outer margin of the hind wing is somewhat strongly produced at the end of the third median nervule.

Egg.— Short, subconical, with from thirty to forty vertical ribs, pale green in color.

Caterpillar.— The caterpillar is cylindrical, with short branching spines arranged in longitudinal rows upon the body, the spines on the first segment being bent forward over the head. The head is somewhat smaller in the mature stage than the body.

Chrysalis.— The chrysalis is suspended, marked upon its dorsal side with a number of small angular eminences, with the head and the ventral side evenly rounded.

The larva of these insects feeds upon the various species of passion-flower. It is also said to feed upon violets. The butterflies frequent open fields, and are sometimes exceedingly abundant in worn-out lands in the Southern States.

There are two species of this genus, both of which are found within the United States, and range southwardly over the greater portion of Central and South America.

(1) **Euptoieta claudia**, Cramer, Plate VIII, Fig. 9, ♂ (The Variegated Fritillary).

Butterfly.—The upper side of both wings is dull ferruginous, darker toward the base, crossed by an irregular black median line, which is darker, broader, and more zigzag on the fore wing than on the hind wing. This line is followed outwardly on both wings by a pair of more or less wavy limbal lines, inclosing between them a series of round blackish spots. The outer margin is black, with the fringes pale fulvous, checkered with black at the end of each nervule. At the end of the cell in the fore wing there are two black lines inclosing paler fulvous spots, and both wings near the base have some curved black lines. On the under side the fore wings are marked somewhat as on the upper side, but paler in color, with a large apical patch of brownish-gray broken by a transverse band of darker brown. The hind wings are dark brown, with the markings of the upper side obscurely repeated; they are mottled with gray and crossed by a broad central band of pale buff.

The species varies very much, according to locality, both in size and in the depth of the markings. Expanse, 1.75–2.75 inches.

Egg.—The egg is conoidal, relatively taller than the eggs of the genus *Argynnis*, which closely resemble it. There is a depression at the apex, surrounded by a serrated rim, formed by the ends of the vertical ribs, of which there are about twenty, some longer and some shorter, about half of them reaching from the apex to the base. Between these vertical ribs there are a multitude of smaller cross-ridges.

Caterpillar.—The caterpillar is cylindrical, reddish-yellow in color, marked with two brown lateral bands and a series of white spots upon the back. There are six rows of short branching spines upon the body, which are black in color; the two upper-most of these spines on the first segment are much elongated and are directed forward. The head is smaller than the body in the mature caterpillar, and is black. On the under side the caterpillar is pale or whitish; the legs are blackish-brown. It feeds upon the passion-flower.

Chrysalis.—The chrysalis is pearly-white, marked with black spots and longitudinal streaks.

99

This species has been taken as far north as Long Island and Connecticut, though it is a very rare visitant in New England; it is quite common in Virginia and thence southward, and occurs not infrequently in southern Illinois and Indiana, ranging westward and southward over the entire continent to the Isthmus of Panama, and thence extending over the South American continent, wherever favorable conditions occur.

(2) **Euptoieta hegesia,** Cramer, Plate VIII, Fig. 8, ♂ (The Mexican Fritillary).

The upper side is marked very much as in the preceding species, but all the lines are finer and somewhat more regular, and the basal and discal areas of the hind wings are without dark spots in most specimens. The under side is less mottled and more uniformly dark rusty-brown than in *E. claudia*. Expanse, about 2 inches.

The life-history of this species has not as yet been thoroughly worked out, but there is every reason to believe that the insect in its early stages very closely approaches the Variegated Fritillary. It is a Southern form, and only occasionally is taken in Arizona and southern California. It is common in Central and South America.

LUTHER'S SADDEST EXPERIENCE

" Luther, he was persecuted,
Excommunicated, hooted,
Disappointed, egged, and booted;
Yelled at by minutest boys,
Waked up by nocturnal noise,
Scratched and torn by fiendish cats,
Highwayed by voracious rats.

" Oft upon his locks so hoary
Water fell from upper story;
Oft a turnip or potato
Struck upon his back or pate, Oh!
And wherever he betook him,
A papal bull was sure to hook him.

" But the saddest of all
I am forced to relate:

Of a diet of worms
He was forced to partake—
Of a *diet of worms*
For the Protestants' sake;
Munching crawling caterpillars,
Beetles mixed with moths and millers;
Instead of butter, on his bread,
A sauce of butterflies was spread.
Was not this a horrid feast
For a Christian and a priest?

" Now, if you do not credit me,
Consult D'Aubigné's history.
You 'll find what I have told you
Most fearfully and sternly true."

Yale Literary Magazine, 1852.

EXPLANATION OF PLATE IX

1. *Argynnis diana*, Cramer, ♂.
2. *Argynnis diana*, Cramer, ♀.
3. *Argynnis cybele*, Fabricius, ♂.
4. *Argynnis cybele*, Fabricius, ♀.
5. *Argynnis leto*, Behr, ♂.
6. *Argynnis leto*, Behr, ♀.

PLATE IX.

Genus ARGYNNIS, Fabricius

(The Fritillaries, the Silver-spots)

"July is the gala-time of butterflies. Most of them have just left the chrysalis, and their wings are perfect and very fresh in color. All the sunny places are bright with them, yellow and red and white and brown, and great gorgeous fellows in rich velvet-like dresses of blue-black, orange, green, and maroon. Some of them have their wings scalloped, some fringed, and some plain; and they are ornamented with brilliant borders and fawn-colored spots and rows of silver crescents. . . . They circle about the flowers, fly across from field to field, and rise swiftly in the air; little ones and big ones, common ones and rare ones, but all bright and airy and joyous—a midsummer carnival of butterflies."—FRANK H. SWEET.

Butterfly.— Butterflies of medium or large size, generally with the upper surface of the wings reddish-fulvous, with well-defined black markings consisting of waved transverse lines, and rounded discal and sagittate black markings near the outer borders. On the under side of the wings the design of the fore wings is generally somewhat indistinctly repeated, and the hind wings are marked more or less profusely with large silvery spots. In a few cases there is wide dissimilarity in color between the male and the female sex; generally the male sex is marked by the brighter red of the upper surface, and the female by the broader black markings, the paler ground-color, and the sometimes almost white lunules, which are arranged outwardly at the base of the sagittate spots along the border.

FIG. 89.— Neuration of the genus *Argynnis*.

The eyes are naked; the palpi strongly developed, heavily clothed with hair rising above the front, with the last joint very small and pointed. The antennæ are moderately long, with a well-defined, flattened club. The abdomen is shorter than the hind wings; the wings are more or less denticulate. The subcostal vein is provided with five nervules, of which the two innermost are invariably given forth before the end of the cell; the third subcostal nervule always is nearer the fourth than the second. The cell of the fore wing is closed by a fine lower discocellular vein, which invariably joins the median vein beyond the origin of the second nervule. The hind wing has a well-defined precostal nervule; the cell in this wing is closed by a moderately thick lower discocellular vein, which joins the median exactly at the origin of the second median nervule. The fore feet of the males are slender, long, and finely clothed with hair. The fore feet of the females are of the same size as those of the males, but thin, covered with scales, and only on the inner side of the tibiæ clothed with moderately long hair.

Egg.— The eggs are conoidal, truncated, and inwardly depressed at the apex, rounded at the base, and ornamented on the sides by parallel raised ridges, not all of which reach the apex. Between these ridges there are a number of small raised cross-ridges.

Caterpillar.— The caterpillar is cylindrical, covered with spines, the first segment always bearing a pair of spines somewhat longer than the others. All of the species in North America, so far as their habits are known, feed upon violets at night. During the daytime the caterpillars lie concealed.

Chrysalis.— The chrysalis is angular, adorned with more or less prominent projections. The head is bifid.

The genus *Argynnis* is one of the largest genera of the brush-footed butterflies. It is well represented in Europe and in the temperate regions of Asia, some magnificent species being found in the Himalayas and in China and Japan. It even extends to Australia, and recently two species have been discovered in the vicinity of the great volcanic peak, Kilima-Njaro, in Africa. But it has found its greatest development upon the continent of North America. The species composing this genus are among our most beautiful butterflies. Owing to the fact that there is a great tendency in many of the forms closely to approximate one another, the accurate distinction of many of the species has troubled naturalists, and it is quite probable that some of the so-called species will ultimately be discovered to be merely local races or varietal forms. The species that are found in the eastern part of the United States have been studied very carefully, and their life-history has been worked out so thoroughly that little difficulty is found in accurately determining them. The greatest perplexity occurs in connection with those species which are found in the region of the Rocky Mountains. While silvery spots are characteristic of the under side of most of the fritillaries, in some species the silvery spots are not found; in others they are more or less evanescent, occurring in the case of some individuals, and being absent in the case of others.

(1) **Argynnis idalia**, Drury, Plate X, Fig. 3, ♀; Plate V, Fig. 4, *chrysalis* (The Regal Fritillary).

Butterfly.—The upper side of the fore wings of the male is bright fulvous, marked very much as in other species of the genus. The upper side of the hind wings is black, glossed with blue, having a marginal row of fulvous and a submarginal row of cream-colored spots. On the under side the fore wings are fulvous, with a marginal row of silver crescents, and some silvery spots on and near the costa. The hind wings are dark olive-brown, marked with three rows of large irregular spots of a dull greenish-silvery color. The female is at once distinguished from the male by having the marginal row of spots on the hind wings cream-colored, like the submarginal row, and by the presence of a similar row of light spots on the fore wings. Expanse, 2.75–4.00 inches.

Egg.—The egg in form is like those of other species of *Argynnis.*

Caterpillar.—The caterpillar moults five times before attaining to maturity. When fully developed it is 1.75 inches long, black, banded and striped with ochreous and orange-red, and adorned with six rows of fleshy spines surmounted by several black bristles. The spines composing the two dorsal rows are white, tipped with black; those on the sides black, tinted with orange at the point where they join the body. The caterpillar feeds on violets, and is nocturnal in its habits.

Chrysalis.—The chrysalis is brown, mottled with yellow and tinted on the wing-cases with pinkish. It is about an inch long, and in outline does not depart from the other species of the genus.

This exceedingly beautiful insect ranges from Maine to Nebraska. It is found in northern New Jersey, the mountainous parts of New York and northern Pennsylvania, and is reported from Arkansas and Nebraska. It is rather local, and frequents open spots on the borders of woodlands. At times it is apparently common, and then for a succession of seasons is scarce. It flies from the end of June to the beginning of September.

(2) **Argynnis diana**, Cramer, Plate IX, Fig. 1, ♂; Fig. 2, ♀ (Diana).

Butterfly.— The male on the upper side has both wings deep rich brown, bordered with fulvous, this border being more or less interrupted by rays of brown along the nervules and two rows of circular brown spots, larger on the fore wings than on the hind wings. The wings on the under side are pale buff, deeply marked with black on the base and middle of the fore wings, and clouded with grayish-fulvous on the inner two thirds of the hind wings. A blue spot is located near the end of the cell in the fore wings, and the hind wings are adorned by a marginal and submarginal row of narrow silvery crescents and a few silvery spots toward the base. The female on the upper side is a rich bluish-black, with the outer border of the fore wings marked by three rows of bluish-white quadrate spots, the outer row being the palest, and often quite white. The hind wings are adorned by three more or less complete rows of bright-blue spots, the inner row composed of large subquadrate spots, each having a circular spot of black at its inner extremity. On the under side the female has the ground-color slaty-brown, paler on the hind wings than on the fore wings, which latter are richly marked with blue and black spots. The silvery crescents found on the under side of the hind wings of the male reappear on the under side of the female, and are most conspicuous on the outer margins. Expanse, 3.25–4.00 inches.

Egg.— The egg is pale greenish-white, and conformed in outline to type.

Caterpillar.— The larva is velvety-black, adorned with six rows of fleshy spines armed with bristles. The spines are orange-red at the base. The head is dull brown.

Chrysalis.— The chrysalis is dusky-brown, with lighter-colored short projections on the dorsal side.

This splendid butterfly, which is the most magnificent species of the genus, is confined to the southern portion of the Appalachian region, occurring in the two Virginias and Carolinas, northern Georgia, Tennessee, and Kentucky, and being occasionally found in the southern portion of Ohio and Indiana, and in Missouri and Arkansas.

(3) **Argynnis nokomis**, Edwards, Plate X, Fig. 1, ♂ ; Fig. 2, ♀ (Nokomis).

Butterfly.— The male on the upper side is bright fulvous, with the characteristic black markings of the genus. On the under

EXPLANATION OF PLATE X

1. *Argynnis nokomis*, Edwards, ♂.
2. *Argynnis nokomis*, Edwards, ♀.
3. *Argynnis idalia*, Drury, ♀.
4. *Argynnis nevadensis*, Edwards, ♂, under side.
5. *Argynnis montivaga*, Behr, ♂, under side.
6. *Argynnis alcestis*, Edwards, ♂, under side.
7. *Argynnis bremneri*, Edwards, ♂.
8. *Argynnis electa*, Edwards, ♂.
9. *Argynnis atlantis*, Edwards, ♂.

PLATE X

side the wings are pale greenish-yellow, with the fore wings laved with bright pink at the base and on the inner margin. The spots of the upper side reappear on the under side as spots of silver bordered narrowly with black. The female has the ground-color of the upper side yellow, shaded outwardly with fulvous. All the dark markings of the male sex reappear in this sex, but are much broader, and tend to fuse and run into one another, so as to leave the yellow ground-color as small subquadrate or circular spots, and wholly to obliterate them at the base of the wings. On the under side this sex is marked like the male, but with all the markings broader. Expanse, 3.40–3.60 inches.

This species, the male of which resembles the male of *A. leto*, and the female the same sex of *A. diana*, is as yet quite rare in collections. It has been taken in Arizona and southern Utah. We have no knowledge of the life-history of the species.

(4) **Argynnis nitocris,** Edwards, Plate XIII, Fig. 4, ♂, *under side* (Nitocris).

Butterfly. — The male is bright reddish-fulvous, marked like *A. nokomis*. The under side of the fore wings is cinnamon-red, ochre-yellow at the tip. The hind wings are deep rusty-red, with a broad yellowish-red submarginal belt. The silver spots are as in *A. nokomis*. The female on the upper side is blackish-brown, darker than *A. nokomis*. The extradiscal spots in the transverse rows are pale yellow, and the submarginal spots whitish. The under side of the fore wings is bright red, with the tip yellow. The hind wings on this side are dark brown, with a submarginal yellow belt. Expanse, 3.25–3.75 inches.

This species, like the preceding, is from Arizona, and nothing is known of its egg, caterpillar, or chrysalis.

(5) **Argynnis leto,** Edwards, Plate IX, Fig. 5, ♂ ; Fig. 6, ♀ (Leto).

Butterfly. — The male on the upper side is marked much as *A. nokomis*, but the ground-color is duller red, and the basal area is much darker. The under side of the fore wings is pale fulvous, upon which the markings of the upper side reappear; but there are no marginal silver crescents. Both wings on the under side are shaded with brown toward the base; the hind wings are traversed by a submarginal band of light straw-yellow. The female is marked as the male, but the ground-color is pale straw-yellow, and all the darker markings are deep blackish-brown, those

105

at the base of both wings being broad and running into one another, so that the inner half of the wings appears to be broadly brownish-black. On the under side this sex is marked as the male, but with the dark portions blacker and the lighter portions pale yellow. Expanse, 2.50–3.25 inches.

The life-history of this insect remains to be worked out. It is one of our most beautiful species, and occurs in California and Oregon.

(6) **Argynnis cybele**, Fabricius, Plate IX, Fig. 3, ♂ ; Fig. 4, ♀ ; Plate XIII, Fig. 1, ♀, *under side;* Plate V, Figs. 1–3, *chrysalis* (The Great Spangled Fritillary).

Butterfly. — The male is much like the male of *A. leto,* but the dark markings of the upper surface are heavier, and the under sides of the hind wings are more heavily silvered. The yellowish-buff submarginal band on the under side of the hind wings is never obliterated by being invaded by the darker ferruginous of the marginal and discal tracts of the wing. The female has the ground-color of the wings paler than the male, and both wings from the base to the angled median band on the upper side are dark chocolate-brown. All the markings of the upper side in this sex are heavier than in the male. On the under side the female is like the male. Expanse, 3.00–4.00 inches.

Egg. — Short, conoidal, ribbed like those of other species, and honey-yellow.

Caterpillar. — The larva in the mature state is black. The head is blackish, shaded with chestnut behind. The body is ornamented with six rows of shining black branching spines, generally marked with orange-red at their base. The caterpillar, which is nocturnal, feeds on violets, hibernating immediately after being hatched from the egg, and feeding to maturity in the following spring.

Chrysalis. — The chrysalis is dark brown, mottled with reddish-brown or slaty-gray.

This species, which ranges over the Atlantic States and the valley of the Mississippi as far as the plains of Nebraska, appears to be single-brooded in the North and double-brooded in Virginia, the Carolinas, and the Western States having the same geographical latitude. A small variety of this species, called *A. carpenteri* by Mr. W. H. Edwards, is found in New Mexico upon the top of Taos Peak, and is believed to be isolated here in a colony, as *Œneis semidea* is isolated upon the summit of Mount Washington. Specimens of *cybele* much like those of this New Mexican variety are found in eastern Maine and Nova Scotia, and on the high mountains of North Carolina.

(7) **Argynnis aphrodite**, Fabricius, Plate XIV, Fig. 11, ♀, *under side;* Plate V, Fig. 5, *chrysalis* (Aphrodite).

Butterfly.—This species closely resembles *cybele*, but is generally smaller, and the yellow submarginal band on the hind wings is narrower than in *cybele*, and often wholly wanting, the hind wings being broadly brown, particularly in the female sex. The under side of the fore wings at the base and on the inner margin is also brighter red.

The caterpillar, chrysalis, and egg of this species closely resemble those of *cybele*. The caterpillar has, however, a velvety-black spot at the base of each spine, the chrysalis has the tubercles on the back shorter than in *cybele*, and the basal segments are party-colored, and not uniformly colored as in *cybele*.

(8) **Argynnis cipris**, Edwards, Plate XII, Fig. 3, ♂ ; Fig. 4, ♀ (The New Mexican Silver-spot).

Butterfly.—This species, which belongs to the Aphrodite-group, may be distinguished by the fact that the fore wings are relatively longer and narrower than in *aphrodite*. The black markings on the upper side of the wings in both sexes are narrower, the dusky clouding at the base of the wings is less pronounced, and the ground-color is brighter reddish-fulvous than in *aphrodite*. On the under side the fore wings lack in the male the pinkish shade at the base and on the inner margin which appears in *aphrodite*, and both the male and the female have the inner two thirds of the hind wings deep cinnamon-red, with only a very narrow buff submarginal band, deeply invaded on the side of the base by rays of the deeper brown color of the inner portion of the wing. Expanse, 2.75–3.15 inches. The insect flies from late June to the end of August.

Caterpillar, etc.—We know nothing of the larval stages of this insect. The specimens contained in the Edwards collection came from Colorado, Utah, and New Mexico, and these localities approximately represent the range of the species.

(9) **Argynnis alcestis**, Edwards, Plate X, Fig. 6, ♂, *under side* (The Ruddy Silver-spot).

Butterfly.—Very much like *aphrodite*, from which it may be most easily distinguished by the fact that the hind wings are uniformly dark cinnamon-brown, without any band of buff on the outer margin. Expanse, 2.50–3.00 inches. The insect flies from late June to the end of August.

Egg.—Greenish, conoidal, with about eighteen vertical ribs.

Caterpillar.—Head black, yellowish behind. The body velvety-black, ornamented with black spines which are yellowish at their basal ends. The caterpillar feeds on violets.

Chrysalis.—Reddish-brown or gray, irregularly mottled and striped with black, the abdominal segments slaty-gray, marked with black on the edges where the short angular projections are located.

This butterfly is found in the Western States, extending from the prairie lands of northwestern Ohio to Montana. It largely replaces *aphrodite* in these regions.

(10) **Argynnis nausicaä**, Edwards, Plate XI, Fig. 9, ♂ (The Arizona Silver-spot).

Butterfly.—The species is related to the foregoing, but is rather smaller in size. The upper side of the wings is dusky reddish-brown, with the characteristic markings of the genus. On the under side the fore wings are pink, laved with buff at the tip. The hind wings on this side are deep cinnamon-brown, mottled with buff on the inner two thirds ; a narrow but clearly defined submarginal band of bright yellowish-buff surrounds them. The silvery spots are clearly marked. The female has the black markings broader and more conspicuous than the male. Expanse, 2.25–2.50 inches.

This insect is quite common in the mountain valleys of Arizona, at an elevation of from six to seven thousand feet above the level of the sea, and flies in July and August. We have no knowledge of the early stages, but it probably does not differ greatly in its larval state from the allied species of the genus.

(11) **Argynnis atlantis**, Edwards, Plate X, Fig. 9, ♂ ; Plate V, Fig. 6, *chrysalis* (The Mountain Silver-spot).

Butterfly.—This insect, which resembles *aphrodite*, is distinguished from that species by its smaller size, its somewhat narrower wings, the deeper brown color of the base of the wings on the upper side, and their darker color on the under side. The submarginal band is pale yellow, narrow, but distinct and always present. Expanse, 2.50 inches.

Explanation of Plate XI

1. *Argynnis callippe,* Boisduval, ♂.
2. *Argynnis callippe,* Boisduval, ♀.
3. *Argynnis callippe,* Boisduval, ♂, under side.
4. *Argynnis edwardsi,* Reakirt, ♂.
5. *Argynnis edwardsi,* Reakirt, ♀.
6. *Argynnis rhodope,* Edwards, ♀, under side.
7. *Argynnis bischoffi,* Edwards, ♂.
8. *Argynnis cornelia,* Edwards, ♂.
9. *Argynnis nausicaä,* Edwards, ♂.
10. *Argynnis coronis,* Behr, ♂.
11. *Argynnis coronis,* Behr, ♀.

PLATE XI

Egg.— Conoidal, with twelve to fourteen ribs, honey-yellow. The caterpillars are hatched in the fall, and hibernate without feeding until the following spring.

Caterpillar.— The head is dark blackish-brown. The body is velvety-purple above, a little paler on the under side. The usual spines occur on the body, and are black, grayish at the base. The larva feeds on violets.

Chrysalis.— The chrysalis is light brown, speckled, except on the abdominal segments, with black.

This species ranges from Maine to the mountains of western Pennsylvania, and thence southward along the central ridges of the Alleghanies into West Virginia. It is also found in Canada, and extends westward into the region of the Rocky Mountains. It is especially common in the White Mountains of New Hampshire and the Adirondacks.

(12) **Argynnis lais,** Edwards, Plate XIV, Fig. 12, ♂ ; Fig. 13, ♀ (The Northwestern Silver-spot).

Butterfly.—The male is bright reddish-fulvous on the upper side, slightly obscured by fuscous at the base. The discal band of spots common to both wings is broken and irregular, and the spots on the hind wings are quite small. The fore wings on the under side are buff at the tips and pale red at the base and on the inner margin, lighter at the inner angle. The under side of the hind wings as far as the outer margin of the discal row of silvery spots is dark brown, mottling a yellowish ground. The submarginal band of the hind wings is pale yellow and moderately broad. The female is marked much as the male, but the discal band of spots on the upper side of the fore wings is confluent and broader, the fringes whitish, and the spots included between the sagittate marginal spots and the marginal lines paler than in the male sex. Expanse, 2.00–2.20 inches.

Caterpillar, etc.— The early stages are unknown.

This species is found in the territories of Alberta and Assiniboia, and in British Columbia among the foot-hills and the lower slopes of the mountain-ranges.

(13) **Argynnis oweni,** Edwards, Plate XII, Fig. 5, ♂ ; Fig. 6, , *under side* (Owen's Silver-spot).

Butterfly, ♂ .—The wings on the upper side are dull reddish-fulvous, not much obscured with brown on the base, the black markings moderately heavy, the two marginal lines tending to flow together. The fore wings on the under side are yellowish-buff from the base to the outer row of spots, or in some specimens with the buff lightly laved with reddish; the nerves reddish-brown. The subapical patch is dark brown, with a small silvered spot; the five submarginal spots are small and obscurely silvered. The hind wings are dark brown on the discal area and outer margin, with a rather narrow grayish-buff submarginal band, strongly invaded by projections of the dark brown of the discal area. The spots of the outer discal row are generally well silvered; the inner spots less so in most cases.

♀.—The female has the wings more or less mottled with yellowish outside of the mesial band. The black markings are very heavy in this sex. On the under side the spots are well silvered.

The dark markings on the upper side of the wings of the male are much heavier than in *A. behrensi*. On the under side of the wings in both sexes it may be distinguished from *behrensi* by the fact that the ground-color toward the base is mottled with yellow, and not solid brown as in *behrensi*. Expanse, 2.25–2.40 inches.

This species abounds on Mount Shasta, in California, at an elevation of seven to eight thousand feet above sea-level.

(14) **Argynnis cornelia**, Edwards, Plate XI, Fig. 8, ♂ (Miss Owen's Fritillary).

Butterfly, ♂.—The upper side of both wings is dark-brown from the base to the mesial band of spots, with the exception of the outer end of the cell. The space beyond the band is reddish-fulvous; the dark markings are not very heavy; the two marginal lines are fine, and confluent at the ends of the nervules. The under side of the fore wings is reddish-brown from the base to the outer margin on the inner half of the wing; the outer spaces toward the apex are yellowish; the subapical patch is reddish-brown, inclosing a small silvery spot; the outer margin is reddish-brown, adorned with five small silvery spots toward the apex. The hind wings on the under side are almost solid reddish-brown to the clear yellow submarginal belt, only slightly mottled on the discal area with buff. The spots are small and well silvered.

♀.—The female on the upper side is duller red, with the dark markings heavier; the marginal spots on the fore wings are pale yellowish, and the marginal lines are confluent on the upper half of these wings. The wings on the under side in this sex are as in the male, but the ground-color on the inner half of the wings is darker, and the spots are more brilliantly silvered. Expanse, 2.30–2.50 inches.

Early Stages.—Unknown.

This pretty species is found with *A. electa* and *A. hesperis* in Colorado. It was originally described from specimens taken at Manitou and Ouray, and named by Edwards in honor of a deceased daughter of Professor Owen of the University of Wisconsin.

(15) **Argynnis electa,** Edwards, Plate X, Fig. 8, ♂ (Electa).

Butterfly.—The male is dull reddish-fulvous on the upper side. The black markings are narrow. The base of both wings is slightly obscured. On the under side the fore wings are pale cinnamon-red, with the tip dark cinnamon-red. The hind wings are broadly dark cinnamon-red, mottled on the disk with a little buff. The submarginal band is buff, quite narrow, and often invaded by the ground-color of the inner area. The silvery spots are usually very well marked and distinct, though in a few instances the silvery color is somewhat obscured. The female has the black markings a little heavier than in the male; otherwise there is but little difference between the sexes. Expanse, 2.00–2.25 inches.

Caterpillar, etc.— The early stages are unknown.

This species has been confounded with *A. atlantis,* from which it is wholly distinct, being much smaller in size, the fore wings relatively broader, and the markings not so dark on the upper surface. It is found in Colorado and Montana, among the mountains.

(16) **Argynnis columbia,** Henry Edwards, Plate XIV, Fig. 3, ♂ (The Columbian Silver-spot).

Butterfly.—The male has the upper side of the fore wings pale reddish-fulvous. In the median band of both wings the spots do not flow together, but are separate and moderately heavy. The under side of the fore wings is pale fulvous, buff at the tip; spots silvered. The hind wings on the under side are light rusty-red, but little mottled with buff on the disk; the submarginal band is narrow, buff, and sometimes almost wholly obscured by the darker ground-color. The spots, which are small, are well silvered.

The female is much lighter than the male, and, as usual, the dark lines are heavier than in that sex. The spots of the median band are bent and partly lanceolate, and the light spots of the outer border are whitish. Expanse, 2.25–2.50 inches.

Caterpillar, etc.—The early stages have not as yet been worked out.

This species, which is related to *electa*, may easily be distinguished from it by the pale marginal series of light spots, in the male, between the sagittate spots and the dark outer marginal lines, which latter are confluent, forming a solid dark outer border to the wing, while in *electa* they are separated by a narrow band of light-brown spots. The female is also much lighter and larger than in *electa*, as has been pointed out. The types which came from the Caribou mining region of British Columbia are in my possession, as are those of most of the other North American species of the genus.

(17) **Argynnis hesperis,** Edwards, Plate XII, Fig. 1, ♂ ; Fig. 2, ♀ (Hesperis).

Butterfly.— The male on the upper side of the wings is fulvous, shaded with dark fuscous for a short distance from the base. The black spots of the median band are rather broad, and seem to coalesce through dark markings along the nervules. The under side of the fore wings is pale ferruginous, tinged with a little buff at the tips, which, together with the outer margin, are somewhat heavily clouded with dark ferruginous. The under side of the hind wings is dark ferruginous, with a narrow buff submarginal band, which in some specimens is almost lost. The female is paler than the male in the ground-color of the upper side, the black markings are heavier, the marginal lines fuse, as do also the sagittate marginal markings, leaving the marginal spots between them, which are quite light in color, deeply bordered on all sides by black. The under side is like that of the male, but darker and richer in color. In neither sex are the light spots marked with silver; they are opaque, yellowish-white. Expanse, 2.25–2.40 inches.

Caterpillar, etc.— The life-history remains to be learned.

This insect is not uncommon among the mountains of Colorado.

(18) **Argynnis hippolyta,** Edwards, Plate XII, Fig. 10, ♂ (Hippolyta).

EXPLANATION OF PLATE XII

1. *Argynnis hesperis*, Edwards, ♂.
2. *Argynnis hesperis*, Edwards, ♀.
3. *Argynnis cypris*, Edwards, ♂.
4. *Argynnis cypris*, Edwards, ♀.
5. *Argynnis oweni*, Edwards, ♂.
6. *Argynnis oweni*, Edwards, ♂, under side.
7. *Argynnis eurynome*, Edwards, ♂.
8. *Argynnis rupestris*, Behr, ♂.
9. *Argynnis rupestris*, Behr, ♂, under side.
10. *Argynnis hippolyta*, Edwards, ♂.
11. *Argynnis laura*, Edwards, ♂.
12. *Argynnis laura*, Edwards, ♀.
13. *Argynnis artonis*, Edwards, ♂, under side.

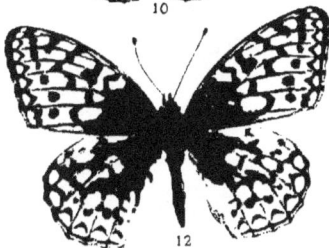

Butterfly.— The male is fulvous upon the upper side, all the dark markings being heavy and black, and the basal areas of the wings clouded with fuscous, this dark clouding on the hind wings reaching down and nearly covering the inner angle. The fore wings on the under side are buff, laved with pale red at the base, marked with ferruginous on the outer margin and about the subapical spots. The submarginal and subapical spots are silvered, especially the latter. The hind wings are deep ferruginous, mottled with buff. The submarginal band is buff, narrow, and dusted with more or less ferruginous. All the spots are well silvered. The female has the basal area of the fore wings bright pinkish-fulvous, and the belt of the secondaries almost lost in the deep ground-color.

(19) **Argynnis bremneri,** Edwards, Plate X, Fig. 7, ♂ (Bremner's Silver-spot).

Butterfly.— The male on the upper side is bright fulvous. The black markings, especially those about the middle of the wing, are heavy. Both wings at the base are clouded with fuscous, the under side of the primaries red toward the base, buff on the apical area; the subapical and the upper marginal spots well silvered; the hind wings with the inner two thirds more or less deeply ferruginous, a little mottled with buff, very rarely encroached upon by the dark color of the inner area, except occasionally near the anal angle. Expanse, ♂, 2.40 inches; ♀, 2.70 inches.

Early Stages.— The early stages have not as yet been described.

This species is found in Oregon, Washington, Montana, and in the southern portions of British Columbia and Vancouver's Island.

(20) **Argynnis zerene,** Boisduval, Plate XIV, Fig. 9, ♂, *under side* (Zerene).

Butterfly.— The male on the upper side is reddish-fulvous, with rather heavy black markings, the mesial band of spots being confluent. The under side of the fore wings is reddish, inclining to pink, with the apex laved with buff. The hind wings have the ground-color purplish-gray, mottled on the inner two thirds with ferruginous. The spots are not silvered, but are a delicate gray color. The female is colored like the male, but the red at the base of the fore wings in this sex is much deeper, and the yellow at the apex of the primaries contrasts much more strongly. The spots on the under side in the female sex are frequently well silvered, though in many specimens they are colored exactly as in the male sex. Expanse of wing, ♂, 2.17 inches; ♀, 2.50 inches.

Early Stages.— The early stages of this species have not as yet been ascertained.

This beautiful butterfly, which is somewhat inclined to variation, is found in northern California, being quite common about Mount Shasta. It is also found in Oregon and Nevada. One of the varietal forms was named *Argynnis purpurescens* by the late Henry Edwards, because of the decided purplish tint which prevails on the under side of the secondaries, extending over the entire surface of the hind wings and covering likewise the apex of the fore wings. This purplish-brown is very marked in specimens collected about the town of Soda Springs, in northern California.

(21) **Argynnis monticola,** Behr, Plate XIII, Fig. 7, ♂, *under side;* Fig. 8, ♂ ; Plate XIV, Fig. 17, ♀ (Behr's Fritillary).

Butterfly.— This species is very closely allied to the preceding in some respects; the upper surface, however, of the wings in both sexes is brighter than in *zerene*, and the dark markings stand forth more clearly upon the lighter ground-color. The wings are not shaded with fuscous toward the base as much as in *A. zerene.* While the markings on the upper side are almost identical with those of Dr. Boisduval's species, they are much brighter and clearer, giving the insects quite a different aspect. On the under side the wings are colored as in *zerene*, the primaries in the male being ferruginous, laved with a little red toward the base, marked with purplish-gray toward the apex, the light spots near the end of the cell on this wing being pale buff. The hind wings are very uniformly purplish-gray, mottled with dark brown, the spots very little, if at all, silvered in the male. In the female the fore wings are bright red at the base, and the hind wings are colored as in the male; but all the spots in both the fore wings and hind wings are broadly and brightly silvered.

Early Stages.— The early stages have not been ascertained, and there remains something here for young entomologists to accomplish.

This species is quite common in the same localities as the last, and some authors are inclined to regard it as being a mere variety, which is a belief that can only be verified by careful breeding from the egg.

(22) **Argynnis rhodope**, Edwards, Plate XI, Fig. 6, ♀, *under side* (Rhodope).

Butterfly.— In the male sex the upper side is bright fulvous, with both wings on the inner half heavily clouded with dark fuscous. The black markings are very heavy and confluent. The outer border is solid black, very slightly, if at all, interrupted by a narrow marginal brown line, in this respect resembling *A. atlantis*. On the under side the fore wings are dark ferruginous, on the outer margin rich dark brown. Between the spots at the end of the cell and the nervules below the apex are some clear, bright straw-yellow spots. The upper spots of the marginal series are silvered. The hind wings are dark reddish-brown, very slightly paler on the line of the marginal band. The spots are pale straw-yellow, except those of the marginal series, which are distinctly silvered. The female on the upper side is of a lighter and brighter red, with the markings dark and heavy as in the male sex. On the under side the markings in the female do not differ from those in the male, except that the primaries on the inner half and at the base are bright pinkish-red. Expanse, ♂, 2.20 inches; ♀, 2.40 inches.

Early Stages.— Unknown.

This striking species has been heretofore only found in British Columbia.

(23) **Argynnis behrensi**, Edwards, Plate XIV, Fig. 10, ♂, *under side* (Behrens' Fritillary).

Butterfly.— The male on the upper side is dull fulvous, clouded with fuscous at the base, the black markings much narrower and lighter than in the preceding species. The primaries on the under side are pale fulvous, clouded with dark brown at the apex. The subapical spots and the upper spots of the marginal series on this wing are well silvered. The hind wings on the under side are deep reddish-brown, with the marginal band only faintly indicated. All the spots are distinctly well silvered. The female does not differ materially from the male, except in the larger size and the somewhat paler ground-color of the upper side of the wings. On the under side the wings are exactly as in the male, with the marginal band even less distinct than in that sex.

Early Stages.— Not yet ascertained.

The type specimens upon which the foregoing description is founded came from Mendocino, in California.

(24) **Argynnis halcyone,** Edwards, Plate XIII, Fig. 5, ♂ ; Fig. 6, ♂, *under side* (Halcyone).

Butterfly, ♂.— The primaries are produced and relatively narrower than in the preceding species, fulvous on the upper side, with the black markings distinct, the mesial band of the secondaries confluent. The fore wings on the under side are pale fulvous, reddish at the base, pale buff at the end of the cell and on the costal margin before the apex. The subapical spots and the pale spots of the marginal series are very little silvered. The hind wings have the inner two thirds deep reddish-brown, slightly mottled with buff. The marginal band is buff, and all the spots are well silvered.

♀.—The female, which is considerably larger than the male, is marked much as in that sex; but all the black markings are heavier, and on the under side of the primaries the base and inner margin are laved with red. The marginal band on the hind wings is not as distinct in this sex as in the male, in many specimens being somewhat obscured by olive-brown. Expanse, ♂, 2.50 inches; ♀, 2.90–3.10 inches.

Early Stages.— Not known.

This species, which is still rare in collections, is found in southern Colorado and the adjacent parts of Utah and Arizona.

(25) **Argynnis chitone,** Edwards, Plate XIV, Fig. 16, ♀ (Chitone).

Butterfly, ♂.—The wings on the upper side are dull fulvous, greatly obscured by brown at the base of the wings. The dark spots and markings are not heavy. The fore wings on the under side are yellowish-fulvous at the base and on the inner half of the wing; the apical patch and the nervules on the apical area are heavy ferruginous; the marginal spots are buff, with no silver. The hind wings on the under side are light ferruginous, mottled with buff; the belt is broad, clear buff; the outer margin is brown. All the spots are small and imperfectly silvered.

♀.—The female is nearly the same shade as the male, with the marginal spots on the under side always silvered, the remainder without silver, or only now and then with a few silvery scales. Expanse, 2.25–2.50 inches.

EXPLANATION OF PLATE XIII

1. *Argynnis cybele*, Fabricius, ♀, under side.
2. *Argynnis semiramis*, Edwards, ♂.
3. *Argynnis semiramis*, Edwards, ♀.
4. *Argynnis inlocris*, Edwards, ♂, under side.
5. *Argynnis halcyone*, Edwards, ♂.
6. *Argynnis halcyone*, Edwards, ♀, under side.
7. *Argynnis monticola*, Behr, ♂, under side.

8. *Argynnis monticola*, Behr, ♂.
9. *Argynnis macaria*, Edwards, ♂.
10. *Argynnis inornata*, Edwards, ♀, under side.
11. *Argynnis liliana*, Henry Edwards, ♂.
12. *Argynnis atossa*, Edwards, ♂.
13. *Argynnis egleis*, Boisduval, ♂.
14. *Argynnis egleis*, Boisduval, ♀, under side.
15. *Argynnis egleis*, Boisduval, ♀.

PLATE XIII

Early Stages.—Not ascertained.

This species occurs in southern Utah and Arizona.

(26) **Argynnis platina**, Skinner, Plate XVIII, Fig. 7, ♂
(Skinner's Fritillary).

Butterfly, ♂.—The original description of this species, contained in the "Canadian Entomologist," vol. xxix, p. 154, is as follows:

"♂.—Expands two and a half inches. Upper side: Rather light tawny or even light buff. Black markings dense and wide, with outer halves of wings looking rather clear or open, with rows of round spots not very large; marginal border light; bases of wings not much obscured. Under side: Superiors have the two subapical silver spots and silver spots on margin well defined; color of inner half of wing rosy. The silver spots on the inferiors are large and well defined, and placed on a very light greenish-gray ground. The intermediate buff band is well defined, comparatively wide, and very light in color. ♀.—The ground-color on the inferiors below is reddish-brown in the female."

Early Stages.— Unknown.

This species occurs in Utah and Idaho, and is possibly a varietal form of *A. coronis*, specimens agreeing very nearly with the type figured in the plate being contained in the Edwards collection under the name of *A. coronis*.

(27) **Argynnis coronis**, Behr, Plate XI, Fig. 10, ♂ ; Fig. 11, ♀ (Coronis).

Butterfly, ♂.—The wings on the upper side are yellowish-brown, with but little brown obscuring the base. The dark markings are not heavy, but distinct. The fore wings on the under side are buff, with the basal area orange-fulvous. The subapical and submarginal spots are more or less imperfectly silvered. The hind wings are brown, mottled with reddish. The discal area is buff, and the belt is pale yellowish-buff. All the spots are large and well silvered on these wings.

♀.—The female is paler than the male, with the markings on the upper side a little heavier. The wings on the under side are much as in the male sex. Expanse, ♂, 2.10–2.50 inches; ♀, 2.50–3.00 inches.

Early Stages.—The early stages remain to be ascertained.

This species ranges from southern California northward to the

117

southern part of British Columbia, and is found as far east as Utah.

(28) **Argynnis snyderi**, Skinner, Plate XVIII, Fig. 6, ♂ (Snyder's Fritillary).

Butterfly, ♂.—The wings on the upper side are light tawny, but little obscured by fuscous at the base. The black markings are moderately heavy and very sharply 'defined against the lighter ground-color. The outer margin is distinctly but not heavily marked. On the under side of the fore wings there are two subapical and five marginal silver spots. The ground-color of the under side of the hind wings is grayish-green, with a narrow pale-buff marginal belt. The spots are large and well silvered.

♀.—The female is much like the male, but on the hind wings the ground-color from the base to the outer belt is brownish. Expanse, ♂, 3.00 inches; ♀, 3.30 inches.

Early Stages.—Unknown.

This species, which is very closely allied to *A. coronis*, is found in Utah.

(29) **Argynnis callippe**, Boisduval, Plate XI, Fig. 1, ♂; Fig. 2, ♀; Fig. 3, ♀, *under side* (Callippe).

Butterfly.—This species may easily be recognized by the general obscuration of the basal area of the wings, the light-buff quadrate spots on the discal area of the fore wings, and the clear oval spots of the same color on the hind wings, as well as by the light triangular marginal spots, all standing out distinctly on the darker ground. The wings on the under side are quite pale buff, with the spots large and well silvered. Expanse, 2.30–3.00 inches.

Early Stages.—Unknown.

Callippe is abundant in California.

(30) **Argynnis nevadensis**, Edwards, Plate X, Fig. 4, ♂, *under side* (The Nevada Fritillary).

Butterfly, ♂.—The ground-color is pale fulvous, but little obscured with fuscous at the base. The outer margins are heavily bordered with black. The dark markings of the discal area are not heavy. The fore wings on the under side are pale buff, the spots well silvered; the hind wings are greenish; the belt is narrow and clear, and the spots are large and well silvered.

♀.—The female is much like the male, but larger and paler. The outer margin of the fore wings in this sex is more heavily marked with black, and the marginal spots are light buff in color. Expanse, ♂, 2.50–3.00 inches; ♀, 3.00–3.50 inches.

Early Stages.—These remain to be discovered.

This species is found in the Rocky Mountains of Utah, Nevada, Montana, and British America.

(31) **Argynnis meadi,** Edwards, Plate XIV, Fig. 1, ♂ ; Fig. 2, ♂, *under side* (Mead's Silver-spot).

Butterfly.—This species is very closely allied to the preceding, of which it may be an extreme variation, characterized by the darker color of the fore wings on the upper side, the nervules being heavily bordered with blackish, and the deeper, more solid green of the under side of the wings. All the specimens I have seen are considerably smaller in size than *A. nevadensis.*

Early Stages.—Wholly unknown.

This species or variety is found from Utah northward to the province of Alberta, in British America.

(32) **Argynnis edwardsi,** Reakirt, Plate XI, Fig. 4, ♂ ; Fig. 5, ♀ (Edwards' Fritillary).

Butterfly.—This beautiful insect is closely related to the Nevada Fritillary, from which it may be distinguished by the brighter color of the upper side, the heavier black borders, especially in the female sex, and the olive-brown color of the under side of the hind wings. The olivaceous of these wings greatly encroaches upon the marginal belt. Expanse, 3.00–3.25 inches.

Early Stages.—These have been carefully and minutely described by Edwards in the "Canadian Entomologist," vol. xx, p. 3. They are not unlike those of *A. atlantis* in many respects.

This species is not uncommon in Colorado and Montana.

(33) **Argynnis liliana,** Henry Edwards, Plate XIII, Fig. 11, ♂ (Liliana).

Butterfly, ♂.—The wings on the upper side are reddish-fulvous. The black markings and the spots are slight. The fore wings on the under side are yellowish-buff; the base and the hind margin to below the cell, brown, with buff on the median interspaces. The outer end of the cell is yellowish-buff. The subapical patch is brown, adorned by two or three well-silvered spots. The five upper marginal spots are well silvered. The hind wings are brown, but little mottled with buff. The spots are well silvered. The marginal belt is narrow, ochreous-brown.

♀.—The female is much paler than the male, and the marginal

119

spots on both wings are much lighter. On the under side the wings are as in the male sex, with the basal area and the nervules of the fore wings red. Expanse, ♂, 2.20 inches; ♀, 2.35 inches.

Egg.—W. H. Edwards gives the following description: "Conoidal, truncated, depressed at summit, marked vertically by twenty-two or twenty-three ribs, which are as in other species of the genus; the outline of this egg is much as in *eurynome*, the base being broad, the top narrow, and the height not much more than the breadth; color yellow."

Caterpillar.—The same author has given us a description of the caterpillar immediately after hatching; but as the young larvæ were lost after being sent to Maine to be kept over winter, we do not yet know the full life-history.

The range of this species is northern California and Utah, so far as is known at present.

(34) **Argynnis rupestris,** Behr, Plate XII, Fig. 8, ♂ ; Fig. 9, ♂, *under side* (The Cliff-dwelling Fritillary).

Butterfly, ♂.—The upper side of the fore wings is deep reddish-fulvous, with the black markings very heavy. The fore wings on the under side are buff, shaded with red at the base and on the inner margin. The spots are buff, without any silver. The hind wings are buff, mottled with cinnamon-red, sometimes dark, sometimes lighter. The marginal belt is narrow, buff, encroached upon by the darker color of the median area at the ends of the oval spots. None of the spots is silvered, except very light y in exceptional cases.

♀.—The female is much like the male on the upper side, with the dark markings much heavier, the ground-color somewhat paler, and the marginal row of spots quite light. The wings on the under side are more brightly tinted than in the male, and the marginal spots are more or less silvered. Expanse, ♂, 2.00 inches; ♀, 2.20 inches.

Early Stages.—Nothing is as yet known about the egg and larva.

This species is quite abundant at a considerable elevation upon Mount Shasta, Mount Bradley, and in the Weber Mountains in Utah.

(35) **Argynnis laura,** Edwards, Plate XII, Fig. 11, ♂ ; Fig. 12, ♀ (Laura).

Butterfly, ♂.—The upper side is deep reddish-fulvous, with both wings somewhat obscured at the base by fuscous. The black markings on the upper side of the wings are heavy; the outer margin is also heavily banded with dark brown, the marginal lines being fulvous. The four spots on the hind wings are lighter in color than the ground. The fore wings on the under side are reddish-orange, with the apex and the hind margin yellowish-buff. The apical and upper marginal spots are more or less well silvered. The hind wings are pale yellow, the marginal belt very broad and clear yellow. All the spots are large and well silvered.

♀.—The female is much paler than the male, but otherwise closely resembles that sex. Expanse, ♂, 2.20 inches; ♀, 2.35 inches.

Early Stages.—Unknown.

This species is found in northern California, Oregon, Washington, and Nevada.

(36) **Argynnis macaria,** Edwards, Plate XIII, Fig. 9, ♂ (Macaria).

Butterfly, ♂.—The upper side of the wings is yellowish-fulvous, the black markings very light. The fore wings on the under side are orange-red, at the apex yellowish-buff. The subapical upper marginal spots are lightly silvered. The hind wings are yellowish-buff on the outer third, mottled with brown on the basal and median areas. The marginal belt is clear buff. The spots are large and well silvered.

♀.—The female is paler than the male. On the upper side of the hind wings the second row of silver spots is indicated by spots much paler than the ground. The black markings are lighter than in the male. Expanse, ♂, 2.00 inches; ♀, 2.20 inches.

Early Stages.—Unknown.

This species, which is somewhat like *A. coronis,* but smaller, and brighter fulvous, is found in California, but is still quite rare in collections.

(37) **Argynnis semiramis,** Edwards, Plate XIII, Fig. 2, ♂, *under side;* Fig. 3, ♀ (Semiramis).

Butterfly, ♂.—The wings are bright fulvous on the upper side, with the black markings much as in *A. adiante,* slight on the fore wings and even slighter on the hind wings. The under side of the fore wings is cinnamon-red at the base and on the inner half of the wing, beyond this buff. The apical patch and the outer

121

margin are brown. The upper marginal spots and two spots on the subapical patch are well silvered. The hind wings are rusty-brown from the base to the second row of spots, mottled with lighter brown. The marginal belt is clear brownish-buff. All the spots are well silvered.

♀.—The female on the upper side is colored like the male, with the dark markings somewhat heavier. On the under side the fore wings are laved over almost their entire surface with red, the upper angle of the cell alone being buff. The hind wings are in many specimens fawn-colored throughout, except that the marginal band is paler. In a few specimens the ground is darker and the band more distinct. All the spots are well silvered. Expanse, ♂, 2.60 inches; ♀, 2.75–3.00 inches.

Early Stages.—The life-history of this butterfly has not been ascertained.

The species appears to be very common at San Bernardino, California, and vicinity, and resembles *A. adiante* on the upper side and *A. coronis* upon the lower side.

(38) **Argynnis inornata**, Edwards, Plate XIII, Fig. 10, ♀, *under side* (The Plain Fritillary).

Butterfly, ♂.—This species resembles *A. rupestris* in its markings, but is somewhat paler, the black margins are heavy and the black markings on the disk comparatively light; the base of the wings is obscured with fuscous. On the under side the fore wings are cinnamon-brown, with the apical area buff. The hind wings are reddish-brown, with the marginal band clear buff. All the spots are buff, and completely devoid of silvery scales.

♀.—Paler than the male on the upper side. The fore wings on the under side are orange-fulvous; the hind wings are pale greenish-brown, mottled with buff. In some specimens a few silver scales are found on the submarginal spots. Expanse, ♂, 2.50 inches; ♀, 2.70 inches.

Early Stages.—Unknown.

This butterfly, which is as yet not very common in collections, is found in California and Nevada.

(39) **Argynnis atossa**, Edwards, Plate XIII, Fig. 12, ♂ (Atossa).

Butterfly, ♂.—The upper side is bright yellowish-fulvous, with the wings at the base slightly dusted with brown. The margins of both wings are bordered by a single line, there being no trace

1. *Argynnis meadi*, Edwards, ♂.
2. *Argynnis meadi*, Edwards, ♂, under side.
3. *Argynnis columbia*, Henry Edwards, ♂.
4. *Argynnis adiante*, Boisduval, ♀.
5. *Argynnis clio*, Edwards, ♂.
6. *Argynnis clio*, Edwards, ♀.
7. *Argynnis clio*, Edwards, ♂, under side.
8. *Argynnis opis*, Edwards, ♂, under side.
9. *Argynnis zerene*, Boisduval, ♂, under side.
10. *Argynnis hebrensi*, Edwards, ♂.
11. *Argynnis aphrodite*, Fabricius, ♀, under side.
12. *Argynnis lais*, Edwards, ♂.
13. *Argynnis lais*, Edwards, ♀.
14. *Argynnis eurynome*, Edwards, ♀.
15. *Argynnis eurynome*, Edwards, ♂, under side.
16. *Argynnis chitone*, Edwards, ♀.
17. *Argynnis monticola*, Behr, ♀.

PLATE XIV.

of the outer line usually found in other species of the genus. The dark markings of the outer margin are almost entirely absent, and those of the discal and basal areas very greatly reduced. On the under side both wings are very pale, the spots entirely without silver, in some specimens even their location being but faintly indicated. The fore wings at the base and on the inner margin are laved with bright red.

♀.—The female resembles the male, except that the red on the under side of the fore wings is in many specimens very bright and fiery. Expanse, ♂, 2.50 inches; ♀, 2.75-3.00 inches.

Early Stages.—Entirely unknown.

This butterfly, which is still rare in collections, has been taken in southern California. It may be an extreme variation of the next species, *A. adiante*, Boisduval.

(40) **Argynnis adiante**, Boisduval, Plate XIV, Fig. 4, ♀ (Adiante).

Butterfly, ♂.—The wings on the upper side are bright fulvous; the black markings are slight. The fore wings on the under side are pale buff, much lighter at the apex, laved with orange-red at the base. The hind wings are pale buff, clouded with fawn-color on the basal and discal areas. All the spots which are generally silvered in other species are in this species wholly devoid of silvery scales.

♀.—The female is like the male, but the black markings on the upper side are heavier, and the basal area and inner half of the primaries are laved with brighter and deeper red. Expanse, ♂, 2.30-2.40 inches; ♀, 2.30-2.60 inches.

Early Stages.—Unknown.

This species is found in southern California, and is somewhat local in its habits, hitherto having been taken only in the Santa Cruz Mountains.

(41) **Argynnis artonis**, Edwards, Plate XII, Fig. 13, ♂, *under side* (Artonis).

Butterfly, ♂.—Closely resembling *A. eurynome*, Edwards, from which species it may be at once distinguished by the entire absence of silvery scales upon the under side of the wings, and also by the fact that the silver spots on the under side of the hind wings are not compressed and elongated as much as in *eurynome*, and by the further fact that all the dark marginal markings of the under side are obliterated.

♀.—The female does not differ materially from the male, except that the dark markings on the upper side are all much heavier, standing out very distinctly upon the paler ground, and the marginal spots within the lunules are very light in color and relatively large. On the under side the fore wings are laved with red, very much as in the female of *A. adiante.* Expanse, ♂, 1.75–2.00 inches ; ♀, 2.00–2.15 inches.

Early Stages.—These still remain to be ascertained.

This interesting butterfly, which seems to indicate a transition between the butterflies of the Adiante-group and those of the Eurynome-group, has been found in Colorado, Nevada, Utah, and Arizona.

(42) **Argynnis clio**, Edwards, Plate XIV, Fig. 5, ♂ ; Fig. 6, ♀ ; Fig. 7, ♂, *under side* (Clio).

Butterfly.—Closely resembling *A. eurynome* and *A. artonis.* Like *artonis*, the spots on the under side of the wing are without silver. The female very closely resembles the female of *artonis*, and in fact I am unable to distinguish the types of the females of the two species by any marks which seem to be satisfactory. Expanse, ♂, 1.75 inch; ♀, 1.75–1.90 inch.

Early Stages.—Unknown.

This species, which is as yet comparatively rare in collections, is found in Montana and the province of Alberta, in British America, at a considerable elevation.

(43) **Argynnis opis**, Plate XIV, Fig. 8, ♂, *under side* (Opis).

Butterfly.—This species, which apparently belongs to the Eurynome-group, appears by the location of its markings to be closely related to *eurynome*, but on the upper side the wings of both the male and female are more heavily obscured with fuscous at the base; the dark markings are heavier than in *eurynome*, and in both sexes it is smaller in size, being the smallest of all the species of the genus thus far found in North America. The spots on the under side of the wings are none of them silvered. Expanse, ♂, 1.50 inch; ♀, 1.60 inch.

Early Stages.—Nothing is known of these.

The types came from Bald Mountain, in the Caribou mining district of British Columbia.

(44) **Argynnis bischoffi**, Edwards, Plate XI, Fig. 7, ♂ (Bischoff's Fritillary).

Butterfly, ♂.—The fore wings on the upper side are bright reddish-fulvous, the base of the primaries and the inner half of the secondaries being heavily obscured by blackish, so as to conceal the markings. Both wings have moderately heavy black marginal borders. The other markings are as in *A. eurynome.* On the under side the fore wings are buff, laved with reddish at the base. The hind wings are pale buff, with the basal and discal areas mottled with green. The marginal belt is clear buff. In some specimens the spots on the under side are not silvered; in others they are well silvered.

♀.—The female on the upper side is very pale buff, slightly laved with fulvous on the outer margin of both wings. All the markings are heavy; the margins of both wings are solid black, the spots within the lunules being pale and almost white. The fore wings at the base and the inner half of the hind wings are almost solid black. On the under side the wings are very much as in the male, and the same variation as to the silvering of the spots is found. Expanse, ♂, 1.80 inch; ♀, 1.90 inch.

Early Stages.—Unknown.

The types of this genus came from Sitka, in Alaska. It may be an extreme boreal variation of *A. eurynome.*

(45) **Argynnis eurynome,** Edwards, Plate XII, Fig. 7, ♂; Plate XIV, Fig. 14, ♀; Fig. 15, ♂. *under side* (Eurynome).

Butterfly, ♂.—The wings on the upper side are bright yellowish-fulvous, but little obscured at the base. The outer margins are edged by two fine lines which are occasionally confluent. The under side of the fore wings is pale buff, laved with cinnamon-brown at the base and along the nervules; the spots on the margin and in the apical area are well silvered. The hind wings on the under side are buff, with the basal and discal areas mottled with pale brown or pale olive-green. The marginal belt is broad and clear buff; all the spots are well silvered.

♀.—The female is like the male, but paler, with the dark markings, especially those of the margin, heavier. The marginal spots inclosed by the lunules are much paler than the ground-color, and in many specimens almost white. On the under side the wings in this sex are like those of the male, but the fore wings are more heavily laved with cinnamon-brown at the base. Expanse, ♂, 1.70–2.00 inches; ♀, 2.00 inches.

Early Stages.—Mr. Edwards, in "The Butterflies of North America," vol. ii, has given us a beautiful figure of the egg of this species. Of the other stages we have no knowledge.

A. eurynome is a very common butterfly in Colorado, Montana, and British America, and is the representative of a considerable group, to which the four preceding species belong, if, indeed, they are not local races or climatic varieties of *eurynome*, a fact which can be demonstrated only by the careful breeding of specimens from various localities. There is a fine field here for study and experiment.

(46) **Argynnis montivaga,** Behr, Plate X, Fig. 5, ♂, *under side* (Montivaga).

Butterfly.—This species in both sexes very closely approximates the foregoing. The main points of distinction consist in the somewhat darker red of the upper side of the wings, the slightly heavier dark markings, and the absence on the under side, especially of the hind wings, of the olive-green shade which is characteristic of typical specimens of *A. eurynome*. The mottling of the basal and median areas on this side is reddish-brown. The spots are more or less silvered on the under side. Expanse, ♂, 1.75 inch; ♀, 1.90 inch.

Early Stages.—Unknown.

This species is found in the Sierras of California and among the mountains of Nevada.

(47) **Argynnis egleis,** Boisduval, Plate XIII, Fig. 13, ♂ ; Fig. 14, ♀, *under side;* Fig. 15, ♀ (Egleis).

Butterfly, ♂.—The ground-color of the wings on the upper side is deep fulvous, with rather heavy black markings. The wings on the under side are pale fulvous, mottled with buff on the subapical interspaces of the fore wings. The basal and discal areas of the hind wings are mottled with brown, which in many specimens is of a distinctly purplish shade. In some specimens the inner half of the primaries is rather heavily laved with red. The spots on the under side are either silvered or without silver, in the latter case being pale buff.

♀.—The female is much like the male, but paler. The red on the under side of the primaries is deeper, and the purplish-brown on the inner surface of the secondaries is also darker. Expanse, ♂, 2.25 inches; ♀, 2.50 inches.

Early Stages.—These remain to be ascertained.

This is a common species in California and Nevada. For many years it has been placed in all catalogues at the end of the list of the species of this genus, where I also leave it, though to my way of thinking its proper location is near *A. rupestris.* It certainly reveals but small affinity to the species of the Eurynome-group.

Besides the species of *Argynnis* enumerated in the foregoing pages and delineated upon the plates, there are several others of more or less doubtful validity credited to our fauna, and a number of varieties which have received names. With all of these the more advanced student will become familiar as he prosecutes his researches, but it is not necessary to speak of them here.

A RACE AFTER A BUTTERFLY

There is much that is pleasing about "first things." I shall never forget the first dollar I earned; the first trout I took with my fly; the first muskalonge I gaffed beside my canoe on a still Canadian lake; the first voyage I made across the Atlantic. So I shall never forget my first capture of a female specimen of *Argynnis diana.*

My home in my boyhood was in North Carolina, in the village of Salem, famous as one of the most successful of the settlements made by the Moravian Brethren under the lead of the good Count Zinzendorf, and well known throughout the Southern States as the seat of an excellent seminary for young ladies. The Civil War broke out, and the hopes cherished of sending me North to be educated were disappointed. I was left to pursue my studies under a tutor, and to roam the neighborhood in quest of insects, of which I gathered a large collection.

One day I spied upon a bed of verbenas a magnificent butterfly with broad expanse of wing and large blue spots upon the secondaries. In breathless haste I rushed into the house and got my net. To the joy of my heart, when I returned to the spot, the beauty was still hovering over the crimson blossoms. But, as I drew near with fell intent, it rose and sailed away. Across the garden, over the fence, across the churchyard, out into the street, with leisurely flight the coveted prize sped its way, while I quickly followed, net in hand. Once upon the dusty street, its flight was accelerated; my rapid walking was converted into a run. Down past the church and — *horribile dictu!* — past the boarding-school that pesky butterfly flew. I would rather have faced a cannonade in those days than a bevy of boarding-school misses, but there was no alternative. There were the dreaded females at the windows (for it was Saturday, and vacation hour), and there was my butterfly. Sweating, blushing, inwardly anathematizing my luck, I rushed past the school, only to be overwhelmed with mortification by the rascally porter of the institution, who was sweeping the pavement, and who bawled out after me: "Oh, it 's no use; you can't catch it! It 's frightened; you 're so ugly!" And now it began to rise in its flight. It was plainly my last chance, for it would in a moment be lost over the housetops. I made an upward leap, and by a fortunate sweep of the net succeeded in capturing my prize.

Many years later, after a long interval in which ornithology and botany had engrossed my mind to the exclusion of entomology, my boyish love for the butterflies was renewed, and I found out the name of the choice thing I had captured on that hot July day on the streets of Salem, and returned to North Carolina for the special purpose of collecting a quantity of these superb insects. My quest was entirely successful, though my specimens were not taken at Salem, but under the shadow of Mount Mitchell, in the flower-spangled valleys which lie at its feet.

Genus BRENTHIS, Hübner

"The garden is fragrant everywhere;
In its lily-bugles the gold bee sups,
And butterflies flutter on winglets fair
Round the tremulous meadow buttercups."
MUNKITTRICK.

Butterfly.—Small or medium-sized butterflies, very closely approximating in form and color the species of the genus *Argynnis*, in which they are included by many writers. The principal structural difference between the two genera is found in the fact that in the genus *Brenthis* only one of the subcostal nervules arises before or at the end of the cell of the primaries, while in *Argynnis* the two innermost subcostal nervules thus arise. In *Brenthis* the palpi are not as stout as in *Argynnis*, and the short basal spur or branch of the median vein of the front wings, which is characteristic of the latter genus, is altogether lacking in *Brenthis*.

Egg.—The eggs are subconical, almost twice as high as wide, truncated at the top, and marked with thirteen or fourteen raised longitudinal ridges connected by a multitude of smaller cross-ridges.

Larva.—The caterpillars are not noticeably different in their general appearance from those of the genus *Argynnis*, except that they are smaller and generally not as dark in color as the larvæ of the latter genus. They feed, like the caterpillars of *Argynnis*, upon violets.

Chrysalis.—The chrysalis is pendant, about six tenths of an inch long, and armed with two rows of sharp conical tubercles on the back.

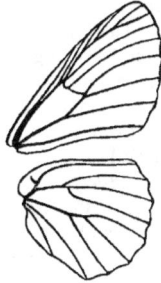

Fig. 90.—Neuration of the genus *Brenthis*, enlarged.

(1) **Brenthis myrina**, Cramer, Plate XV, Fig. 1, ♂ ; Fig. 2, ♂, *under side;* Plate V, Figs. 12–14, *chrysalis* (The Silver-bordered Fritillary).

Butterfly.—The upper side of the wings is fulvous; the black markings are light, the borders heavy. The fore wings on the under side are yellowish-fulvous, ferruginous at the tip, with the marginal spots lightly silvered. The hind wings are ferruginous, mottled with buff. The spots, which are small, are well silvered. Expanse, ♂, 1.40 inch; ♀, 1.70 inch.

Egg.—The egg is conoidal, about one third higher than wide, marked by sixteen or seventeen vertical ribs, between which are a number of delicate cross-lines. It is pale greenish-yellow in color.

Caterpillar.—The caterpillar has been carefully studied, and its various stages are fully described in "The Butterflies of New England," by Dr. Scudder. In its final stage it is about seven eighths of an inch long, dark olive-brown, marked with green, the segments being adorned with fleshy tubercles armed with needle-shaped projections, the tubercles on the side of the first thoracic segment being four times as long as the others, cylindrical in form, and blunt at the upper end, the spines projecting upward at an angle of forty-five degrees to the axis of the tubercle.

129

Chrysalis.—The chrysalis is yellowish-brown, spotted with darker brown spots, those of the thoracic and first and second abdominal segments having the lustre of mother-of-pearl.

This very pretty little species has a wide range, extending from New England to Montana, from Nova Scotia to Alaska, and southward along the ridges of the Alleghanies into Virginia and the mountains of North Carolina.

(2) **Brenthis triclaris,** Hübner, Plate XV, Fig. 3, ♂ (Hübner's Fritillary).

Butterfly, ♂.—The male above is bright fulvous, with the base of the fore wings and the inner margin of the hind wings heavily obscured with blackish scales. The usual dark markings are finer than in the preceding species; the black marginal borders are not so heavy. The submarginal spots are relatively large and distinct in most specimens, and uniform in size. The light spots of the under side of the median band of the hind wings show through from below on the upper side lighter than the ground-color of the wings. On the under side the fore wings are fulvous, tipped with ferruginous. The hind wings are broadly ferruginous, with a couple of bright-yellow spots near the base and a curved band of yellow spots crossing the median area. The outer margin about the middle is marked with pale fulvous. The spots on the under side are none of them silvered.

♀.—The female is much paler than the male in most cases, and the marginal spots within the lunules are very pale, almost white. The submarginal row of round black spots is relatively large and distinct, quite uniform in size. On the under side the wings are much more conspicuously marked on the secondaries than in the male sex, being crossed by three conspicuous bands of irregularly shaped yellow spots, one at the base and one on either side of the discal area. The submarginal round spots of the upper side reappear on the under side as small, slightly silvered, yellow spots. The marginal spots are bright yellow, slightly glossed with silver. Expanse, ♂, 1.50 inch; ♀, 1.60 inch.

Early Stages.—Unknown.

This extremely beautiful little species is found throughout arctic America, is not uncommon in Labrador, and also occurs upon the loftier summits of the Rocky Mountains in Colorado and elsewhere. It is, as most species of the genus, essentially arctic in its habits.

EXPLANATION OF PLATE XV

1. *Brenthis myrina*, Cramer, ♂.
2. *Brenthis myrina*, Cramer, ♂, under side.
3. *Brenthis triclaris*, Hübner.
4. *Brenthis chariclea* Schneider,
5. *Brenthis boisduvali* Duponchel.
6. *Brenthis boisduvali* Duponchel. under side.
7. *Brenthis montinus*, Scudder, ♂.
8. *Brenthis montinus*, Scudder, ♂, under side.
9. *Brenthis freija*, Thunberg, ♂.
10. *Brenthis freija*, Thunberg, ♀, under side.
11. *Brenthis polaris*, Boisduval. ♂.
12. *Brenthis polaris*, Boisduval, ♂, under side.
13. *Brenthis frigga*, Thunberg, ♂.
14. *Brenthis frigga*, Thunberg, ♀, under side.
15. *Brenthis alberta*, Edwards, ♂.
16. *Brenthis bellona*, Fabricius. ♂.
17. *Brenthis epithore*, Boisduval, ♂.
18. *Brenthis epithore*, Boisduval, ♂, under side.

(3) **Brenthis helena**, Edwards, Plate XVIII, Fig. 16, ♂, *under side:* Fig. 17, ♂ (Helena).

Butterfly, ♂.—The wings on the upper side are fulvous, greatly obscured by brown at the base of the fore wings and along the inner margin of the hind wings. The usual black markings are light, and the marginal border is also not so heavily marked as in *B. myrina*. The fore wings on the under side are pale fulvous, laved with ferruginous at the tip. The hind wings are brightly ferruginous, with small yellow marginal spots, and paler spots inclining to buff on the costal border and at the end of the cell, about the region of the median nervules.

♀.—The female is very much like the male on the upper side, but the ground-color is paler. On the under side the wings are somewhat paler, and all the spots and light markings, especially on the secondaries, are far more conspicuous, being bright yellow, and standing out very prominently upon the dark ferruginous ground. Expanse, 1.40 inch.

Early Stages.—The early stages of this insect are not as yet known.

Helena appears to be a common species in Colorado, Montana, and New Mexico. It is subject to considerable variation, both in the intensity of the coloring of the under side of the wings, and in the distinctness of the maculation.

(4) **Brenthis montinus**, Scudder, Plate XV, Fig. 7, ♂ ; Fig. 8, ♀, *under side* (The White Mountain Fritillary).

Butterfly, ♂.—The upper side is fulvous, closely resembling *B. chariclea*, but the ground-color is darker. The under side of the hind wings is deep ferruginous, mottled with white, the most conspicuous of the white spots being a white bar occurring at the end of the cell, and a small round white spot at the base of the wing. The hind wings have also a marginal row of slightly silvered white spots.

♀.—The female is very much like the male, but the ground-color of the upper side is paler. Expanse, ♂, 1.50 inch; ♀, 1.75 inch.

This interesting butterfly is found on the barren summits of Mount Washington, New Hampshire. It represents the survival of the arctic fauna on these desolate peaks, and, like the arctic flora of the spot where it is found, is a souvenir of the ice-age, which once shrouded the northeastern regions of the United States with glaciers.

(5) **Brenthis chariclea,** Schneider, Plate XV, Fig. 4, ♂ (Chariclea).

Butterfly, ♂.—Fulvous on the upper side, with heavy black markings, both wings greatly obscured at the base by fuscous. On the under side the fore wings are pale yellowish-fulvous, mottled with ferruginous at the tip and on the outer margin. The hind wings on the under side are dark purplish-ferruginous, mottled with yellow, crossed by a central row of conspicuous yellow spots. The row of marginal spots and two or three small spots at the base are white, slightly silvered.

♀.—The female differs from the male in having the markings of the upper side darker and heavier, and the outer margins more heavily marked with black, and having all the spots on the under side more distinctly defined against the dark ground. Expanse, ♂, 1.50 inch; ♀,1.75 inch.

Early Stages.—Undescribed.

This species, like *B. freija,* is circumpolar, being found in Lapland, Greenland, and throughout arctic America. It also occurs within the limits of the United States, in the Yellowstone Park at considerable elevations, and is not uncommon on the high mountains in British Columbia, numerous specimens having been captured in recent years about Banff and Laggan, in Alberta.

(6) **Brenthis boisduvali,** Duponchel, Plate XV, Fig. 5, ♂ ; Fig. 6, ♀, *under side* (Boisduval's Fritillary).

Butterfly.—Somewhat closely resembling *B. chariclea,* but with the markings much heavier on the outer margin, and the base of the wings generally more deeply obscured with dark brown. The wings on the under side in color and marking closely approximate those of *B. chariclea,* and I have been unable to distinguish the specimens marked as *boisduvali,* and contained in the Edwards collection, from the specimens designated as *B. chariclea* in the same collection, so far as the color and maculation of the under sides of these specimens are concerned. Expanse, ♂, 1.50 inch; ♀, 1.75 inch.

Early Stages.—Unknown.

This species, originally described from Labrador, is found throughout boreal America and British Columbia.

(7) **Brenthis freija,** Thunberg, Plate XV, Fig. 9, ♂ ; Fig. 10, ♀, *under side* (The Lapland Fritillary).

Butterfly.— The wings are pale fulvous, the fore wings at the base and the hind wings on the inner half being deeply obscured with fuscous. The markings are quite heavy. The fore wings on the under side are very pale fulvous, yellowish at the tip, mottled with ferruginous. The hind wings are ferruginous on the under side, mottled with yellow. The spots are quite large, consisting of lines and dashes, and a marginal row of small lunulate spots, pale yellow or white, slightly silvered. Expanse, 1.50 inch.

This butterfly is circumpolar, being found in Norway, Lapland, northern Russia, and Siberia, through Alaska, British America, and Labrador, occurring also upon the highest peaks of the Rocky Mountains as far south as Colorado.

(8) **Brenthis polaris,** Boisduval, Plate XV, Fig. 11, ♂ ; Fig. 12, ♂, *under side* (The Polar Fritillary).

Butterfly. — The upper side dull fulvous; the markings on the inner half of the wings are confluent, and lost in the brownish vestiture which obscures this portion of the wing. The outer median area is defined by irregular zigzag spots which flow together. Beyond these the submarginal row of small black spots stands out distinctly upon the lighter ground-color of the wings. The outer margin is marked by black spots at the end of the nervules, on the fore wings somewhat widely separated, on the hind wings narrowly separated by the lighter ground-color. On the under side the wings are fulvous, with a marginal row of white checkerings on both wings. The hind wing is deeply mottled with ferruginous, on which the lighter white markings stand forth very conspicuously. Expanse, ♂, 1.50 inch; ♀, 1.50–2.00 inches.

Early Stages. — Unknown.

This butterfly has been found in Labrador, Greenland, and other portions of arctic America, as far north as latitude 81° 52'.

(9) **Brenthis frigga,** Thunberg, Plate XV, Fig. 13, ♂ ; Fig. 14, ♀, *lower side* (Frigga).

Butterfly, ♂. — On the upper side this butterfly somewhat closely resembles *polaris,* but the markings are not so compact — more diffuse. The fore wings at the base and the hind wings on the inner two thirds are heavily obscured with brown. The outer margins are more heavily shaded with blackish-brown than in *B. polaris.* On the under side the wings are quite differently marked. The fore wings are fulvous, shaded with brown at the tips, and marked with light yellow on the interspaces beyond the end of the cell. The hind wings are dark ferruginous, shading into purplish-gray on the outer margin, with a whitish quadrate spot on the costa near the base, marked with two dark spots, and a bar of pale, somewhat obscured spots, forming an irregular band across the middle of the hind wings.

♀.—The female does not differ greatly from the male, except that the spots on the under side of the hind wings stand forth more conspicuously, being lighter in color and better defined. Expanse, 1.65–2.00 inches.

This pretty little butterfly occurs in Labrador, across the continent as far west as northern Alaska, and is also occasionally taken upon the alpine summits of the Rocky Mountains as far south as Colorado.

(10) **Brenthis bellona**, Fabricius, Plate XV, Fig. 16, ♂ ; Plate V, Fig. 10, *chrysalis, side view;* Fig. 11, *chrysalis, side view* (Meadow Fritillary).

Butterfly.—Pale fulvous on the upper side, with the dark markings on the inner half of the wing narrow, but more or less confluent. The dark markings on the outer part of the wing are slighter. The fore wings are a little angled on the outer margin below the apex. On the under side the fore wings are pale fulvous, mottled with purple at the tip and on the outer margin. The hind wings on this side are ferruginous, mottled with purple. Expanse, 1.65–1.80 inch.

Egg.—The egg of this species is similar in form, size, color, and markings to the egg of *B. myrina*.

Caterpillar.—The caterpillar also in its early stages closely resembles *myrina*, but in its mature form it differs in not having the spines on the second segment of the body lengthened as in that species.

Chrysalis.—The chrysalis, which is represented in Plate V, is bluish-gray in color, marked with dark spots. The life-history has been given us by several authors.

This butterfly is very common in the whole of the northern United States, as far south as the mountain-ranges of Virginia, and occurs throughout Quebec, Ontario, and British America, as far west as the foot-hills of the Rocky Mountains. It flies commonly with *B. myrina*, the only other species of the genus found in the densely populated portions of our territory, from which it may be at once distinguished by the entire absence of the silvered markings which make *B. myrina* so bright and attractive.

(11) **Brenthis epithore**, Boisduval, Plate XV, Fig. 17, ♂; Fig. 18, ♂, *under side* (Epithore).

Butterfly.—This species on the upper side is pale fulvous, with the markings slighter than in *B. bellona*, and the inner half of the hind wings much more heavily clouded with fuscous. On the under side the wings are somewhat like those of *B. bellona*, but less purple and mottled more distinctly with yellow. Expanse, ♂, 1.50 inch; ♀, 1.85 inch.

Early Stages.—Undescribed.

This species appears to replace *B. bellona*, its close ally, in California, Oregon, and the States eastward as far as parts of Colorado.

(12) **Brenthis alberta**, Edwards, Plate XV, Fig. 15, ♂ (Alberta).

Butterfly.—This, the least attractive in appearance of the species composing the genus, has pale wings with a "washed-out" appearance on the upper side, almost all the dark markings being greatly reduced or obliterated. On the under side the wings are even more obscurely marked than on the upper side. The female is darker than the male, and specimens have a greasy look. Expanse, ♂,1.55 inch; ♀,1.65–1.75 inch.

Early Stages.—Unknown, except the egg and the young caterpillar, which have been most beautifully figured by Edwards in vol. iii of "The Butterflies of North America." The only locality from which specimens have as yet been received by collectors is Laggan, in Alberta, where the species apparently is not uncommon at lofty elevations above sea-level.

(13) **Brenthis astarte**, Doubleday and Hewitson, Plate XVIII, Fig. 14, ♂; Fig. 15, ♀, *under side* (Astarte).

Butterfly.—This rare insect, the largest of the genus, may at once be distinguished from all others by the very beautiful markings of the under side of the hind wings, crossed by a band of irregular, bright-yellow spots, which are narrowly edged with black, and beyond the black bordered by red. Expanse, ♂, 2.00 inches; ♀, 2.15 inches.

Early Stages.—Unknown.

The first description and figure of this insect were given by Doubleday and Hewitson in their large and now very valuable work on "The Genera of Diurnal Lepidoptera." They correctly attributed it to the Rocky Mountains, but Kirby afterward gave Jamaica as its habitat, and this led to its subsequent redescription by Edwards under the name *Victoria*. It is a rare species still, having been received only from Laggan, Alberta, where it was rediscovered by that most indefatigable collector and observer, Mr. T. E. Bean. It frequents the highest summits of the lofty mountains about this desolate locality. Mr. Bean says: "*Astarte* seems always on the lookout for an entomologist, whose advent is carefully noted, and at any approach of such a monster nearer than about fifteen feet, its wings rise to half-mast, vibrate there a doubtful instant, and away goes the butterfly."

In addition to the thirteen species figured in our plates there are two other species of the genus, *B. butleri*, Edwards, from Grinnell Land, and *B. improba*, Butler, from near the arctic circle. It is not likely that many of the readers of this book will encounter these insects in their rambles, and if they should, they will be able to ascertain their names quickly, by conferring with the author.

SUSPICIOUS CONDUCT

The entomologist must not expect to be always thoroughly understood. The ways of scientific men sometimes appear strange, mysterious, bordering even upon the insane, to those who are uninitiated. A celebrated American naturalist relates that on one occasion, when chasing butterflies through a meadow belonging to a farmer, the latter came out and viewed him with manifest anxiety. But when the nature of the efforts of the man of science had been finally explained, the farmer heaved a sigh of relief, remarking, in Pennsylvania Dutch, that "he had surely thought, when he first saw him, that he had just escaped from a lunatic asylum." The writer, a number of years ago, after having despatched a very comfortable lunch, sallied forth one afternoon, in quest of insects, and in the course of his wanderings came upon a refuse-heap by the roadside, opposite a substantial house, and on this heap discovered an ancient ham, which was surrounded by a multitude of beetles of various species known to be partial to decomposed, or semi-decomposed, animal matter. He proceeded immediately to bottle a number of the specimens.

136

While engaged in so doing, the window of the house across the way was thrown up, and an elderly female thrust her head out, and in strident voice exclaimed: "Hey, there! What are you doin' with that ham? I say, don't you know that that ham is spiled?" As he paid no attention to her, she presently appeared at the door, came across the street, and remarked: "See here, mister; that ham 's spiled; Lucy and me throwed it out, knowin' it was no good. If you want a good meal of wittles, come into the house, and we will feed you, but for mercy's sake leave that spiled ham alone." It took considerable effort to assure her that no designs upon the ham were cherished, and she went away, evidently completely mystified at the wild conduct of the well-dressed man who was grubbing in the rubbish-pile.

Genus MELITÆA, Fabricius

(The Checker-spots)

" The fresh young Flie, . . .
. . . joy'd to range abroad in fresh attire,
Through the wide compass of the ayrie coast;
And, with unwearied wings, each part t' inquire
Of the wide rule of his renowned sire."

SPENSER.

Butterfly.—Small. The tibiæ and the tarsi of the mesothoracic and metathoracic legs are more lightly armed with spines than in the genera *Argynnis* and *Brenthis.* The palpi are not swollen. They are clothed with long hairs and have the third joint finely pointed. The antennæ are about half as long as the costa of the fore wings, and are provided with a short, heavy, excavated, or spoon-shaped club. The subcostal of the fore wings is five-branched, the first nervule always arising before the end of the cell, the second at the end or just beyond it. The cell of the primaries is closed, of the secondaries open. The markings upon the wings are altogether different from those in the two preceding genera, and the spots on the under side of the wings are not silvered, as in the genus *Brenthis.*

Egg.—The egg is rounded at the base, subconical, truncated, and depressed at the upper end and fluted by light raised ridges (see p. 4, Fig. 8).

137

Caterpillar.—The larvæ are cylindrical, armed in the mature form on each segment with comparatively short spines thickly

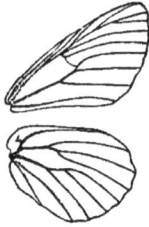

covered with diverging hairs, or needle-shaped spines. They are known in some species to be gregarious in their early stages, and then to separate before maturity. They feed upon the *Scrophulariaceæ*, upon *Castileja, Diplopappus*, and other plants.

Chrysalis.— The chrysalis is pendant, rounded at the head, provided with more or less sharply pointed tubercles on the dorsal surface, and generally white or some shade of light gray, blotched with brown or black, and marked with reddish or orange spots on the dorsal side.

Fig. 91.—Neuration of the genus *Melitæa*.

This genus is very large and is distributed widely over all the colder portions of the north temperate zone. There are many species found in Europe, in Siberia, in China, and in the northern islands of Japan. On the upper slopes of the Himalayas it is also represented by a few species. In North America the genus is well represented, the most of the species being found upon the mountain-slopes and in the valleys of the Pacific coast region. Only two species occur in the Eastern States.

(1) **Melitæa phaëton**, Drury, Plate XVI, Fig. 1, ♂ ; Plate V, Figs. 15, 16, *chrysalis* (The Baltimore).

Butterfly, ♂.—The upper side is black, with a marginal row of red spots, followed by three rows of pale-yellow spots on the fore wings and two on the hind wings. Besides these there are some large red spots on the cells of both wings, a large red spot about the middle of the costa of the hind wing, and a few scattering yellow spots, forming an incomplete fourth row on the fore wing and an incomplete third row on the hind wing. On the under side all the spots of the upper side reappear, but heavier and more distinct, and on the hind wings there are two additional rows of yellow spots, and a number of irregular patches of red and yellow at the base of both wings.

♀.—The female is much like the male. Expanse, ♂, 1.75–2.00 inches; ♀, 2.00–2.60 inches.

Egg.—The egg which is outlined upon p. 4, **Fig. 8**, is brownish-yellow when first laid, then changes to crimson, and

EXPLANATION OF PLATE XVI

1. *Melitæa phaëton*, Drury, ♂.
2. *Melitæa chalcedon*, Doubleday and Hewitson, ♂.
3. *Melitæa macglashani*, Rivers, ♀.
4. *Melitæa augusta*, Edwards, ♂.
5. *Melitæa colon*, Edwards, ♂.
6. *Melitæa nubigena*, Behr, ♂.
7. *Melitæa baroni*, Henry Edwards. ♂.
8. *Melitæa editha*, Boisduval, ♂.
9. *Melitæa nubigena*, var. *wheeleri*, Henry Edwards, ♂.
10. *Melitæa rubicunda*, Henry Edwards, ♂.
11. *Melitæa acastus*, Edwards. ♀.
12. *Melitæa acastus*, Edwards, ♂, under side.
13. *Melitæa palla*, Boisduval ♀.
14. *Melitæa palla*, Boisduval, ♂, under side.
15. *Melitæa gabbi*, Behr, ♂.
16. *Melitæa taylori*, Edwards, ♂.
17. *Melitæa fulvia*, Edwards, ♂.
18. *Melitæa dymas*, Edwards, ♀.
19. *Melitæa perse*, Edwards, ♀.
20. *Melitæa leanira*, Boisduval, ♂.
21. *Melitæa nympha*, Edwards, ♂.
22. *Melitæa arachne*, Edwards, ♀.

PLATE XVI.

becomes black just before hatching. The eggs are laid by the female in large clusters on the under side of the leaf of the food-plant.

Caterpillar.—The life-history in all the stages will be found minutely described by Edwards in "The Butterflies of North America," vol. ii, and·by Scudder in "The Butterflies of New England," vol. i. The mature larva is black, banded with orange-red, and beset with short, bristly, black spines. Before and during hibernation, which takes place after the third moult, the caterpillars are gregarious, and construct for themselves a web in which they pass the winter. After the rigors of winter are past, and the food-plant, which is commonly *Chelone glabra*, begins to send up fresh shoots, they recover animation, scatter, and fall to feeding again, and after the fifth moult reach maturity.

Chrysalis.—The chrysalis is pendant, formed generally at a considerable distance from the spot where the caterpillar feeds, for the larvæ wander off widely just before pupation. It is pearly-gray, blotched with dark brown in stripes and spots, with some orange markings.

This very beautiful butterfly is quite local, found in colonies in swampy places where the food-plant grows, but in these spots sometimes appearing in swarms. It occurs in the northern portions of the United States and in Canada, extending as far north as the Lake of the Woods, and as far south as West Virginia. It does not occur west of the Rocky Mountains.

(2) **Melitæa chalcedon,** Doubleday and Hewitson, Plate XVI, Fig. 2, ♂ (Chalcedon).

Butterfly.—The male and female are much alike. The wings are black, spotted with red and ochreous-yellow. On the under side they are brick-red, with the spots of the upper side repeated, and in addition at the base a number of large and distinct yellow spots. Expanse, ♂, 1.75-2.00 inches; ♀, 2.50 inches.

Early Stages.—For a knowledge of these the reader may consult Edwards, "The Butterflies of North America," vol. i, and "Papilio," vol. iv, p. 63; Wright, "Papilio," vol. iii, p. 123, and other authorities. The egg is pale yellowish when first laid, pitted at the base, and ribbed vertically above. The caterpillar is black, with the bristling processes on the segments longer than in the preceding species. The chrysalis is pale gray, blotched with brown. The food-plants are *Mimulus* and *Castileja*.

This very pretty species is apparently quite common in northern California about Mount Shasta. It is subject to variation, and I possess a dozen remarkable aberrations, in one of which the fore wings are solid black without spots, and the hind wings marked by only one central band of large yellow spots; another representing the opposite color extreme, in which yellow has almost wholly replaced the black and red. The majority of these aberrant forms are females. They are very striking.

(3) **Melitæa macglashani**, Rivers, Plate XVI, Fig. 3, ♂ (Macglashan's Checker-spot).

Butterfly.—Larger than the preceding species, with the red spots on the outer margin bigger, the yellow spots generally larger and paler. Expanse, ♂, 1.85–2.00 inches; ♀, 2.25–3.00 inches.

Early Stages.—Unknown.

This insect is represented in the Edwards collection by a considerable series. They come from Truckee, California.

(4) **Melitæa colon**, Edwards, Plate XVI, Fig. 5, ♂ (Colon).

Butterfly.—Of the same size and general appearance as *M. chalcedon*, with which I believe it to be identical, the only possible satisfactory mark of distinction which I am able to discover on comparing the types with a long series of *chalcedon* being the reduced size of the marginal row of yellow spots on the upper side of the primaries, which in one of the types figured in the plate are almost obsolete. They appear, however, in other specimens labeled "Type." The learned author of the species lays stress, in his original description, upon the shape of the spots composing the band of spots second from the margin on the under side of the hind wings; but I find that the same points he dwells upon as diacritic of this species are apparent in many specimens of what undoubtedly are *chalcedon*. Expanse, 1.75–2.50 inches.

Early Stages.—These have not been recorded.

The types came from the region of the Columbia River, in Washington and Oregon.

(5) **Melitæa anicia**, Doubleday and Hewitson, var. **beani**, Skinner, Plate XVIII, Fig. 13, ♂ (Bean's Checker-spot).

Butterfly.—*M. anicia* is a well-known Californian species, smaller than *M. chalcedon*, and with a great deal of red on the basal and discal areas of both wings upon the upper side. An extremely small and dark form of this species, found on the bleak, inhospitable mountain-tops about Laggan, in Alberta, has been named by Dr. Skinner in honor of Mr. Bean, its discoverer. The figure in our plate, which is taken from Dr. Skinner's original type, sufficiently defines the characteristics of the upper surface. Expanse, 1.50 inch.

Early Stages.—The early stages of *M. anicia* and its varietal forms are quite unknown.

M. anicia is found in Colorado, Montana, Washington, and British America.

(6) **Melitæa nubigena,** Behr, Plate XVI, Fig. 6, ♂ ; var. **wheeleri,** Edwards, Plate XVI, Fig. 9, ♂ (The Clouded Checker-spot).

Butterfly.—Smaller than any of the foregoing species, and characterized by the much redder ground-color of the upper side of the wings, an extreme form being the variety *M. wheeleri*, in which the black ground-color is greatly reduced and almost wholly obliterated on parts of the primaries. There are other marks of distinction given in the figures in the plate which will enable the student easily to recognize this species, which is subject to much variation, especially in the female sex. Expanse, 1.20-1.50 inch.

Early Stages.—Mead, in the "Report upon the Lepidoptera of the Wheeler Survey," has described the caterpillar and chrysalis.

The species is common in Nevada.

(7) **Melitæa augusta,** Edwards, Plate XVI, Fig. 4, ♂ (Augusta).

Butterfly.—This is another species in which red predominates as the color of the upper side, but it may at once be distinguished by the broad, clear red band on the secondaries, on either side of which are the marginal and outer median rows of yellow spots, and by the bands of yellow spots on the primaries, which are not so well marked in *M. nubigena.* Expanse, ♂, 1.50-1.75 inch; ♀, 1.75-2.00 inches.

Early Stages.—Unknown.

The habitat of this species is southern California.

(8) **Melitæa baroni,** Henry Edwards, Plate XVI, Fig. 7, ♂ (Baron's Checker-spot).

Butterfly.—This species closely resembles *chalcedon* upon the upper side, but is smaller and much more heavily spotted with deep red on the upper side toward the base and on the median area of the wings. The bands of light spots on the under side are paler than in *chalcedon*, being white or very pale yellow, narrow, and more regular. Expanse, ♂, 1.50–1.80 inch; ♀, 1.60–1.90 inch.

Early Stages.—These are in part given by Edwards, "The Butterflies of North America," vol. iii. The food-plant is *Castileja*. The young larvæ have the same habit as those of *M. phaëton* in the matter of spinning a common web in which to hibernate.

The species is found in northern California.

(9) **Melitæa rubicunda**, Henry Edwards, Plate XVI, Fig. 10, ♂ (The Ruddy Checker-spot).

Butterfly.—Of the same size as *M. baroni*, from which it is most easily distinguished, among other things, by the tendency of the outer row of small yellow spots near the margin of the hind wings on the upper side to become greatly reduced, and in a majority of specimens to be altogether wanting, as in the specimen figured in our plate. Expanse, ♂, 1.50–1.60 inch; ♀, 1.80 inch.

Early Stages.—For a knowledge of what is thus far known of these the reader may consult the "Canadian Entomologist," vol. xvii, p. 155. The caterpillar feeds on *Scrophularia*.

The range of this species is in northern California.

(10) **Melitæa taylori**, Plate XVI, Fig. 16, ♂ (Taylor's Checker-spot).

Butterfly.—This insect resembles *M. baroni*, but is smaller, the red spots on the wings are larger and more conspicuous, and the light bands of pale spots more regular and paler in color, in many specimens being quite white. It looks at first sight like a diminutive edition of Baron's Checker-spot, and possibly is only a northern race of this species. Expanse, ♂, 1.25–1.50 inch; ♀, 1.50–1.75 inch.

Early Stages.—Mr. W. H. Danby of Victoria, B. C., informs us in the "Canadian Entomologist," vol. xxi, p. 121. that the food-plant of this species is the ribwort-plantain (*Plantago lanceolata*, Linn.).

It is found on Vancouver's Island.

(11) **Melitæa editha**, Boisduval, Plate XVI, Fig. 8, ♂ (Editha).

Butterfly.—Characterized by the considerable enlargement and the disposition in regular bands of the pale spots on the upper side of the primaries, and by the tendency to a grayish cast in the darker markings of the upper side, some specimens, especially females, being quite gray. Expanse, ♂, 1.50 inch; ♀, 2.00 inches.

Early Stages.—The food-plants, according to Henry Edwards, who described the caterpillar and chrysalis in the "Canadian Entomologist," vol. v, p. 167, are *Erodium cicutarium*, clover, and violets.

The habitat of this species is southern California.

(12) **Melitæa acastus,** Edwards, Plate XVI, Fig. 11, ♂ ; Fig. 12, ♂, *under side* (Acastus).

Butterfly.—With thinner and less robust wings than any of the species of the genus hitherto mentioned. It is prevalently fulvous upon the upper side, and on the under side of the hind wings heavily and somewhat regularly banded with yellowish-white spots, possessing some pearly luster. Expanse, ♂, 1.50 inch; ♀, 1.60 inch.

Early Stages.—Unknown.

Common in Nevada, Utah, and Montana.

(13) **Melitæa palla,** Boisduval, Plate XVI, Fig. 13, ♂ ; Fig. 14, ♂, *under side* (The Northern Checker-spot).

Butterfly.—On the upper side resembling the preceding species, but with the median band of spots on the hind wings paler. On the under side the markings are different, as is shown in the plate. Expanse, ♂, 1.50 inch; ♀, 1.75 inch.

Early Stages.—The larva and chrysalis were described by Henry Edwards, the actor naturalist, in the "Proceedings of the California Academy of Sciences," vol. v, p. 167. The food-plant is *Castileja*.

The species ranges from California to Colorado, and northward into British Columbia.

(14) **Melitæa whitneyi,** Behr, Plate XVII, Fig. 7, ♂ ; Fig. 8, ♂, *under side* (Whitney's Checker-spot).

Butterfly.—The markings are much as in *M. palla*, the spots are lighter fulvous and larger than in that species, the yellow bands on the under side are more prominent, and the marginal spots have a silvery luster which is lacking in *M. palla*. The female has the yellow of the under side more prominent than is the case in the male sex. Expanse, ♂, 1.50 inch; ♀, 1.70 inch.

Early Stages.—Altogether unknown.

Whitney's Checker-spot ranges from California into Nevada.

(15) **Melitæa hoffmanni,** Behr, Plate XVII, Fig. 13, ♂ ; Fig. 14, ♀, *aberration* (Hoffmann's Checker-spot).

Butterfly, ♂.—General style of marking much as in the two

preceding species, but with the basal area black, and the black markings toward the outer margin not so heavy, giving it here a more fulvous appearance. The median bands on both wings are broader and paler than in *M. palla.* The under side is much as in the last-mentioned species, but the yellow markings are more prominent.

♀.—Much like the male. Expanse, ♂, 1.35 inch; ♀, 1.45 inch.

Early Stages.—Unknown.

This species, which is found in California and Nevada, is subject to extreme variation, and I have placed upon the plate one out of many beautiful and singular aberrations which I possess.

(16) **Melitæa gabbi,** Behr, Plate XVI, Fig. 15, ♂ (Gabb's Checker-spot).

Butterfly.—In the style of its markings on the upper side it almost completely resembles *M. acastus,* but the dark markings are slighter, giving the wings a more fulvous appearance. On the under side the bands are narrower, defined more sharply with black, and pearly, almost silvery white, whereas in *acastus* they are pale yellowish-white, and not so lustrous. Expanse, ♂, 1.20 inch; ♀, 1.50 inch.

Early Stages.—Unknown.

The habitat of this species is southern California.

(17) **Melitæa harrisi,** Scudder, Plate XVII, Fig. 5, ♂; Fig. 6, ♀, *under side;* Plate V, Figs. 17–18, *chrysalis* (Harris' Checker-spot).

Butterfly, ♂.—Wings fulvous, black at the base and on the outer margin, with five fulvous spots in the cell of the fore wing, two below the cell; and three in the cell of the hind wing. The black border is widest at the apex of the fore wing, and below this runs inwardly on the veins. There are two white spots near the apex. At the anal angle on the hind wing the border is somewhat divided so as to present the appearance of two indistinct lines. On the under side the wings are fulvous, marked with black bands and spots, and crossed by bands and crescents of pale yellow, as is shown in the figure on the plate.

♀.—The female is much like the male. Expanse, ♂, 1.50 inch; ♀, 1.75 inch.

Egg.—The eggs are lemon-yellow, in the form of a truncated cone, with fifteen or sixteen vertical ribs, which are highest about the middle.

Caterpillar.—The matured caterpillar is reddish-fulvous, with a black stripe on the back. Each segment is marked with one black ring before and two black rings behind the sets of spiny tubercles with which the segments are adorned. There are nine rows of spines, those above the feet being quite small. The spines are black, tapering, and set with diverging black hairs. The food-plants are aster and *Diplopappus umbellatus.*

Chrysalis.—The chrysalis is pearly-gray or white, blotched with dark brown or black.

This choice little butterfly ranges from Nova Scotia to Wisconsin, extending as far south as northern Illinois, and northward to Ottawa.

(18) **Melitæa elada**, Hewitson, Plate XVII, Fig. 2, ♂ (Hewitson's Checker-spot).

Butterfly, ♂.—The wings on the upper side are black, crossed by numerous bands of small fulvous spots, the one crossing the middle of the median area being composed of the largest spots. The fore wings on the under side are fulvous, shading outwardly into ferruginous. The spots and bands of the upper side reappear upon the under side, but are lighter, and the submarginal row of crescents is pale yellow and very distinct, the spot between the second and third median nervules being the largest, and the spot between the fourth and fifth subcostals being only a little smaller. The under side of the hind wings is deep ferruginous, crossed by bands of pearly pale-yellow spots, those of the outer margin being the largest.

♀.—The female is much like the male, with the ground-color a little paler. Expanse, ♂, .90 inch; ♀, 1.00–1.10 inch.

Early Stages.—Unknown.

This little species is found in western Texas, Arizona, and northern Mexico.

(19) **Melitæa dymas**, Edwards, Plate XVI, Fig. 18, ♀ (Dymas).

Butterfly.—This species is closely related in size and the style of some of the markings to the foregoing species, but may be at once distinguished by the lighter ground-color, which is pale fulvous, and the totally different style of the marginal markings on the under side of the wings. The female represented in the plate is a trifle paler than the male. Expanse, ♂, .85 inch; ♀, 1.00 inch.

Early Stages.—Unknown.

The habitat of this species is southwestern Texas.

(20) **Melitæa perse,** Edwards, Plate XVI, Fig. 19, ♂ (Perse).

Butterfly.—This is nearly related to the two foregoing species, but the ground-color is darker fulvous than in *dymas*, the markings are slight as in that species, and the arrangement of the spots and bands on the under side is similar. The marginal crescents on the under side of the primaries are largest at the apex and rapidly diminish in size, vanishing altogether about the middle of the wing. Expanse, ♂, 1.00 inch; ♀, 1.10 inch.

Early Stages.—These remain to be discovered.

The only specimens so far found have come from Arizona.

(21) **Melitæa chara,** Edwards, Plate XVII, Fig. 3, ♂ ; Fig. 4, ♂, *under side* (Chara).

Butterfly.—No lengthy description of this pretty little species is required, as the plate, which gives both sides of the wings, shows their peculiarities with sufficient accuracy to enable an exact determination to be made. The whitish spot on the costa before the apex on the upper side, and the chalky-white markings and spots on the under side, serve at once to distinguish this form from its near allies. Expanse, ♂, 1.00 inch; ♀, 1.25 inch.

Early Stages.—Unknown.

I have a large series of this species, all from Arizona, where it appears to be common.

(22) **Melitæa leanira,** Boisduval, Plate XVI, Fig. 20, ♀ (Leanira).

Butterfly, ♂.—Ground-color brownish-black, fulvous on the costa, with submarginal, median, and basal rows of yellow spots. Both the primaries and secondaries have a marginal row of red spots, and the former have in addition a submarginal row of such spots. The under side of the primaries is reddish-fulvous, with the markings of the upper side reproduced. The secondaries have a marginal row of yellow crescents, then a black band inclosing yellow spots, then a median band of long yellow crescents. The remainder of the wing to its insertion is black, spotted with yellow.

146

♀.—Much like the male. Expanse, ♂, 1.50 inch; ♀, 1.75 inch.
Early Stages.—Unknown.

This pretty insect ranges from southern California and Arizona to Nevada, Montana, and British America.

(23) **Melitæa wrighti,** Edwards, Plate XVII, Fig. 9, ♂ ; Fig. 10, ♀, *under side* (Wright's Checker-spot).

Butterfly.—Much like *M. leanira,* but with more fulvous upon the upper side of the wings, and the under side yellow. The black bands on the secondaries are reduced, and the dividing-lines between the spots are confined to the nervules, which are narrowly black. This is probably only a varietal form of the preceding species. I figure the types. Expanse, ♂, 1.30 inch; ♀, 1.80 inch.

Early Stages.— Unknown.

Habitat, southern California.

(24) **Melitæa alma,** Strecker, Plate XVII, Fig. 1, ♂ (Strecker's Checker-spot).

Butterfly, ♂.—The upper side of the wings is bright fulvous, with the margins and veins black. There are three rows of transverse spots paler than the ground-color. The fore wings on the under side are pale fulvous, with pale-yellow spots and a sub-marginal and marginal row of yellow spots separated by a narrow black line. The hind wings on this side are yellow, with the veins and margins black, and a transverse double band of black on the outer margin of the median area.

♀.—Much like the male, but larger, and redder on the upper side. Expanse, ♂, 1.25 inch; ♀, 1.50 inch.

Early Stages.—Unknown.

The specimens I have came from the Death Valley. The species occurs in southern Utah and Arizona.

(25) **Melitæa thekla,** Edwards, Plate XVII, Fig. 15, ♂, *under side;* Fig. 16, ♂ (Thekla).

Butterfly, ♂.—The upper side of the wings is fulvous, black toward the base and on the outer margin. The primaries are adorned with a large oval pale-fulvous spot at the end of the cell, a small one on the middle of the upper side of the cell, and another small one below the cell, at the origin of the first median nervule. The discal area is defined outwardly by a very irregular fine black transverse line, beyond which is a transverse band of pale-fulvous oblong spots, an incomplete series of spots of the ground-color

147

sharply defined upon the black outer shade, followed by a row of irregular white submarginal spots. The transverse bands of spots on the primaries are repeated upon the secondaries, where they are more regular and the spots more even in size. On the under side both wings are pale red, with the light spots of the upper side reappearing as pale-yellow sharply defined spots. The fringes are checkered black and white.

♀.—Much like the male, but larger. Expanse, ♂, 1.35–1.50 inch; ♀, 1.50–1.75 inch.

Early Stages.—Unknown.

This species is common in Texas. It is identical, as an examination of the type shows, with *M. bolli*, Edwards, and the latter name as a synonym falls into disuse.

(26) **Melitæa minuta**, Edwards, Plate XVII, Fig. 11, ♂, *under side;* Fig. 12, ♂ (The Smaller Checker-spot).

Butterfly, ♂.—This species is fulvous on the upper side, rather regularly banded with black lines. The veins are also black. The result is that the wings appear to be more regularly checkered than in any other species which is closely allied to this. The markings of the under side are white edged with black, and are shown very well in the plate, so that a lengthy description is unnecessary. Expanse, ♂, 1.25–1.35 inch; ♀, 1.50–1.60 inch.

Early Stages.—Unknown.

The specific name, *minuta*, is not altogether appropriate. There are many smaller species of the genus. It is found rather commonly in Colorado, Arizona, and New Mexico.

(27) **Melitæa arachne**, Edwards, Plate XVI, Fig. 22, ♀ (Arachne).

Butterfly.—I have given in the plate a figure of a female bearing this name in the Edwards collection. It is remarkably pale on the upper side. There is a large series of types and paratypes in the collection, but all of them vary on the upper side of the wings in the intensity of the fulvous ground-color and the width of the black markings. Underneath they are absolutely like *M. minuta.* I think *M. arachne* is without much doubt a synonym for *M. minuta.* The species varies very greatly. The types are from Colorado and western Texas. Expanse as in *M. minuta.*

Early Stages.—Unknown.

(28) **Melitæa nympha**, Edwards, Plate XVI, Fig. 21, ♂ (Nympha).

Butterfly.—This species differs from *M. minuta* only in having
the black markings darker and the outer median bands of spots
on the upper side yellow. On the under side the pattern of the
markings is exactly as in *M. minuta.*   It seems to me to be a
dark, aberrant form of *M. minuta,* but is very well marked, and
constant in a large series of specimens, so that we cannot be sure
until some one breeds these creatures from the egg. Expanse,
the same as that of *M. minuta.*

Early Stages.—Unknown.

Habitat, Arizona.

In addition to the species of the genus *Melitæa* illustrated in our
plates there are a few others which are credited to our fauna, some
of these correctly and some erroneously, and a number of so-called
species have been described which are not true species, but varieties or aberrations.

COLLECTING IN JAPAN

I was tired of the Seiyo-ken, the only hotel at which foreigners
could be entertained without the discomfort of sleeping upon the
floor. There is a better hotel in Tokyo now.   I had looked out
for five days from my window upon the stinking canal through
which the tide ebbs and flows in Tsukiji.   I felt if I stayed longer
in the lowlands that I would contract malarial fever or some other
uncomfortable ailment, and resolved to betake myself to the mountains, the glorious mountains, which rise all through the interior
of the country, wrapped in verdure, their giant summits capped
with clouds, many of them the abode of volcanic thunder.   So I
went by rail to the terminus of the road, got together the coolies
to pull and push my jinrikishas, and, accompanied by a troop of
native collectors, made my way up the Usui-toge, the pass over
which travelers going from western Japan into eastern Japan
laboriously crept twelve years ago.

What a sunset when we reached an elevation of three thousand feet above the paddy-fields which stretch across the Kwanto
to the Gulf of Yeddo! What a furious thunder-storm came on just
as night closed in!   Then at half-past nine the moon struggled
out from behind the clouds, and we pushed on up over the muddy
roads, until at last a cold breath of night air sweeping from the
west began to fan our faces, and we realized that we were at the

top of the pass, and before us in the dim moonlight loomed the huge form of Asama-yama, that furious volcano, which more than once has laid the land waste for leagues around, and compared with which Vesuvius is a pygmy. We slept on Japanese mats, and in the morning, the drops glittering on every leaf, we started out to walk through the fields to Oiwake, our baggage going forward, we intending to loiter all day amid the charms of nature. Seven species of lilies bloomed about us in the hedges and the fields; a hundred plants, graceful and beautiful in blossom, scented the air with their aroma, and everywhere were butterflies and bees. Above us hung in the sky a banner, the great cloud which by day and by night issues from the crater of Asama-yama. Five species of fritillaries flashed their silvery wings by copse and stream; great black papilios soared across the meadows; blue lycænas, bright chrysophani, and a dozen species of wood-nymphs gamboled over the low herbage and among the grass. Torosan, my chief collector, was in his element. "Dana-san" (*my lord*, or *my master*), "this kind Yokohama no have got." "Dana-san, this kind me no catchee Tokyo side." And so we wandered down the mountain-slope, taking species new alike to American and Japanese, until the sun was sinking in the west. The cloud-banner had grown crimson and purple in the sunset when we wandered into the hospitable doorway of the wayside inn at Oiwake. There we made our headquarters for the week, and thence we carried away a thousand butterflies and moths and two thousand beetles as the guerdon of our chase.

Genus PHYCIODES, Doubleday

(The Crescent-spots)

" Flusheth the rise with her purple favor,
Gloweth the cleft with her golden ring.
'Twixt the two brown butterflies waver,
Lightly settle, and sleepily swing."
JEAN INGELOW.

Butterfly.—The butterflies composing this genus are generally quite small. Their wings on the upper side are fulvous, or brown, with black margins, spots, and lines upon the upper side of the wings, and with the under side of the wings reproducing the spots of the upper side in paler tints. Of the spots of the under side of the wings one of the most characteristic is the pale crescent situated on the outer margin of the hind wings, between the ends of the second and third median nervules. This spot is frequently pearly-white or silvered. Structurally the butterflies of this genus may be distinguished from the preceding genus by the enlarged second joint of the palpi and the very fine, extremely pointed third joint. In the neuration of the wings and in their habits these butterflies closely approximate *Melitæa*.

FIG. 92.— Neuration of the genus *Phyciodes*.

Eggs.—The eggs are always higher than broad, with the surface at the base more or less pitted, giving them a thimble-like appearance. On the upper end in some species they have a few short, vertical ridges, radiating from the micropyle.

Caterpillar.—The caterpillar is cylindrical, marked with pale longitudinal stripes upon a darker ground, and adorned with tubercles arranged in regular rows. These tubercles are generally much shorter than in the genus *Melitæa*. The caterpillars do not, so far as is known, weave webs at any time.

Chrysalis.—The chrysalis is pendant, with the head slightly bifid. The dorsal region of the abdomen is provided with slight tubercles. The color is generally some shade of pale gray, blotched with black or dark brown.

This genus finds its principal development in South and Central America, which are very rich in species, some of them mimicking in a most marvelous manner the butterflies of the protected genus *Heliconius* and its allies. The species found in the United States and Canada are for the most part not very gaily colored insects, chaste shades of brown, or yellow, and black predominating.

(1) **Phyciodes nycteis,** Doubleday and Hewitson, Plate XVII, Fig. 28, ♂, *under side;* Fig. 29, ♂ ; Fig. 30, ♀ ; Plate V, Fig. 19, *chrysalis* (Nycteis).

Butterfly.—On the upper side very closely resembling *Melitæa harrisi,* for which it may easily be mistaken upon the wing. The under side of the hind wings is very different, and may at once be distinguished by the lighter color of the base of the wing,

151

and the pale, silvery crescent on the outer margin. Expanse, ♂, 1.25–1.65 inch; ♀, 1.65–2.00 inches.

Egg.—The egg is half as high again as broad, marked with sixteen or seventeen vertical ribs above, and pitted about the middle by hexagonal cells. It is pale green in color.

Caterpillar.—The caterpillar undergoes four moults after hatching. In the mature stage it is velvety-black, with a dull orange stripe along the back, and purplish streaks on the sides. The body is studded with whitish spots, each giving rise to a delicate black hair, and is further beset with rather short, black, hairy spines.

Chrysalis.—The chrysalis is pearly-gray, blotched with dark brown.

The life-history of this species has been carefully worked out, and all the details may be found described in the most minute manner by Edwards and by Scudder.

The insect ranges from Maine to North Carolina, and thence westward to the foot-hills of the Rocky Mountains.

(2) **Phyciodes ismeria**, Boisduval and Leconte, Plate XVII, Fig. 24, ♂ ; Fig. 25, ♂, *under side* (Ismeria).

Butterfly, ♂.—Easily distinguished from all other allied species by the double row of small light spots on the dark margin of the fore wings on the upper side, and by the silvery, narrow, and greatly bent line of bright silvery spots crossing the middle of the hind wings on the under side.

♀.—The female is like the male, but larger and paler, and all the spots on the upper side are pale fulvous, and not as distinctly white on the outer margin as in the male sex. Expanse, ♂, 1.15–1.35 inch; ♀, 1.35–2.00 inches.

Caterpillar.—The caterpillar, according to Boisduval and Leconte, is yellowish, with blackish spines and three longitudinal blackish stripes. The head, the thoracic legs, and the under side are black; the other legs are yellow.

Chrysalis.—According to the same authors, the chrysalis is pale gray, with paler light spots and nearly white dorsal tubercles.

This insect ranges over a wide territory from Canada to the Southern and Western States east of the Rocky Mountains.

(3) **Phyciodes vesta**, Edwards, Plate XVII, Fig. 17, ♂ ; Fig. 18, ♀ ; Fig. 19, ♀, *under side* (Vesta).

Butterfly, ♂.—On the upper side it closely resembles the win-

1. *Melitæa alma*, Strecker, ♂.
2. *Melitæa clada*, Hewitson, ♂.
3. *Melitæa chara*, Edwards, ♂.
4. *Melitæa chara*, Edwards, ♂, *under side*.
5. *Melitæa harrisi*, Scudder, ♂.
6. *Melitæa harrisi*, Scudder, ♀, *under side*.
7. *Melitæa whitneyi*, Behr, ♂.
8. *Melitæa whitneyi*, Behr, ♂, *under side*.
9. *Melitæa wrighti*, Edwards, ♂.
10. *Melitæa wrighti*, Edwards, ♀, *under side*.
11. *Melitæa minuta*, Edwards, ♂, *under side*.
12. *Melitæa minuta*, Edwards, ♂.
13. *Melitæa boffmanni*, Behr, ♂.
14. *Melitæa boffmanni*, Behr, ♀, *aberration*.
15. *Melitæa thekla*, Edwards, ♂, *under side*.
16. *Melitæa thekla*, Edwards, ♂.
17. *Phyciodes vesta*, Edwards, ♂.
18. *Phyciodes vesta*, Edwards, ♀.
19. *Phyciodes vesta*, Edwards, ♀, *under side*.
20. *Phyciodes picta*, Edwards, ♀, *under side*.
21. *Phyciodes picta*, Edwards, ♂.
22. *Phyciodes phaon*, Edwards, ♂.

23. *Phyciodes phaon*, Edwards, ♀, *under side*.
24. *Phyciodes ismeria*, Boisduval and Leconte, ♂.
25. *Phyciodes ismeria*, Boisduval and Leconte, ♂, *under side*.
26. *Phyciodes montana*, Behr, ♀, *under side*.
27. *Phyciodes montana*, Behr, ♂.
28. *Phyciodes nycteis*, Doubleday and Hewitson, ♂, *under side*.
29. *Phyciodes nycteis*, Doubleday and Hewitson, ♂.
30. *Phyciodes nycteis*, Doubleday and Hewitson, ♀.
31. *Phyciodes orseis*, Edwards, ♂.
32. *Phyciodes camillus*, Edwards, ♂.
33. *Phyciodes camillus*, Edwards, ♀.
34. *Phyciodes camillus*, Edwards, ♂, *under side*.
35. *Phyciodes batesi*, Reakirt, ♂.
36. *Phyciodes batesi*, Reakirt, ♂, *under side*.
37. *Phyciodes pratensis*, Behr, ♂.
38. *Phyciodes pratensis*, Behr, ♀, *under side*.
39. *Eresia punctata*, Edwards, ♂.
40. *Phyciodes mylitta*, Edwards, ♂, *under side*.
41. *Phyciodes mylitta*, Edwards, ♂.
42. *Eresia frisia*, Poey, ♂.

PLATE XVII.

ter form *marcia* of *Phyciodes tharos*, Drury; but the black markings are more evenly distributed. The under side is a pale yellowish-fulvous, and the black markings are slight.

♀.—The female is like the male, but paler. Expanse, ♂, 1.15 inch; ♀, 1.25 inch.

Early Stages.—The chrysalis has been described by Edwards in the "Canadian Entomologist," vol. xi, p. 129. This is all we know of the early life of the insect.

It is found in Texas and Mexico.

(4) **Phyciodes phaon**, Edwards, Plate XVII, Fig. 22, ♂ ; Fig. 23, ♀, *under side* (Phaon).

Butterfly, ♂.—The ground-color of the male is paler on the upper side than in *Phyciodes tharos*, and the black markings are much heavier. The median band on the fore wings is yellowish. The wings on the under side are yellow, shaded with fulvous on the primaries, on which the dark markings are heavy.

♀.—Like the male. Expanse, ♂, .90 inch; ♀, 1.25 inch.

Early Stages.—Unknown.

This insect inhabits the Gulf States, and has been occasionally taken in Kansas.

(5) **Phyciodes tharos**, Drury, Plate XVIII, Fig. 1, ♂ ; Fig. 2, ♀ ; var. **marcia**, Edwards, Plate XVIII, Fig. 3, ♂ ; Fig. 4. ♀ ; Plate V, Figs. 20-22, *chrysalis* (The Pearl Crescent).

Butterfly.—This very common and well-known little insect scarcely needs to be described. The upper side is bright fulvous, with heavy black borders; all the other dark markings are slight. The wings on the under side are paler, with the dark markings of the upper side showing through, and there are additional markings of brown on the hind wings. Expanse, ♂, 1.25 inch; ♀, 1.65 inch.

Early Stages.—The early stages of this insect have been worked out with the most extreme care by Mr. Edwards, and the reader who is curious to know about them should consult "The Butterflies of North America." Dr. Scudder also has minutely and laboriously described the early stages in "The Butterflies of New England." The egg is light greenish-yellow. The caterpillar, which feeds upon various species of aster and allied *Compositæ*, is dark brown after the third moult, its back dotted with yellow, adorned with short, black, bristly spines, which are yellow at the base. The chrysalis is grayish-white, mottled with dark spots and lines.

This species is one of many dimorphic species, the winter form *marcia*, which emerges in spring, having the under side brighter, and the light markings more conspicuous on that side than in the summer form, which has been called *morpheus*. Concerning all of this, and the way in which cold affects the color of butterflies, the reader will do well to consult the splendid pages of Edwards and of Scudder.

The pretty little Pearl Crescent ranges from southern Labrador to Florida; in fact, all over North America north of Texas and south of the region of Hudson Bay, except the Pacific coast of California.

(6) **Phyciodes batesi,** Reakirt, Plate XVII, Fig. 35, ♂ ; Fig. 36, ♀, *under side* (Bates' Crescent-spot).

Butterfly, ♂.—On the upper side much like *P. tharos*, with the black markings very heavy. The under side of the hind wings is uniformly pale fulvous or yellow, with a row of faint submarginal brown spots.

♀.—Like the male. Expanse, ♂, 1.25 inch; ♀, 1.50–1.65 inch.

Early Stages.—Unknown.

This species ranges from New York to Virginia, and westward to Ohio.

(7) **Phyciodes pratensis,** Behr, Plate XVII, Fig. 37, ♂ ; Fig. 38, ♀, *under side* (The Meadow Crescent-spot).

Butterfly, ♂.—The butterfly resembles the preceding species on the upper side, but the ground-color is much paler and the black markings are not so heavy. The under side of the wings is pale fulvous, spotted with yellow.

♀.—The female has the black markings of the upper side heavier than the male, and all the spots pale yellow. The markings on the under side are heavier than in the male sex. Expanse, ♂, 1.15 inch; ♀, 1.40 inch.

Early Stages.—Unknown.

The range of this species is the Pacific coast from Oregon to Arizona.

(8) **Phyciodes orseis,** Edwards, Plate XVII, Fig. 31, ♂ (Orseis).

Butterfly, ♂.—The dark markings on the upper side are much heavier than in either of the two preceding species, and the fulvous spots are smaller, the marginal crescents more regular and distinct. The markings on the under side are also much heavier than in *P. batesi* or *P. pratensis*.

♀.—The female is like the male, but all the dark markings are heavier and the pale markings lighter. Expanse, ♂, 1.35 inch; ♀, 1.60 inch.

Early Stages.—These remain to be described.

Phyciodes orseis ranges from Washington Territory in the north to Mexico in the south.

(9) **Phyciodes camillus,** Edwards, Plate XVII, Fig. 32, ♂; Fig. 33, ♀; Fig. 34, ♂, *under side* (The Camillus Crescent).

Butterfly, ♂.—The male is more like *P. pratensis*, but the light spots on the primaries are paler, on the secondaries brighter, fulvous. The dark markings on the under side are less pronounced than in *pratensis*.

♀.—The female is much like the male. Expanse, ♂, 1.30 inch; ♀, 1.50 inch.

Early Stages.—These are wholly unknown.

The species is reported from British Columbia, Colorado, Montana, Kansas, and Texas.

(10) **Phyciodes mylitta,** Edwards, Plate XVII, Fig. 40, ♂, *under side;* Fig. 41, ♂ (The Mylitta Crescent).

Butterfly, ♂.—Broadly bright fulvous on the upper side, with the dark markings slight; on the under side closely resembling *P. tharos*, var. *marcia*, Edwards.

♀.—The female is like the male, but paler. Expanse, ♂, 1.15 inch; ♀, 1.25–1.50 inch.

Early Stages.—These have been described by Mr. Harrison G. Dyar in the "Canadian Entomologist," vol. xxiii, p. 203. The eggs are laid in clusters upon the thistle (*Carduus*). The caterpillar in its final stage after the fourth moult is black, yellowish below, with a faint twinned yellow dorsal line and faint lines of the same color on the sides. The spines, which are arranged in six rows, are black; those of segments four, five, and six, yellow. The chrysalis is dull wood-brown.

This species has a wide range in the region of the Rocky Mountains, extending from Washington to Arizona, and eastward to Colorado.

(11) **Phyciodes barnesi,** Skinner, Plate XVIII, Fig. 5, ♂ (Barnes' Crescent-spot).

Butterfly, ♂.—Very like the following species, with the light

155

fulvous of the upper side of the wings more widely extended, causing the dark markings to be greatly restricted. The figure in the plate is, in this species as in most others, that of the type, and I am under obligations to Dr. Skinner for kind permission to have the use of the specimen. Expanse, 1.75 inch.

The type came from Colorado Springs.

(12) **Phyciodes montana**, Behr, Plate XVII, Fig. 26, ♀, *under side ;* Fig. 27, ♀ (The Mountain Crescent-spot).

Butterfly.—Upon the upper side the wings are marked much as in P. *camillus*, but are prevalently bright fulvous, with the dark markings quite slight in most specimens. On the under side the wings are pale yellowish-fulvous. The female usually has the secondaries crossed by a broad median band of very pale spots. Expanse, ♂, 1.25 inch; ♀, 1.50 inch.

Early Stages.—Unknown.

The habitat of this species is the Sierras of California and Nevada.

(13) **Phyciodes picta**, Edwards, Plate XVII, Fig. 20, ♀, *under side ;* Fig. 21, ♂ (The Painted Crescent-spot).

Butterfly.—The butterfly in both sexes somewhat closely resembles P. *phaon* on the upper side. On the under side the fore wings are red on the median area, with the base, the costa, the apex, and the outer margin pale yellow; the black markings very prominent. The hind wings on the under side are nearly immaculate yellow. Expanse, ♂, .80–1.10 inch; ♀, 1.10–1.25 inch.

Early Stages.—These may be found described with minute exactness by Mr. W. H. Edwards in the pages of the "Canadian Entomologist," vol. xvi, pp. 163–167. The egg is yellowish-green. The caterpillar moults five times. When mature it is about six tenths of an inch long, armed with seven principal rows of short spines, which appear to vary in color in the spring and fall broods, being light brown in the June brood and greenish-yellow in the October brood. The prevalent color of the caterpillar is some shade of yellowish- or greenish-brown, mottled with lighter and darker tints. The chrysalis is yellowish-brown. The food-plants of the caterpillar are various species of aster.

This species is found as far north as Nebraska, and is abundant in Colorado and New Mexico, ranging southward through Arizona into Mexico.

Explanation of Plate XVIII

1. *Ph...des Tharos*, Drury, ?.
2. *Phr...des Tharos*, Drury, ...
3. *Phrc...des Tharos*, var. *marcia*, Edwards, ♂.
4. *Ph...odes Tharos*, var. *marcia*, Edwards, ♀.
5. *Phr...odes barnesi*, Skinner, ♂.
6. *...tegennis snyderi*, Skinner, ♀.
7. *tegennis platina*, Skinner, ♂.
8. *Fresia texana*, Edwards, ♀.
9. *Fresia texana*, Edwards, ♂, under side.
10. *...gar...hloë janais*, Drury, ♀.
11. *Synchloë latinia*, Hubner, ♂.

2. *Fresia ianthe*, Fabricius, ♀.
3. *Melitaea anicia*, var. ...taut, Skinner, ♂.
4. *Brenthis astarte*, Doubleday and Hewitson, ♂.
5. *Brenthis astarte*, Doubleday and Hewitson, ♀, under side.
6. *Brenthis helena*, Edwards, ♀, under side.
7. *Brenthis helena*, Edwards, ♂.
8. *Debis creola*, Skinner, ♀.
9. *Debis creola*, Skinner, ♀.
10. *Debis portlandia*, Fabricius, ♀.
11. *Geirocheilus tritonia*, Edwards, ♂.

PLATE XVIII

Genus ERESIA, Doubleday

Butterfly.—Small butterflies, closely resembling the species of the genus *Phyciodes* in the neuration of the wings, and only differing from them in the outline of the outer margin of the primaries, which are more or less excavated about the middle. In the style of the markings they differ somewhat widely from the butterflies of the genus *Phyciodes*, notably in the absence of the crescents on the margins of the wings. The wings on the upper side are generally some shade of deep brown or black, marked with spots and bands of white or fulvous, the median band on the hind wings being generally more or less conspicuous. In the pattern of their markings they illustrate a transition from the genus *Phyciodes* to the genus *Synchloë*.

Fig. 93.—Neuration of the genus *Eresia*, slightly enlarged.

Egg.—Hitherto undescribed.

Caterpillar.—Cylindrical, with seven rows of spines, one dorsal, and three lateral on each side; the spines are short, blunt, and armed with short bristles. The head is subcordate, with the vertices rounded. It moults four times.

Chrysalis.—Cylindrical, abdomen stout, head-case short, beveled, nearly square at top, the vertices pyramidal. There are three rows of small tubercles on the dorsal side of the abdomen.

The caterpillars so far as known feed upon various *Compositæ*, as *Diclippa* and *Actinomeris*.

The genus, which is somewhat doubtfully separable from *Phyciodes*, and probably possesses only subgeneric value, is well represented in Central and South America. But three species are found in the faunal region of which this book treats.

(1) **Eresia frisia**, Poey, Plate XVII, Fig. 42, ♂ (Frisia).

Butterfly.—Upper side reddish-fulvous, clouded with fuscous at the base. On the basal area are waved black lines, separate on the hind wings, more or less blended on the fore wings. The outer border is broadly black. Between this border and the basal third the wing is crossed by irregular black bands, the spaces between which are paler fulvous than the base and the hind wings, those near the outer margin being whitish. These

157

bands are continued broadly across the hind wings. The wings on the under side are fulvous, mottled with dark brown and white, and spotted with conspicuous white spots. The male and the female closely resemble each other. Expanse, 1.40 inch.

The early stages are wholly unknown.

The only locality within the limits of the United States in which this insect has been found is Key West, in Florida. It is abundant in the Antilles, Mexico, Central and South America.

(2) **Eresia texana,** Edwards, Plate XVIII, Fig. 8, ♀ ; Fig. 9, ♂, *under side* (The Texan Eresia).

Butterfly.—Black on the upper side of the wings, shading into reddish-brown on the basal area. The fore wings are spotted on the median and limbal areas with white, and the hind wings are adorned by a conspicuous median band of small white spots. On the under side the fore wings are fulvous at the base, broadly dark brown beyond the middle. The light spots of the upper side reappear on the lower side. The hind wings on the under side are marbled wood-brown on the basal area and the inner margin, darker brown externally. The white macular band of the upper side reappears on this side, but less distinct than above. Expanse, ♂, 1.25–1.50 inch; ♀, 1.60–1.75 inch.

Early Stages.—For the only account of the life-history of this species the reader is referred to the "Canadian Entomologist," vol. xi, p. 127, where the indefatigable Edwards gives us an interesting account of his original observations.

This insect ranges from Texas into Mexico. It has been confounded by some with a closely allied insect, **Eresia ianthe,** Fabricius, and to show the difference we have given in Plate XVIII, Fig. 12, a representation of that species, by means of which the reader will be enabled to mark the difference on the upper surfaces of the two species.

(3) **Eresia punctata,** Edwards, Plate XVII, Fig. 39, ♂ (The Dotted Eresia).

Butterfly.—A lengthy description of this little species is scarcely necessary, as the figure in the plate will suffice for its accurate determination. Nothing is known of its early stages. Expanse, 1.10 inch. It is found in New Mexico, Texas, Arizona, and Mexico. It has been recently declared to be identical with *E. tulcis*, Bates, an opinion I am not quite prepared to accept, but which, if correct, will force us, according to the law of priority, to substitute the name given by Bates for that given by Edwards.

Genus SYNCHLOË, Boisduval
(The Patched Butterflies)

Butterfly.—Medium-sized or small butterflies, rather gaily colored, although the species found in the United States are not very brilliant. They may be distinguished structurally from the butterflies of the two preceding genera not only by their larger size and the spindle-formed third article of the palpi, which in the genera *Eresia* and *Phyciodes* is thin and pointed like a needle, but also by the fact that the lower discocellular vein of the fore wings is generally quite straight and not bowed or angled as in the before-mentioned genera.

Egg.—Similar in appearance to the eggs of the genus *Phyciodes:* obovoid, truncated and slightly depressed at top, rounded at the bottom; the lower three fifths with shallow depressions ; the upper part with about twenty-four light blunt-edged ribs. The eggs are laid in clusters upon the leaves of *Helianthus.*

FIG. 94.—Neuration of the genus *Synchloë,* enlarged.

Caterpillar.—Varying in color, generally black or some shade of red or brown, covered with spines which are arranged as in the genus *Melitæa* and are thickly beset with diverging bristles. The caterpillar moults four times.

Chrysalis.—Shaped as in the genus *Melitæa*, light in color, blotched with dark-brown or black spots and lines.

The genus is well represented in Central and South America. Some of the species are polymorphic, many varieties being produced from a single batch of eggs. The result has been considerable confusion in the specific nomenclature.

(1) **Synchloë janais,** Drury, Plate XVIII, Fig. 10, ♂ (The Crimson-patch).

Butterfly.—Fore wings black above, spotted with white; hind wings black above, marked in the center with a broad band

of crimson. On the under side the markings of the upper side of the fore wings are reproduced. The hind wings on the under side are black at the base and on the outer third; immediately at the base is a yellow bar; across the middle is a broad yellow band laved outwardly by red, upon which are numerous black spots. There is a marginal row of yellow spots and an inner row of smaller white spots on the limbal area. Expanse, 2.50-3.00 inches.

Early Stages.—What is known of these is contained in articles published by Mr. William Schaus, "Papilio," vol. iii, p. 188; and by Henry Edwards, "Entomologica Americana," vol. iii, p. 161, to which the reader may refer.

The habitat of the species is Texas, Mexico, and Central America. The insect is very variable in the markings both of the upper and under sides, and several so-called species are only varietal forms of this.

(2) **Synchloë lacinia,** Hübner, Plate XVIII, Fig. 11, ♂; form **crocale,** Edwards, Plate XXIV, Fig. 8, ♂, *under side;* Fig. 9, ♂ (Lacinia).

Butterfly.—This is a protean species, a dozen or more well-marked varietal forms being produced, many of them from a single batch of eggs. The wings on the upper side are black; both primaries and secondaries are crossed about the middle by a band of spots, generally broken on the primaries and continuous on the secondaries. These spots in the typical form *lacinia* are fulvous, and the bands are broad. In the form *crocale* the spots are white, the bands narrow. A great variety of intergrading forms are known and are represented in the author's collection, most of them bred specimens reared from the egg. On the under side the fore wings are marked as on the upper side. The hind wings on the under side are black, with a marginal row of spots, a transverse straight median band, a short basal band, and a costal edging, all bright straw-yellow; in addition there is a submarginal row of small white spots and a crimson patch of variable size at the anal angle. Expanse, ♂, 1.50-2.00 inches; ♀, 1.75-2.75 inches.

Early Stages.—These are described fully by Edwards in the "Canadian Entomologist," vol. xxv, p. 286.

Lacinia ranges from Texas and New Mexico to Bolivia.

FAUNAL REGIONS

That branch of zoölogical science which treats of the geographical distribution of animals is known as zoögeography. None of the zoölogical sciences has contributed more to a knowledge of the facts with which zoögeography deals than the science of entomology. Various divisions of the surface of the earth, based upon the character of the living beings which inhabit them, have been suggested. At the present time, however, it is agreed that in a general way five major subdivisions are sufficient for the purposes of the science, and we therefore recognize five faunal regions, namely, the *Palæarctic*, which includes the temperate regions of the eastern hemisphere; the *Indo-Malayan*, covering the tropics of Asia and the islands lying south of that great continent, including Australia; the *Ethiopian*, covering the continent of Africa south of the lands bordering on the Mediterranean, and extending northward into the southern part of Arabia; the *Neotropical*, covering the continent of South America and the islands of the Caribbean Sea and the Gulf of Mexico; and, finally, the *Nearctic*, covering the temperate and polar regions of North America. The butterflies with which this volume deals are mostly nearctic species, only a few species representing the neotropical region being found as stragglers into the extreme southern portion of the United States.

These five faunal regions are characterized by the presence of certain groups of insects which are more or less peculiar to them. In the Palæarctic Region, for instance, we find a very great development of the *Satyrinæ*, of the genera *Argynnis*, *Melitæa*, and *Lycæna*, and of the genus *Colias*. The genus *Papilio* is but poorly represented, there being only three species found on the entire continent of Europe, and comparatively few in Asia north of the Himalayan mountain-ranges.

As soon as we pass from the boundaries of the Palæarctic Region into India there is discovered a great number of species of the genus *Papilio*. The *Euplœinæ*, of various genera, swarm, and splendid creatures, magnificent in color, present themselves, replacing among the *Nymphalinæ* the small and obscurely colored forms which are found among the mountains of Europe and on the great Asiatic steppes. In the Indo-Malayan Region one

161

of the most gorgeous of the papilionine genera is known as *Orni-thoptera*. These great "bird-wing" butterflies are most brilliant in color in the male, and in the female attain an expanse of wing reaching in some species eight and even nine inches, so that it would be impossible to represent them in their natural dimensions upon a page such as that which is before the reader. One of these giants of the butterfly family, named *Victoria* after her Majesty the Queen of England, is found in the Solomon Islands, and is probably the largest of all known butterflies. One specimen, belonging to the author, has an expanse of wing exceeding nine inches. Among the strangest of recent discoveries is *Orni-thoptera paradisea*, which is found in New Guinea. The male has the hind wings produced in the form of a very delicate and slender tail; the upper surfaces of the wing are broadly marked with shining green and lustrous orange upon a velvety-black ground. The female is black with white spots, slightly marked with yellow, being obscure in color, as is for the most part characteristic of this sex among butterflies, as well as other animals.

The Ethiopian Region is rich in beautiful butterflies of the genus *Callosune*, which are white or yellow, having the tips of the anterior wings marked with crimson or purple. There are many scores of species of these which are found on the grassy park-like lands of southeastern Africa, and they range northward through Abyssinia into Arabia, and a few species even invade the hot lands of the Indian peninsula. In the great forests of the Congo, and in fact throughout tropical Africa, the genus *Acræa*, composed of beautiful insects with long, narrow wings like the genus *Heliconius*, but for the most part yellow, rich brown, and red, spotted with black, abound. And here, too, are found some of the noblest species belonging to the great genus *Papilio*, among them that most singular and, until recently, rarest of the genus, *Papilio antimachus* of Drury, one specimen of which, among a dozen or more in the author's possession, has wings which exceed in expanse even those of *Ornithoptera victoria*, though this butterfly, which seems to mimic the genus *Acræa*, has comparatively narrow wings, and they, therefore, do not cover so large an area as is covered in the case of the genus *Ornithoptera*.

In the Neotropical Region we are confronted by swarms of butterflies belonging to the *Ithomiinæ*, the *Heliconiinæ*, and the *Acræinæ*, all of which are known to be protected species, and which are mimicked by other species among the butterflies and moths of the region which they frequent. A naturalist familiar with the characteristics of the butterfly fauna of South America can at a glance determine whether a collection placed before him is from that country or not, merely by his knowledge of the peculiar coloration which is characteristic of the lepidoptera of the region. The most brilliant butterflies of the neotropical fauna are the *Morphos*, glorious insects, the under side of their wings marked with eye-like spots, the upper side resplendent in varying tints of iridescent blue.

In the Nearctic Region there is a remarkable development of the genera *Argynnis, Melitæa,* and *Phyciodes.* There are also a great many species of the *Satyrinæ* and of the *Hesperiidæ,* or "skippers." The genus *Colias* is also well represented. The Nearctic Region extends southwardly into northern Mexico, at high elevations, and is even continued along the chain of the Andes, and there are species which are found in the vicinity of San Francisco which occur in Chili and Patagonia. In fact, when we get to the southern extremity both of Africa and of South America we find certain genera characteristic of the north temperate zone, or closely allied to them, well represented.

Genus GRAPTA, Kirby
(The Angle-Wings)

Butterfly.—Medium-sized or small, characterized by the more or less deeply excavated inner and outer margins of the fore wings, the tail-like projection of the hind wings at the extremity of the third median nervule, the closed cell of the same wings, and the thick squamation of the palpi on the under side, while on the sides and tops of the palpi there are but few scales. They are tawny on the upper side, spotted and bordered with black; on the under side mimicking the bark of trees and dead leaves, often with a *c*-shaped silvery spot on the hind wings. The insects hibernate in the butterfly form in hollow trees and other hiding-places.

FIG. 95.—Neuration of the genus *Grapta.*

Egg.—The eggs are taller than broad, taper-

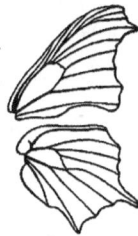

163

ing upward from the base. The summit is broad and flat. The sides are marked by a few equidistant narrow longitudinal ribs, which increase in height to the top. A few delicate cross-lines are interwoven between these ribs. They are laid in clusters or in short string-like series (see p. 5, Fig. 10).

Caterpillar.—The head is somewhat quadrate in outline, the body cylindrical, adorned with rows of branching spines (see Plate III, Figs. 23, 27, 31-33, 38).

Chrysalis.—The chrysalids have the head more or less bifid. There is a prominent thoracic tubercle, and a double row of dorsal tubercles on the abdomen. Viewed from the back they are more or less excavated on the sides of the thorax. In color they are generally some shade of wood-brown or greenish.

The caterpillars feed for the most part upon the *Urticaceæ*, plants of the nettle tribe, such as the stinging-nettle, the elm, and the hop-vine, though the azalea and wild currants furnish the food of some species.

The genus is confined mainly to the north temperate zone.

(1) **Grapta interrogationis,** Fabricius, Plate I, Fig. 3, ♂, *under side;* form **fabricii,** Edwards, Plate XIX, Fig. 1, ♂ ; form **umbrosa,** Lintner, Plate XIX, Fig. 2, ♀ ; Plate III, Fig. 23, *larva,* from a blown specimen; Fig. 27, *larva,* copied from a drawing by Abbot; Plate IV, Figs. 21, 22, 24-26, 40, *chrysalis* (The Question-sign).

Butterfly.—Easily distinguished by its large size, being the largest species of the genus in our fauna. The fore wings are decidedly falcate, or sickle-shaped, bright fulvous on the upper side, spotted and bordered with dark brown and edged with pale blue. On the under side they are mottled brown, shaded with pale purplish, and have a silvery mark shaped like a semicolon on the hind wings. The dimorphic variety *umbrosa,* Lintner, has the upper side of the hind wings almost entirely black, except at the base. Expanse, 2.50 inches.

Early Stages.—These have been frequently described, and the reader who wishes to know all about the minute details of the life-history will do well to consult the pages of Edwards and Scudder, who have written voluminously upon the subject. The food-plants are the elm, the hop-vine, and various species of nettles.

This is one of our commonest butterflies. It is double-brooded in the Middle States. It hibernates in the imago form,

EXPLANATION OF PLATE XIX

1. *Grapta interrogationis*, Fabricius, var. *fabricii*, Edwards, ♂.
2. *Grapta interrogationis*, Fabricius, var. *umbrosa*, Lintner, ♀.
3. *Grapta comma*, Harris, var. *dryas*, Edwards, ♀.
4. *Grapta comma*, Harris, var. *harrisi*, Edwards, ♂.
5. *Grapta silenus*, Edwards, ♂.
6. *Grapta silenus*, Edwards, ♂, under side.
7. *Grapta hylas*, Edwards, ♂.
8. *Grapta hylas*, Edwards, ♂, under side.
9. *Vanessa j-album*, Boisduval and Leconte, ♀.
10. *Grapta gracilis*, Grote and Robinson, ♂.
11. *Grapta gracilis*, Grote and Robinson, ♀, under side.
12. *Grapta faunus*, Edwards, ♂.
13. *Grapta faunus*, Edwards, ♂, under side.
14. *Grapta satyrus*, Edwards, var. *marsyras*, Edwards, ♂.
15. *Grapta satyrus*, Edwards, var. *marsyras*, Edwards, ♀, under side.

PLATE XIX

and when the first warm winds of spring begin to blow, it may be found at the sap-pans in the sugar-camps, sipping the sweets which drip from the wounded trunks of the maples. It ranges all over the United States, except the Pacific coast, and is common throughout Canada and Nova Scotia.

(2) **Grapta comma,** Harris, form **dryas,** Plate XIX, Fig. 3, ♂ ; form **harrisi,** Edwards, Fig. 4, ♂ ; Plate III, Fig. 38, *larva ;* Plate IV, Figs. 27, 29, 30, 39, 46-48, *chrysalis* (The Comma Butterfly).

Butterfly.—Dimorphic, in the form *dryas* with the hind wings heavily suffused with black, in the form *harrisi* predominantly fulvous. Expanse, 1.75-2.00 inches.

The caterpillars feed upon the *Urticaceæ,* and are very common upon the nettle. They vary greatly in color, some being almost snow-white. This species is found throughout Canada and the adjacent provinces, and ranges south to the Carolinas and Texas and over the Northwestern States.

(3) **Grapta satyrus,** Edwards, Plate XX, Fig. 1, ♀ ; Fig. 2, ♀, *under side ;* form **marsyas,** Edwards, Plate XIX, Fig. 14, ♂ ; Fig. 15, ♂, *under side*; Plate III, Fig. 33, *larva ;* Plate IV, Figs. 41, 42, *chrysalis* (The Satyr).

Butterfly.—The species is so accurately depicted in the plates that a description is hardly necessary. The form *marsyas* is smaller, brighter, and with the dark spots on the upper side of the hind wings reduced in size. Expanse, 1.75-2.00 inches.

The food-plant of the caterpillar is the nettle. It occurs occasionally in Ontario, and thence ranges west, being not uncommon from Colorado to California and Oregon.

(4) **Grapta hylas,** Edwards, Plate XIX, Fig. 7, ♂ ; Fig. 8, ♂, *under side* (The Colorado Angle-wing).

Butterfly.—The butterfly closely resembles *G. silenus* on the upper side, but may easily be distinguished by the uniform pale purplish-gray of the lower side of the wings. Expanse, 2.00 inches.

The early stages are unknown. The insect has thus far been found only in Colorado, but no doubt occurs in other States of the Rocky Mountain region.

(5) **Grapta faunus,** Edwards, Plate XIX, Fig. 12, ♂ ; Fig. 13, ♂, *under side ;* Plate III, Fig. 32, *larva ;* Plate IV, Figs. 31, 33-35, *chrysalis* (The Faun).

Butterfly.—This species is readily recognized by the deep indentations of the hind wings, the heavy black border, and the

165

dark tints of the under side mottled with paler shades. Expanse, 2.00–2.15 inches.

The caterpillar feeds on willows. It is found from New England to the Carolinas, and thence westward to the Pacific.

(6) **Grapta zephyrus,** Edwards, Plate XX, Fig. 5, ♂ ; Fig. 6, ♂, *under side* (The Zephyr).

Butterfly.—Fulvous, marked with yellowish toward the outer margins, the dark markings upon which are not as heavy as in the other species of the genus. On the under side the wings are paler than is the case in other species, reddish-brown, marbled with darker brown lines and frecklings. Expanse, 1.75–2.00 inches.

The caterpillar, which feeds upon *Azalea occidentalis,* is described and figured by Edwards in "The Butterflies of North America," vol. i. *Zephyrus* is found throughout the region of the Rocky Mountains, from Colorado to California, and from Oregon to New Mexico.

(7) **Grapta gracilis,** Grote and Robinson, Plate XIX, Fig. 10, ♂ ; Fig. 11, ♀, *under side* (The Graceful Angle-wing).

Butterfly.—A small species, rather heavily marked with dark brown or blackish on the upper side. The wings on the under side are very dark, crossed about the middle by a pale-gray or white band shading off toward the outer margins. This light band serves as a means of easily identifying the species. Expanse, 1.75 inch.

The early stages are unknown.

The species has been found on the White Mountains in New Hampshire, in Maine, Canada, and British America, as far west as Alaska.

(8) **Grapta silenus,** Edwards, Plate XIX, Fig. 5, ♂ ; Fig. 6, ♀, *under side* (Silenus).

Butterfly.—Larger than *gracilis,* and the wings more deeply excised, as in *faunus.* On the under side the wings are very dark, with lighter irrorations, especially on the fore wings. Expanse, 2.00–2.30 inches.

The early stages have never been studied. This species appears to be found ·only in Oregon, Washington, and British Columbia.

(9) **Grapta progne,** Cramer, Plate XX, Fig. 3, ♂ ; Fig. 4, ♂, *under side;* Plate III, Fig. 31, *larva;* Plate IV, Figs. 32, 37, 38, *chrysalis* (Progne).

Butterfly.—A rather small species, with light-fulvous fore wings, shading into yellow toward the outer margins; the dark markings slight, but deep in color. The secondaries are heavily bordered with black on the outer margin. On the under side the wings are very dark, variegated with paler shades, somewhat as in *G. gracilis.* Expanse, 1.85-2.00 inches.

The early stages have been quite fully described by various authors, and the reader may consult "The Butterflies of New England," vol. i, pp. 266-268, for a full account. The caterpillar feeds on the elm, but more commonly on various species of the *Grossulaceæ*, or currant tribe, wild or domesticated. It ranges from Siberia to Nova Scotia, and southward as far as Pennsylvania.

There are several other species of *Grapta* found in our fauna, which are not delineated in this book; but they are rare species, of which little is as yet known. The types are in the collection of the writer, and if the reader finds any species which he cannot identify by means of this book the author will be pleased to help him to the full extent of his ability.

Genus VANESSA, Fabricius
(The Tortoise-shells)

Butterfly.—Medium-sized insects, the wings on the upper side generally some shade of black or brown, marked with red, yellow, or orange. The head is moderately large, the eyes hairy, the palpi more or less heavily scaled, the prothoracic legs feeble and hairy. The lower discocellular vein of the fore wings, when present, unites with the third median nervule, not at its origin, but beyond on the curve. The cell of the primaries may or may not be closed. The cell of the secondaries is open. The fore wings have the outer margin more or less deeply excavated between the extremities of the upper radial and the first median, at which points the wings are rather strongly produced.

FIG. 96.—Neuration of the genus *Vanessa*.

The hind wings have the outer margin denticulate, strongly produced at the extremity of the third median nervule.

Egg.—Short, ovoid, broad at the base, tapering toward the summit, which is broad and adorned with a few narrow, quite high longitudinal ridges, increasing in height toward the apex. Between these ribs are a few delicate cross-lines. They are generally laid in large clusters upon twigs of the food-plant.

Caterpillar.—The caterpillar moults four times. In the mature form it is cylindrical, the segments adorned with long, branching spines arranged in longitudinal rows; the spines much longer, and branching rather than beset with bristles, as in the genus *Grapta*. It lives upon elms, willows, and poplars.

Chrysalis.—The chrysalis in general appearance is not unlike the chrysalis of *Grapta*.

The genus is mainly restricted to the north temperate zone and the mountain regions of tropical lands adjacent thereto. The insects hibernate in the imago form, and are among the first butterflies to take wing in the springtime.

(1) **Vanessa j-album,** Boisduval and Leconte, Plate XIX, Fig. 9, ♀ (The Compton Tortoise).

Butterfly.—No description is required, as the figure in the plate will enable it to be immediately recognized. On the under side of the wings it resembles in color the species of the genus *Grapta*, from which the straight edge of the inner margin of the primaries at once distinguishes it. It is a very close ally of the European *V. vau-album*. Expanse, 2.60–2.75 inches.

The caterpillar feeds upon various species of willow. It is a Northern form, being found in Pennsylvania upon the summits of the Alleghanies, and thence north to Labrador on the east and Alaska on the west. It is always a rather scarce insect.

(2) **Vanessa californica,** Boisduval, Plate XX, Fig. 11, ♂ (The California Tortoise-shell).

Butterfly.—On the upper side deep fulvous, mottled with yellow, spotted and bordered with black. On the under side dark brown; pale on the outer half of the primaries, the entire surface marked with dark lines and fine striæ. Expanse, 2.00–2.25 inches.

Early Stages.—The larva and chrysalis have been described by Henry Edwards in the "Proceedings of the California Academy of Sciences," vol. v, p. 171. The caterpillar feeds upon *Ceanothus thyrsiflorus*.

This insect is a close ally of the European *V. xanthomelas*. It ranges from Colorado to California and as far north as Oregon.

Explanation of Plate XX

1. *Grapta satyrus*, Edwards, ♀.
2. *Grapta satyrus*, Edwards, ♂, under side.
3. *Grapta progne*, Cramer, ♀.
4. *Grapta progne*, Cramer, ♂, under side.
5. *Grapta zephyrus*, Edwards, ♂.
6. *Grapta zephyrus*, Edwards ♂, under side.

7. *Junonia cænia*, Hübner, ♀
8. *Junonia lavinia*, Cramer, ♂.
9. *Junonia genoveva*, Cramer, ♂
10. *Vanessa milberti*, Godart, ♂.
11. *Vanessa californica*, Boisduval, ♂.
12. *Pyrameis carye*, Hübner, ♂.
13. *Anartia jatrophæ*, Linnæus, ♂.

.

(*3*) **Vanessa milberti,** Godart, Plate XX, Fig. 10, ♂ ; Plate III, Fig. 36, *larva;* Plate IV, Figs. 43, 49, 50, *chrysalis* (Milbert's Tortoise-shell).

Butterfly.—Easily distinguished by the broad yellow submarginal band on both wings, shaded outwardly by red. It is nearly related to the European *V. urticæ.* Expanse, 1.75 inch.

The life-history has been worked out and described by numerous writers. The caterpillars feed upon the nettle (*Urtica*).

This pretty little fly ranges from the mountains of West Virginia northward to Nova Scotia and Newfoundland, thence westward to the Pacific.

(4) **Vanessa antiopa,** Linnæus, Plate I, Fig. 6, ♀ ; Plate III, Fig. 28, *larva;* Plate IV, Figs. 51, 58, 59, *chrysalis* (The Mourning-cloak; The Camberwell Beauty).

Butterfly.—This familiar insect needs no description. It is well known to every boy in the north temperate zone. It is one of the commonest as well as one of the most beautiful species of the tribe. A rare aberration in which the yellow border invades the wing nearly to the middle, obliterating the blue spots, is sometimes found. The author has a fine example of this " freak."

The eggs are laid in clusters upon the twigs of the food-plant in spring (see p. 5, Fig. 11). There are at least two broods in the Northern States. The caterpillars feed on willows, elms, and various species of the genus *Populus.*

Genus PYRAMEIS, Doubleday

Butterfly.—The wings in their neuration approach closely to the preceding genus, but are not angulate, and the ornamentation of the under side tends to become ocellate, or marked by eye-like spots, and in many of the species is ocellate.

Egg.—The egg is broadly ovoid, being much like the egg of the genus *Vanessa.*

Caterpillar.—The caterpillar in its mature form is covered with spines, but these are not relatively as large as in *Vanessa*, and are not as distinctly branching.

Chrysalis.—The chrysalis approaches in outline the chrysalis of the preceding genus, and is only differentiated by minor structural peculiarities.

The genus includes only a few species, but some of them have a wide range, *Pyrameis cardui* being almost cosmopolitan, and having a wider distribution than any other known butterfly.

(1) **Pyrameis atalanta,** Linnæus, Plate XLIII, Fig. 4, ♂ ; Plate III, Fig. 35, *larva;* Plate IV, Figs. 52, 53, 55, *chrysalis* (The Red Admiral).

This familiar butterfly, which is found throughout North America, Europe, northern Asia, and Africa, needs no description beyond what is furnished in the plates. Expanse, 2.00 inches. The food-plants are *Humulus, Bœhmeria,* and *Urtica.*

(2) **Pyrameis huntera,** Plate I, Fig. 2, ♂ ; Plate XXXIII, Fig. 6, ♂, *under side;* Plate III, Fig. 34, *larva;* Plate IV, Figs. 54, 63, 64, *chrysalis* (Hunter's Butterfly).

Fig. 97.—Neuration of the genus *Pyrameis.*

Butterfly.—Marked much like the following species, but easily distinguished at a glance by the two large eye-like spots on the under side of the hind wings. Expanse, 2.00 inches.

Early Stages.—These have been frequently described, and are in part well depicted in Plates III and IV. The food-plants are cudweed (*Gnaphalium*) and *Antennaria.*

Hunter's Butterfly ranges from Nova Scotia to Mexico and Central America east of the Sierras.

(3) **Pyrameis cardui,** Linnæus, Plate I, Fig. 1, ♂ ; Plate III, Fig. 37, *larva;* Plate IV, Figs. 60-62, *chrysalis* (The Painted Lady; The Thistle-butterfly).

Butterfly.—This is undoubtedly the most widely distributed of all known butterflies, being found in almost all parts of the temperate regions of the earth and in many tropical lands in both hemispheres. It is easily distinguished from the preceding species by the more numerous and much smaller eye-like spots on the under side of the hind wings. Expanse, 2.00-2.25 inches.

Early Stages.— These have been again and again described at great length and with minute particularity by a score of authors. The food-plants of the caterpillar are thistles (*Carduus*), *Urtica, Cnicus,* and *Althæa.*

(4) **Pyrameis caryæ,** Hübner, Plate XX, Fig. 12, ♂ (The West Coast Lady).

Butterfly.—This species is easily distinguished from *P. cardui,* its nearest ally, by the absence of the roseate tint peculiar to that species, the tawnier ground-color of the upper surfaces, and the complete black band which crosses the middle of the cell of the primaries. Expanse, 2.00 inches.

Early Stages.—These have not all been thoroughly described, but we have an account of the larva and chrysalis from the pen of Henry Edwards, in the "Proceedings of the California Academy of Sciences," vol. v, p. 329. The food-plant of the caterpillar is *Lavatera assurgentiflora.* This species ranges from Vancouver's Island to Argentina, and is found as far east as Utah.

WIDELY DISTRIBUTED BUTTERFLIES

The primal curse declared that the earth, because of man's sin, should bring forth thorns and thistles, and thistles are almost everywhere. Wherever thistles grow, there is found the thistle-butterfly, or the "Painted Lady," as English collectors are in the habit of calling it, *Pyrameis cardui.* All over Europe, all over North America, in Africa,—save in the dense jungles of the Congo,—throughout South America, in far-off Australia, and in many of the islands of the sea this beautiful butterfly is found. At some times it is scarce, and then again there are seasons when it fairly swarms, every thistle-top having one of the gaily colored creatures seated upon its head, and among the thorny environment of the leaves being found the web which the caterpillar weaves. Another butterfly which bids fair ultimately to take possession of the earth is our own *Anosia plexippus,* the wanderings of which have already been alluded to.

Many species are found in the arctic regions both of the Old World and the New. Obscure forms are these, and lowly in their organization, survivors of the ice-age, hovering on the border-line of eternal frost, and pointing to the long-distant time when the great land-masses about the northern pole were knit together, as geologists teach us.

One of the curious phenomena in the distribution of butterflies is the fact that in Florida we find *Hypolimnas misippus,* a species which is exceedingly common in Africa and in the Indo-Malayan subregion. Another curious phenomenon of a like character is the presence in the Canary Islands of a *Pyrameis,* which appears to be only a subvariety of the well-known *Pyrameis indica*, which is common in India, southern China, and Japan. Away off in southeastern Africa, upon the peaks and foot-hills which surround the huge volcanic masses of Kilima-Njaro, Kenia, and Ruwenzori, was discovered by the martyred Bishop Hannington a beautiful species of *Argynnis*, representing a genus nowhere else found upon the continent of Africa south of Mediterranean lands. Strange isolation this for a butterfly claiming kin to the fritillaries that sip the sweets from clover-blossoms in the Bernese Oberland, in the valleys of Thibet, and on the prairies of the United States.

Genus JUNONIA, Hübner.
(Peacock Butterflies)

Butterfly.—Medium-sized butterflies, with eye-like spots upon the upper wings. Their neuration is very much like that of the butterflies belonging to the genus *Pyrameis*, to which they are closely

allied. The eyes are naked, the fore feet are scantily clothed with hair, and the lower discocellular vein of the fore wing, when present, does not terminate on the arch of the third median nervule before its origin, as in the genus *Vanessa*, but immediately at the origin of the third median nervule.

Egg.—Broader than high, the top flattened, marked by ten vertical ribs, very narrow, but not very high. Between the ribs are a few delicate cross-lines.

Fig. 98.—Neuration of the genus *Junonia*.

Caterpillar.—The caterpillar is cylindrical, the segments being adorned with rows of branching spines and longitudinally striped.

Chrysalis.—The chrysalis is arched on the dorsal surface and marked by two rows of dorsal tubercles, concave on the ventral side. The head is slightly bifid, with the vertices rounded.

There are eighteen or more species which belong to this genus, of which some are neotropical, but the greater number are found in the tropical regions of the Old World. Three forms occur within the limits of the United States, which have by some authors been reckoned as distinct species, and by others are regarded merely as varietal forms.

(1) **Junonia cœnia,** Hübner, Plate XX, Fig. 7, ♀ ; Plate III, Figs. 29, 30, *larva;* Plate IV, Figs. 56, 57, 65–67, *chrysalis* (The Buckeye).

Butterfly.—The figure in the plate is far better than any verbal description. On the under side the eye-like spots of the upper side are reproduced, but are much smaller, especially on the hind wings. There is much variety in the ground-color of the wings on the under side. Some specimens are reddish-gray, and some are quite heavily and solidly pinkish-red on the secondaries. Expanse, 2.00–2.25 inches.

Egg.—The egg is dark green.

Caterpillar.—The caterpillar is dark in color, longitudinally striped, and adorned with branching spines, two of which are on the head and point forward.

Chrysalis. — The chrysalis is generally pale wood-brown, strongly arched on the dorsal and concave on the ventral side. It always hangs at less than a right angle to the surface from which it depends.

This is a very common butterfly in the Southern States, ranging northward as far as New England, westward to the Pacific, and southward to Colombia. The caterpillar feeds on various species of plantain (*Plantago*), also *Gerardia* and *Antirrhinum*. When I was a lad in western North Carolina these insects fairly swarmed one summer; thousands of the caterpillars could be found in worn-out fields, feeding on the narrow-leaved plantain, and every fence-rail had one or more of their chrysalids hanging from the under side. I have never seen such multitudes of this species since then. The butterflies are quite pugnacious, and will fight with other passing butterflies, dashing forth upon them, and chasing them away.

(2) **Junonia lavinia,** Cramer, Plate XX, Fig. 8, ♂ (Lavinia).

Butterfly.—This species may be distinguished by the more rounded apex and the more deeply excavated outer margin of the fore wings, and also by the decided elongation of the outer margin of the hind wings at the end of the submedian vein. The wings are paler on the upper side than in the preceding species, and the eye-like spots much smaller. Expanse, 2.00 inches.

The early stages are not accurately known. The insect is common in the Antilles and South America, but is only now and then taken in the extreme southern parts of Texas.

(3) **Junonia genoveva,** Cramer, Plate XX, Fig. 9, ♂ (Genoveva).

Butterfly.—Much darker above than either of the two preceding species. The transverse subapical band is pale yellow, almost white; the ocelli of the wings are more as in *lavinia* than in *cœnia.* Expanse, about 2.00 inches.

This form, if found at all in our fauna, is confined to the extreme South. I have seen and possess some specimens reputed to have come from Texas. The specimen figured in the plate was taken in Jamaica, where this form is prevalent.

Genus ANARTIA, Doubleday

Butterfly.—The head is small; the eyes are round and prominent; the tongue is long; the antennæ are relatively long, having the club short, compressed, and pointed. The palpi have the second joint thick, the third joint gradually tapering and lightly clothed with scales. The fore wings are rounded at the apex, and have the outer and inner margins somewhat excavated. The outer margin of the hind wings is sinuous, produced at the end of the third median nervule. The cell of the hind wing is open. The subcostal nervules in the fore wing are remarkable because of the tendency of the first and second to fuse with the costal vein. The prothoracic feet of the male are small and weak; of the female, stronger.

FIG. 99.—Neuration of the genus *Anartia.*

Early Stages.—These, so far as is known to the writer, await description.

There are four species belonging to this genus, only one of which is found within the limits of the United States. The others are found in Central and South America.

(1) **Anartia jatrophæ,** Linnæus, Plate XX, Fig. 13, ♂ (The White Peacock).

Butterfly.—There can be no mistake made in the identification of this species if the figure we give is consulted. The male and female are much alike. Expanse, 1.75-2.00 inches.

Early Stages.—So far as is known to me, these have never been described. The butterfly is common throughout the tropics of the New World, and is occasionally found in southern Texas and Florida.

Genus HYPANARTIA, Hübner
(The Banded Reds)

Butterfly.—The palpi of medium size, well clothed with scales; the second joint moderately thick; the third very little thinner, blunt at the tip. The antennæ have a distinct, short, well-rounded club. The fore wings have the first two subcostal nervules arising before the end of the cell, close to each other. The third subcostal arises midway between the end of the cell and the origin of the fourth subcostal. The cell of the fore wing is closed by a stout lower discocellular vein which is more or less continuous with the third median nervule. The hind wing has the cell open or only partially closed.

FIG. 100.—Neuration of the genus *Hypanartia.*

Early Stages.—But little is known of the early stages of this genus.

The species reckoned as belonging to *Hypanartia* number less than a dozen, most of which are found in tropical America, but, singularly enough, two species occur in tropical and southern Africa, and another has been described from Madagascar.

(1) **Hypanartia lethe**, Fabricius, Plate XXIV, Fig. 10, ♂ (Lethe).

This very handsome insect, which is quite common in tropical America, is another straggler into our fauna, being occasionally found in southern Texas. But little is known of its early life-history. Expanse, 2.00 inches.

Genus EUNICA, Hübner
(The Violet-wings)

Butterfly.—The head is narrow, hairy; the eyes prominent. The antennæ are long and slender, having a greatly enlarged club marked with two grooves. The palpi have the third joint in the case of the female longer than in the case of the male. They are relatively short, thickly clothed with hairs and scales lying closely

175

appressed to the surface. The fore wing has the costal and median vein enlarged and swollen at the base. The subcostal has five nervules, the first two of which arise before the end of the cell, the third midway between the end of the cell and the fourth nervule. The upper discocellular vein is wanting; the middle discocellular vein is bent inwardly; the lower discocellular vein is somewhat weak and joins the median vein exactly at the origin of the second median nervule. The cell of the hind wing is lightly closed.

Early Stages.—Very little is known of the early stages of this genus.

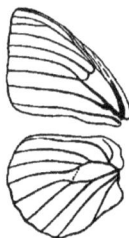

Fig. 101.—Neuration of the genus *Eunica*.

The butterflies are characterized by the dark-brown or black ground-color of the upper side, generally glossed with rich blue or purple. On the under side the markings are exceedingly variable and in most cases very beautiful. The genus is characteristic of the neotropical fauna, and there are over sixty species which have been described. The males are said by Bates, to whom we are indebted for most of our knowledge of these insects, to have the habit of congregating about noon and in the early afternoon in moist places by the banks of streams, returning toward nightfall to the haunts of the females. In this respect they resemble club-men, who at the same hours are generally to be found congregating where there is something to drink. Only two species are found in our region, and are confined to the hottest parts of Texas and Florida, ranging thence southward over the Antilles and Central America as far as Bolivia.

(1) **Eunica monima,** Cramer, Plate XXI, Fig. 7, ♂ ; Fig. 8, ♀ (The Dingy Purple-wing).

Butterfly.—This obscure little butterfly represents in Florida and Texas the great genus to which it belongs, and gives but a feeble idea of the splendid character of its congeners, among which are some exceedingly beautiful insects. Nothing is known of its life-history. It is common in the Antilles and Mexico.

Another species of the genus, *Eunica tatila*, has recently been reported from the extreme southern portion of Florida.

Genus CYSTINEURA, Boisduval

" And here and yonder a flaky butterfly
Was doubting in the air."
McDonald.

Butterfly.—Small butterflies, with elongated fore wings, the hind wings with the outer margin rounded, slightly crenulate. The head is small; the palpi are very delicate and thin, scantily clothed with scales. The costal vein of the fore wing is much swollen near the base. The subcostal vein of this wing sends forth two branches before the end of the cell. The upper discocellular vein is lacking; the middle discocellular is short and bent inwardly; the lower discocellular is almost obliterated, and reaches the median vein at the origin of the second median nervule. In the hind wing the cell is open, and the two radial veins spring from the same point.

FIG. 102.—Neuration of the genus *Cystineura.*

Early Stages.—Very little is as yet definitely ascertained as to these.

But one species is found within the limits covered by this work. Seven species have been described, all of them inhabiting Central or South America.

(1) **Cystineura amymone**, Ménétries, Plate XXIV, Fig. 7, ♂ (Amymone).

Butterfly.— The fore wings are white on the upper side, dusted with gray at the base, on the costa, the apex, and the outer margin. The hind wings are gray on the basal area, pale yellowish-brown on the limbal area, with a narrow fuscous margin. On the under side the markings of the upper side reappear, the gray tints being replaced by yellow. The hind wings are yellowish, with a white transverse band near the base and an incomplete series of white spots on the limbal area. Expanse, 1.50 inch.

The early stages await description. The insect is found about Brownsville, Texas, and throughout Mexico and Central America.

Genus CALLICORE, Hübner
(The Leopard-spots)

Butterfly.—Small-sized butterflies, with the upper side of the wings dark in color, marked with bands of shining metallic blue or silvery-green, the under side of the wings generally more or less brilliantly colored, carmine upon the primaries and silvery-white upon the secondaries, with the apex of the primaries marked with black transverse bands and the body of the secondaries traversed by curiously arranged bands of deep black, these bands inclosing about the middle of the wing circular or pear-shaped spots. All of the subcostal nervules in this genus arise beyond the end of the cell. The costal and the median veins are swollen near the base. The cell in both the fore and hind wings is open.

Fig. 103.—Neuration of the genus *Callicore.*

Early Stages.—Very little is known of these.

This genus numbers about thirty species, almost all of which are found in South America, only one being known to inhabit the United States, being found in the extreme southern portion of Florida, and there only rarely.

(1) **Callicore clymena,** Hübner, Plate XXI, Fig. 5, ♂ ; Fig. 6, ♂, *under side* (The Leopard-spot).

Butterfly.—The wings on the upper side are black, the primaries crossed by an oblique iridescent bluish-green band, and the secondaries marked by a similarly colored marginal band. On the under side the primaries are crimson from the base to the outer third, which is white, margined with black, and crossed by an outer narrow black band and an inner broad black band. The secondaries on this side are white, marked about the middle by two large coalescing black spots, and nearer the costa a large pear-shaped spot, both ringed about with black lines. Beyond these black rings are two black bands conformed to the outline of the inner and outer margins of the wing, and, in addition, a fine black marginal line. The costa is edged with crimson. Expanse, 1.75 inch.

178

EXPLANATION OF PLATE XXI

1. *Timetes coresia*, Godart, ♂.
2. *Timetes coresia*, Godart, ♂, under side.
3. *Timetes petreus*, Cramer, ♂.
4. *Timetes chiron*, Fabricius, ♂.
5. *Callicore clymena*, Cramer, ♂.
6. *Callicore clymena*, Cramer, ♂, under side.
7. *Eunica monima*. Cramer, ♂.
8. *Eunica monima*, Cramer, ♀.
9. *Hypolimnas misippus*, Linnæus, ♂.
10. *Hypolimnas misippus*, Linnæus, ♀.

PLATE XXI.

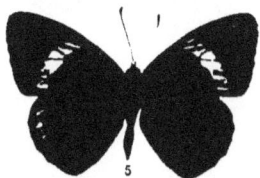

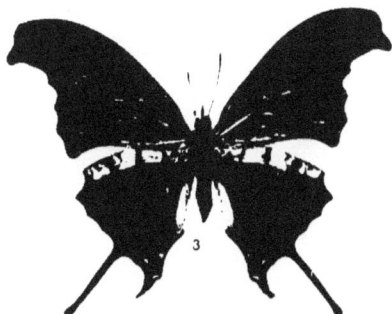

Early Stages.—Unknown.

The Leopard-spot is found occasionally in Florida, but quite ommonly in the Antilles, Mexico, and Central America.

Genus TIMETES, Boisduval
(The Dagger-wings)

Butterfly.—The palpi are moderately long, thickly clothed with scales, the last joint elongated and pointed. The antennæ have a well-developed club. The fore wings and the hind wings have the cell open. In the fore wing the sub-costal vein, which has five branches, emits the first nervule well before the end of the cell, the second a little be-yond it, and the third and fourth near together, before the apex of the wing. The third median nervule of the hind wing is greatly produced and forms the support of the long tail which adorns this wing. Between the end of the sub-median vein and the first median nervule is another lobe-like prolongation of the outer margin of the wing. The butter-flies are characterized for the most part by dark upper surfaces, with light under surfaces marked with broad bands and lines of varying intensity of color. They are easily distinguished from the butter-

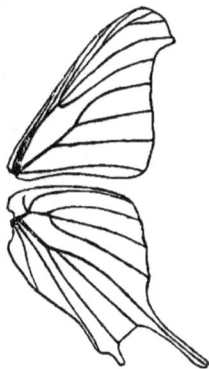

Fig. 104.—Neuration of the genus *Timetes.*

flies of all other genera of the Nymphalidæ by the remarkable tail-like appendage of the hind wing, giving them somewhat the appearance of miniature Papilionidæ.

Early Stages.—Nothing of note has been recorded of their early stages which may be accepted as reliable, and there is an opportunity here for study and research.

There are about twenty-five species belonging to the genus, all found within the tropical regions of America. Four species are occasionally taken in the extreme southern portions of Florida and Texas. They are all, however, very common in the An-tilles, Mexico, and more southern lands.

179

(1) **Timetes coresia,** Godart, Plate XXI, Fig. 1, ♂ ; Fig. 2, ♂, *under side* (The Waiter).

Butterfly.— Easily recognized by means of our figures, which show that this creature deserves the trivial name I have bestowed upon it. In its dark coat and white vest it gracefully attends the feasts of Flora. Expanse, 2.50 inches.

So far as I am aware, nothing reliable has been recorded as to the early stages of this insect. It is occasionally found in Texas.

(2) **Timetes petreus,** Cramer, Plate XXI, Fig. 3, ♂ (The Ruddy Dagger-wing).

Butterfly.—The upper side of the wings is accurately delineated in the plate. On the under side the wings are pale, with the dark bands of the upper side reproduced. Expanse, 2.60 inches. It occurs in southern Florida and Texas, and elsewhere in tropical America.

(3) **Timetes chiron,** Fabricius, Plate XXI, Fig. 4, ♂ (The Many-banded Dagger-wing).

Butterfly.—Easily recognized by means of the figure in the plate. Like the preceding species, this is occasionally found in Texas. It is very common in Mexico, South America, and the Antilles.

Genus HYPOLIMNAS, Hübner

(The Tropic Queens)

Butterfly.— Eyes naked. The palpi are produced, rising above the head, heavily scaled. The antennæ have a well-developed, finely pointed club. The fore wings have stout costal and median veins. The subcostal throws out five nervules, the first two before the end of the cell, the third midway between the end of the cell and the outer border; the fourth and the fifth diverge from each other midway between the third and the outer border, and both terminate below the apex. The upper discocellular vein is wanting; the middle discocellular vein is bent inwardly; the lower discocellular is very weak, and, in some species, wanting. The cell of the hind wing is lightly closed.

Caterpillar.— The caterpillar is cylindrical, thickest toward the middle. The head is adorned with two erect rugose spines; the segments have dorsal rows of branching spines, and three lateral

180

rows on either side of the shorter spines. It feeds on various species of malvaceous plants and also on the common portulaca.

Chrysalis.—The chrysalis is thick, with the head obtusely pointed; the abdominal segments adorned with a double row of tubercles. The thorax is convex.

This genus, which includes a large number of species, reaches its fullest development in the tropics of the Old World, and includes some of the most beautiful, as well as the most singular, forms, which mimic the protected species of the Euplœinæ, or milkweed butterflies, of the Indo-Malayan and Ethiopian regions. In some way one of the most widely spread of these species, which is found throughout the tropics of Asia and Africa, has obtained lodgment upon the soil of the New World, and is occasionally found in Florida, where it is by no means common. It

Fig. 105.—Neuration of the genus *Hypolimnas.*

may be that it was introduced from Africa in the time of the slave-trade, having been accidentally brought over by ship. That this is not impossible is shown by the fact that the writer has, on several occasions, obtained in the city of Pittsburgh specimens of rare and beautiful tropical insects which emerged from chrysalids that were found attached to bunches of bananas brought from Honduras.

(1) **Hypolimnas misippus**, Linnæus, Plate XXI, Fig. 9, ♂ ; Fig. 10, ♀ (The Mimic).

Butterfly, ♂ .—On the upper side the wings are velvety-black, with two conspicuous white spots on the fore wing, and a larger one on the middle of the hind wing, the margins of these spots reflecting iridescent purple. On the under side the wings are white, intricately marked with black lines, and black and reddish-ochraceous spots and shades.

♀ .—The female mimics two or three forms of an Oriental milkweed butterfly, the pattern of the upper side of the wings conforming to that of the variety of the protected species which

181

is most common in the region where the insect is found. The species mimicked is *Danais chrysippus*, of which at least three varietal forms or local races are known. The American butterfly conforms in the female sex to the typical *D. chrysippus*, to which it presents upon the upper side a startling likeness. On the under side it is marked much as the male. Expanse, ♂, 2.50 inches; ♀, 3.00 inches.

Early Stages.—What has been said as to the early stages in the description of the genus must suffice for the species. But little is as yet accurately known upon the subject.

The range of *H. misippus* is southern Florida, the Antilles, and the northern parts of South America. It is not common on this side of the Atlantic, but very common in Africa, tropical Asia, and the islands south as far as northern Australia.

Genus BASILARCHIA, Scudder
(The White Admirals)

Butterfly.—Head large; the eyes are large, naked; the antennæ are moderately long, with a distinct club; the palpi are compact, stout, produced, densely scaled. The fore wings are subtriangular, the apex well rounded, the lower two thirds of the outer margin slightly excavated. The first two subcostal nervules arise before the end of the cell. The hind wings are rounded, crenulate.

Egg.—Nearly spherical, with the surface pitted with large hexagonal cells (see p. 3, Fig. 1).

Caterpillar.—The caterpillar in its mature state is cylindrical, somewhat thicker before than behind, with the second segment adorned with two prominent rugose club-shaped tubercles. The fifth segment, and the ninth and tenth segments also, are ornamented with dorsal prominences (see p. 8, Fig. 20).

Fig. 106.— Neuration of the genus *Basilarchia*.

Chrysalis.—The chrysalis is suspended by a stout cremaster; the abdominal segments are rounded. On the middle of the dorsum is a prominent projecting boss. The thorax is rounded. The head is rounded or slightly bifid.

The caterpillars feed upon the leaves of various species of oak, birch, willow, and linden. The eggs are laid upon the extreme tip of the leaves, and the infant caterpillar, feeding upon the leaf in immediate proximity to the point where it has been hatched, attaches bits of bitten leaf by strands of silk to the midrib, thus stiffening its perch and preventing its curling as the rib dries. Out of bits of leaves thus detached it constructs a packet of material, which it moves forward along the midrib until it has completed its second moult. By this time winter begins

Fig. 107. — Leaf cut away at end by caterpillar of *Basilarchia* (Riley).

to come on, and it cuts away for itself the material of the leaf on either side of the rib, from the tip toward the base, glues the rib of the leaf to the stem by means of silk, draws together

the edges of the remaining portions of the leaf, and constructs a tube-like hibernaculum, or winter quarters, exactly fitting the body, in which it passes the winter.

Fig. 108.—Hibernaculum, or winter quarters, of larva of *Basilarchia*.

There are a number of species of the genus found in the United States, the habits of which have been carefully studied, and they are among our most interesting butterflies, several species being mimics of protected species.

(1) **Basilarchia astyanax**, Fabricius, Plate XXII, Fig. 1, ♂ ; Plate III, Figs. 17, 21, 25, *larva;* Plate IV, Figs. 12, 13, *chrysalis* (The Red-spotted Purple).

Butterfly.—This common but most beautiful species is sufficiently characterized by the plate so far as the upper surface is concerned. On the under side the wings are brownish, banded with black on the margins; the lunules are on this side as above, but the inner band of spots is red. There are two red spots at the base of the fore wings, and four at the base of the hind wings. The palpi are white below, and the abdomen is marked with a lateral white line on each side. Expanse, 3.00–3.25 inches.

Egg.—The egg, which resembles somewhat closely that of B. *disippus* (see p. 3, Fig. 1), is yellowish-green, gradually turning dark brown as the time for the emergence of the caterpillar approaches.

Caterpillar.—The caterpillar is so well delineated in Plate III,

183

Fig. 17, as to obviate the necessity for a lengthy verbal description.

Chrysalis.—What has been said of the caterpillar is also true of the chrysalis (see Plate IV).

The larva feeds upon the willow, cherry, apple, linden (*Tilia*), huckleberry, currant, and other allied shrubs and trees. The butterfly is somewhat variable, and a number of varietal forms have been described. It ranges generally over the United States and southern Canada as far as the Rocky Mountain ranges in the West, and is even said to occur at high elevations in Mexico.

(2) **Basilarchia arthemis,** Drury, Plate XXII, Fig. 4, ♂, form **lamina,** Fabricius; Fig. 5, ♂, form **proserpina,** Edwards, Plate III, Fig. 26, *larva;* Plate IV, Figs. 14, 23, *chrysalis* (The Banded Purple).

Butterfly.—Easily distinguished in the form *lamina* from *astyanax,* which in other respects it somewhat closely resembles, by the broad white bands crossing both the fore wings and the hind wings, and followed on the secondaries by a submarginal row of red spots shading inwardly into blue. In the form *proserpina* there is a tendency on the part of the white bands to become obsolete, and in some specimens they do entirely disappear. The likeness to *astyanax* in such cases is striking, and the main point by which the forms may then be discriminated is the persistence of the red spots on the upper side of the secondaries; but even these frequently are obsolete. Expanse, 2.50 inches.

Egg.—The egg is grayish-green, with "kite-shaped" cells.

Caterpillar.—Greenish- or olive-brown, blotched with white in its mature form, which is well represented in Plate III. It feeds upon the willow, the hawthorn (*Cratægus*), and probably other plants.

Chrysalis.—The figure in Plate IV is sufficiently exact to obviate the necessity for further description.

This beautiful insect ranges through northern New England and New York, Quebec, Ontario, and the watershed of the Great Lakes, spreading southward at suitable elevations into Pennsylvania. I have taken it about Cresson, Pennsylvania, at an elevation of twenty-five hundred feet above sea-level. It is not uncommon about Meadville, Pennsylvania. The species appears to be, like all the others of the genus, somewhat unstable and plastic, or else hybridization is very frequent in this genus. Probably all the species have arisen from a common stock.

1. *Basilarchia astyanax*, Fabricius. ♂. 5. *Basilarchia arthemis*, Drury, var.
2. *Heterochroa californica*. Butler.   *proserpina*, Edwards, ♂.
3. *Basilarchia lorquini*, Boisduval. ♂. 6. *Basilarchia weidemeyeri*, Edwards,
4. *Basilarchia arthemis*, Drury. ♂.   ♂.

(3) **Basilarchia weidemeyeri,** Edwards, Plate XXII, Fig. 6, ♂ (Weidemeyer's Admiral).

Butterfly.—Superficially like *arthemis*, but easily distinguished by the absence of the lunulate marginal bands of blue on the margins of the hind wings and by the presence of a submarginal series of white spots on both wings. Expanse, 3.00 inches.

Early Stages.—These have been described by W. H. Edwards in the "Canadian Entomologist," vol. xxiv, p. 107, and show great likeness to the following species, *B. disippus.* The caterpillar feeds upon cottonwood (*Populus*).

The insect is found on the Pacific slope and eastward to Montana, Nebraska, and New Mexico.

(4) **Basilarchia disippus,** Godart, Plate VII, Fig. 4, ♂ ; Plate III, Figs. 19, 22, 24, *larva;* Plate IV, Figs. 18–20, *chrysalis* (The Viceroy).

Butterfly.—This species mimics *Anosia plexippus* in a remarkable manner, as may be seen by referring to Plate VII. An aberration in which the mesial dark transverse band on the secondaries has disappeared was named *pseudodorippus* by Dr. Strecker. The type is in the Mead collection, now belonging to the writer. Expanse, 2.50–2.75 inches.

Early Stages.—These have all been carefully studied by numerous writers. The egg is depicted on p. 3, Fig. 1. The caterpillar is shown on p. 8, as well as in Plate III.

The species ranges everywhere from southern Canada and British America into the Gulf States.

(5) **Basilarchia hulsti,** Edwards, Plate VII, Fig. 5, ♂ (Hulst's Admiral).

Butterfly.—This form is apparently a mimic of *Anosia berenice.* The ground-color of the wings is not so bright as in *B. disippus,* and the mesial band of the secondaries on the upper side is relieved by a series of small whitish spots, one on each interspace. The perfect insect can easily be distinguished by its markings. Expanse, 2.50–2.60 inches. Thus far it is only known from Utah and Arizona. The early stages have not been described.

(6) **Basilarchia lorquini,** Boisduval, Plate XXII, Fig. 3, ♂ (Lorquin's Admiral).

Butterfly.—Easily distinguished from all the other species of the genus by the yellowish-white bar near the end of the cell of the fore wings and the reddish color of the apex and upper margin of the same wings. Expanse, 2.25-2.75 inches.

Early Stages.—These have been partially described by Henry Edwards, and minutely worked out by Dr. Dyar, for whose description the reader may consult the "Canadian Entomologist," vol. xxiii, p. 172. The food-plant of the caterpillar is *Populus*, willows, and the choke-cherry (*Prunus demissa*).

Besides the forms figured in our plates there is a species in Florida named *floridensis* by Strecker, and subsequently *eros* by Edwards, which is generally larger and much darker than *B. disippus*, which it otherwise closely approximates.

THE BUTTERFLIES' FAD

"I happened one night in my travels
   To stray into Butterfly Vale,
Where my wondering eyes beheld butterflies
   With wings that were wide as a sail.
They lived in such houses of grandeur,
   Their days were successions of joys,
And the very last fad these butterflies had
   Was making collections of boys.

"There were boys of all sizes and ages
   Pinned up on their walls. When I said
'Twas a terrible sight to see boys in that plight,
   I was answered: '*Oh, well, they are dead.*
We catch them alive, but we kill them
   With ether—a very nice way:
Just look at this fellow—his hair is so yellow,
   And his eyes such a beautiful gray.

"'Then there is a droll little darky,
   As black as the clay at our feet;
He sets off that blond that is pinned just beyond
   In a way most artistic and neat.
And now let me show you the latest,—
   A specimen really select,
A boy with a head that is carroty-red
   And a face that is funnily specked.

"'We cannot decide where to place him;
   Those spots bar him out of each class;
We think him a treasure to study at leisure
   And analyze under a glass.'

I seemed to grow cold as I listened
To the words that these butterflies spoke;
With fear overcome, I was speechless and dumb,
And then with a start — I awoke ! "

ELLA WHEELER WILCOX.

Genus ADELPHA, Hübner
(The Sisters)

Butterfly.—This genus is very closely allied to the preceding, and is the South American representative of *Basilarchia.* The only difference which is noticeable structurally is in the fact that the eyes are hairy, the palpi not so densely clothed with scales. The prothoracic legs of the males are smaller than in *Basilarchia.* The cell of the primaries is very slightly closed by the lower discocellular vein, which reaches the median a little beyond the origin of the second median nervule. The outer margin of the fore wing is rarely excavated, as in *Basilarchia,* and the lower extremity of the hind wing near the anal angle is generally more produced than in the last-mentioned genus.

Early Stages.—The life-history of the genus has not been carefully worked out, but an account has been published recently of the caterpillar of the only species found within our fauna, which shows that, while in general resembling the caterpillars of the genus *Basil-*

Fig. 100.—Neuration of the genus *Adelpha.*

archia, the segments are adorned with more branching spines and with short fleshy tubercles, giving rise to small clusters of hairs.

The chrysalids are of a peculiar form, with bifid heads and broad wing-cases. They are generally brown in color, with metallic spots. The only species in our fauna is confined to southern California, Arizona, and Mexico.

(1) **Adelpha californica,** Butler, Plate XXII, Fig. 2, ♀ (The Californian Sister).

Butterfly.—Easily recognized by the large subtriangular patch of orange-red at the apex of the primaries. In its habits and

187

manner of flight it closely resembles the species of the genus *Basilarchia*. Expanse, 2.50–3.00 inches.

Early Stages.—So far as is known to the writer, these have not been described, except partially by Henry Edwards in the "Proceedings of the California Academy of Sciences," vol. v, p. 171. The caterpillar feeds upon oaks.

The insect is found in California, Nevada, Arizona, and Mexico.

Genus CHLORIPPE, Boisduval
(The Hackberry Butterflies)

Butterfly.—Small butterflies, generally some shade of fulvous, marked with eye-like spots on the posterior margin of the secondaries, and occasionally upon the outer margin of the primaries, the fore wings as well as the hind wings being in addition more or less strongly spotted and banded with black. The eyes are naked; the antennæ are straight, provided with a stout, oval club; the palpi are porrect, the second joint heavily clothed with hairs, the third joint short, likewise covered with scales. The costal vein of the fore wing is stout. The first subcostal vein alone arises before the end of the cell. The cell is open in both wings.

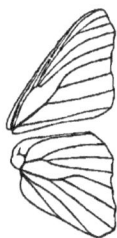

Fig. 110.—Neuration of the genus *Chlorippe*, ♂.

Egg.—The eggs, which are deposited in clusters, are nearly globular, the summit broad and convex. The egg is ornamented by from eighteen to twenty rather broad vertical ribs, having no great elevation, between which are numerous faint and delicate cross-lines.

Caterpillar. — The head is subquadrate, with the summit crowned by a pair of diverging stout coronal spines which have upon them a number of radiating spinules. Back of the head, on the sides, is a frill of curved spines. The body is cylindrical, thickest at the middle, tapering forward and backward from this point. The anal prolegs are widely divergent and elongated, as in many genera of the *Satyrinæ*.

Chrysalis. — The chrysalis is compressed laterally and keeled on the dorsal side, concave on the ventral side, the head distinctly bifid. The cremaster is very remarkable, presenting the appearance of a flattened disk, the sides studded with hooks, by means of which the chrysalis is attached to the surface, from which it depends in such a manner that the ventral surface is parallel to the plane of support.

The caterpillars feed upon the *Celtis*, or hackberry.

There are a number of species, mainly confined to the south-western portion of the United States, though some of them range southward into Mexico. Two only are known in the Middle States. The species are double-brooded in the more northern parts of the country, and the caterpillars produced from eggs laid by the second brood hibernate.

(1) **Chlorippe celtis**, Boisduval and Leconte, Plate XXIII, Fig. 3, ♂ ; Fig. 4, ♀ ; Fig. 11, ♂, *under side* (The Hackberry Butterfly).

Butterfly, ♂.—The primaries at the base and the secondaries except at the outer angle pale olive-brown, the rest of the wings black. The dark apical tract of the primaries is marked by two irregular, somewhat broken bands of white spots. There is a red-ringed eye-spot between the first and second median nervules, near the margin of the fore wing, and there are six such spots on each hind wing. On the under side the ground-color is grayish-purple; the spots and markings of the upper side reappear on this side.

♀.—The female has the wings, as is always the case in this genus, much broader and not so pointed at the apex of the primaries as in the male sex, and the color is much paler. Expanse, ♂, 1.80 inch; ♀, 2.10 inches.

Early Stages.—These are beautifully described and delineated by Edwards in "The Butterflies of North America," vol. ii. The caterpillar feeds on the hackberry (*Celtis occidentalis*).

This species is found generally from southern Pennsylvania, Ohio, Indiana, and Illinois to the Gulf of Mexico. It is not, so far as is known, found on the Pacific coast.

(2) **Chlorippe antonia**, Edwards, Plate XXIII, Fig. 12, ♂ (Antonia).

Butterfly. — Bright yellowish-fulvous on the upper side. Easily distinguished from *celtis* by the two eye-spots near the margin of the primaries. Expanse, 1.75–2.00 inches.

Early Stages.—Unknown.

Antonia is found in Texas.

189

(3) **Chlorippe montis,** Edwards, Plate XXII, Fig. 7, ♂ ; Fig. 8, ♀ (The Mountain Emperor).

Butterfly.— Very closely allied to *C. antonia* in the style and location of the markings, but tinted with pale ashen-gray on the upper side of the wings, and not yellowish-fulvous as in the last-named species. Expanse, ♂, 1.75 inch; ♀, 2.15 inches.

The early stages are unknown.

Montis occurs in Arizona and Colorado, and by some writers is regarded as a varietal form of *antonia*, in which opinion they may be correct.

(4) **Chlorippe leilia,** Edwards, Plate XXIII, Fig. 11, ♂ (Leilia).

Butterfly.— Like *antonia*, this species has two extra-median eye-spots on the primaries, and thus may be distinguished from *celtis*. From *antonia* it may be separated by its larger size and the deeper reddish-brown color of the upper surfaces. Expanse, 2.10–2.50 inches.

Early Stages.—Unknown.

So far we have received this butterfly only from Arizona.

(5) **Chlorippe alicia,** Plate XXIII, Fig. 9, ♂ ; Fig. 10, ♀ (Alicia).

Butterfly.—Very bright fawn at the base of the wings, shading into pale buff outwardly. There is but one eye-spot on the primaries. The six eye-spots on the secondaries are black and very conspicuous. The marginal bands are darker and heavier than in any other species of the genus. Expanse, ♂, 2.00 inches; ♀, 2.50 inches.

The early stages are only partially known.

Alicia ranges through the Gulf States from Florida to Texas.

(6) **Chlorippe clyton,** Boisduval and Leconte, Plate XXIII, Fig. 5, ♂ ; Fig. 6, ♀ ; Plate III, Fig. 20, *larva;* Plate IV, Figs. 15–17, *chrysalis* (The Tawny Emperor).

Butterfly, ♂.—The fore wings without an extra-median eye-spot, and the secondaries broadly obscured with dark brown or blackish, especially on the outer borders, so that the eye-spots are scarcely, if at all, visible.

♀.—Much larger and paler in color than the male, the eye-spots on the secondaries conspicuous. Expanse, ♂, 2.00 inches; ♀, 2.50–2.65 inches.

Early Stages.—The life-history has been carefully worked out,

190

EXPLANATION OF PLATE XXIII

1. *Chlorippe flora*, Edwards, ♂.
2. *Chlorippe flora*, Edwards, ♀.
3. *Chlorippe celtis*, Boisd.-Lec., ♂.
4. *Chlorippe celtis*, Boisd.-Lec., ♀.
5. *Chlorippe clyton*, Boisd.-Lec., ♂.
6. *Chlorippe clyton*, Boisd.-Lec., ♀.
7. *Chlorippe montis*, Edwards, ♂.
8. *Chlorippe montis*, Edwards, ♀.
9. *Chlorippe alicia*, Edwards, ♂.
10. *Chlorippe alicia*, Edwards, ♀.
11. *Chlorippe leilia*, Edwards, ♂.
12. *Chlorippe antonia*, Edwards, ♂.
13. *Chlorippe celtis*, Boisd.-Lec., ♂, under side.

PLATE XXIII.

and the reader who wishes to know all about it should consult the writings of Edwards and Scudder.

This species is occasionally found in New England, and ranges thence westward to Michigan, and southward to the Gulf States. It is quite common in the valley of the Ohio.

(7) **Chlorippe flora**, Edwards, Plate XXIII, Fig. 1, ♂ ; Fig. 2, ♀ (Flora).

Butterfly, ♂.—The ground-color is bright reddish-fulvous on the upper side. The usual markings occur, but there is no eye-spot, or ocellus, on the primaries. The hind wings are not heavily obscured with dark brown, as in *clyton*, and the six ocelli stand forth conspicuously upon the reddish ground. The hind wings are more strongly angulated than in any other species. The borders are quite solidly black.

♀.—The female is much larger than the male, and looks like a very pale female of *clyton*. Expanse, ♂, 1.75 inch; ♀, 2.35 inches.

Early Stages.—The life-history has been described by Edwards in the "Canadian Entomologist," vol. xiii, p. 81. The habits of the insect in its early stages and the appearance of the larva and chrysalis do not differ widely from those of *C. clyton*, its nearest ally.

Flora is found in Florida and on the borders of the Gulf to Texas.

Genus PYRRHANÆA, Schatz
(The Leaf-wings)

Butterfly.—Medium-sized butterflies, on the upper side of the wings for the most part red or fulvous, on the under side of the wings obscurely mottled on the secondaries and the costal and apical tracts of the primaries in such a manner as to cause them to appear on this side like rusty and faded leaves. Structurally they are characterized by the somewhat falcate shape of the primaries and the strongly produced outer margin of the secondaries about the termination of the third median nervule. The first and second subcostal nervules coalesce with one another and with the costal vein. The costal margin of the fore wing at the base is strongly angulated, and the posterior margin of the primaries is straight. The cell of the secondaries is very feebly closed.

Egg.—Spherical, flattened at the base and somewhat depressed at the apex, with a few parallel horizontal series of raised points about the summit.

191

Caterpillar.— Head somewhat globular in appearance; the anterior portion of the first thoracic segment of the body is much smaller in diameter than the head; the body is cylindrical, tapering to a point.

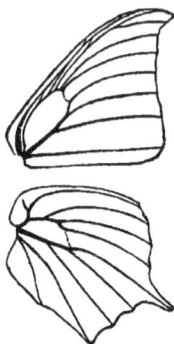

Chrysalis.— Short, stout, with transverse ridges above the wings on the middle of the abdomen, keeled on the sides. The cremaster is small and furnished with a globular tip, the face of which is on the same plane as the ventral surface of the body, causing the chrysalis to hang somewhat obliquely from the surface which supports it.

This is a large genus of mostly tropical species, possessed of rather singular habits. The caterpillars in the early stages of their existence have much the same habits as the caterpillars of the genus *Basilarchia*, which have been already described. After passing the third moult they construct for themselves nests by weaving the edges of a leaf together, and thus conceal themselves from sight, emerging in the dusk to feed upon the food-plant. They live upon the *Euphorbiaceæ*, the *Lauraceæ*, and the *Piperaceæ*. The insects are double-brooded in the cooler regions of the North, and are probably many-brooded in the tropics.

Fig. 111.—Neuration of the genus *Pyrrhanæa.*

(1) **Pyrrhanæa andria**, Scudder, Plate XXIV, Fig. 1, ♀ (The Goatweed Butterfly).

Butterfly, ♂.— Solidly bright red above, the outer margins narrowly dusky on the borders. On the under side the wings are gray, dusted with brown scales, causing them to resemble the surface of a dried leaf.

♀.— The female has the upper side paler and marked by pale fulvous bands, as shown in the plate. Expanse, ♂, 2.50 inches; ♀, 3.00 inches.

Early Stages.— In Fig. 21, on p. 9, is a good representation of the mature caterpillar, the nest which it constructs for itself, and the chrysalis. A full account of the life-history may be found in the " Fifth Missouri Report " from the pen of the late C. V. Riley. The caterpillar feeds on *Croton capitatum.*

The insect ranges from Illinois and Nebraska to Texas.

(2) **Pyrrhanæa morrisoni,** Edwards, Plate XXIV, Fig. 2, ♀ (Morrison's Goatweed Butterfly).

Butterfly, ♂.—Much like *P. andria,* but more brilliantly and lustrously red on the upper side, and marked with paler macular bands like the female.

♀.—Differing from the female of *P. andria* in the more macular, or spotted, arrangement of the light bands on the wings, as is well shown in the plate. Expanse, 2.25–2.50 inches.

Early Stages.—Unknown.

This species occurs in Arizona and Mexico.

(3) **Pyrrhanæa portia,** Fabricius, Plate XXIV, Fig. 3, ♂ (Portia).

Butterfly.—Splendid purplish-red on the upper side. On the under side the fore wings are laved with bright yellow on the basal and inner marginal tracts, and the secondaries are dark brown, irrorated with blackish scales arranged in spots and striæ. Expanse, 2.75–3.00 inches.

Early Stages.—Unknown.

Portia occurs in the extreme southern part of Florida and in the Antilles.

Genus AGERONIA, Hübner
(The Calicoes)

Butterfly.—The antennæ moderately long, delicate, terminated in a gradually thickened club. The eyes are naked; the palpi are compressed, only slightly porrect, not densely covered with scales. The neuration is alike in both sexes, the costal and the median veins greatly thickened toward the base. The first and second subcostals arise from before the end of the cell; the fourth and fifth subcostals arise from a common stem emitted from the third subcostal beyond the end of the cell. The cells in both the fore and hind wings are closed. The butterflies are of medium or large size, curiously marked with checkered spots, blue and white, with broad paler shades on the under side of the secondaries. They are rapid fliers and are said to alight on the trunks of trees with their wings expanded and their heads

Fig. 112.—Neuration of the genus *Ageronia.*

193

down. When flying they emit a clicking sound with their wings.

Early Stages.—Very little is known of these.

The chrysalids are slender and have two ear-like tubercles on the head.

This genus is, strictly speaking, neotropical. About twenty-five species have been described from Central and South America, some of them being exceedingly beautiful and rich in color. The two species credited to our fauna are reported as being occasionally found in Texas. I have specimens of one of the species which certainly came from Texas. I cannot be so sure of the other.

(1) **Ageronia feronia,** Linnæus, Plate XXIV, Fig. 4, ♂ (The White-skirted Calico).

Butterfly.—Easily distinguished from the only other species of the genus found in our fauna by the white ground-color of the under side of the hind wings. Expanse, 2.50 inches.

Early Stages.—Unknown.

This remarkable insect is said to be occasionally found in Texas.

(2) **Ageronia fornax,** Hübner, Plate XXIV, Fig. 5, ♂, *under side* (The Orange-skirted Calico).

Butterfly.— Closely resembling the preceding species on the upper side, but at once distinguished by the orange-yellow ground-color of the under side of the hind wing. Expanse, 2.60 inches.

Early Stages.—Unknown.

Like its congener, *A. fornax* is reported only from the hotter parts of Texas.

Genus VICTORINA, Blanchard
(The Malachites)

Butterfly. — Large butterflies, curiously and conspicuously marked with light-greenish spots upon a darker ground; wings upon the under side marbled with brown about the spots and having a satiny luster. The third median nervule of the fore wing is very strongly bowed upward. The cells of both wings are open. The hind wing is tailed at the end of the third median nervule. The two first subcostals arise before the end of the

194

cell; the fourth and fifth spring from a common stem which is emitted from the third beyond the end of the cell, as the cut shows.

Early Stages.—We know nothing of these.

This genus, in which are reckoned five species, all found in the tropics of the New World, is represented by but a single species in our fauna, which occurs in southwestern Texas and in Florida. It is very common in the West Indies and Central America.

(1) **Victorina steneles,** Linnæus, Plate XXIV, Fig. 6, ♂ (The Pearly Malachite).

This splendid insect is occasionally found in southern Florida and the extreme southern part of Texas. It is common throughout tropical America. Nothing has ever been written upon its early stages.

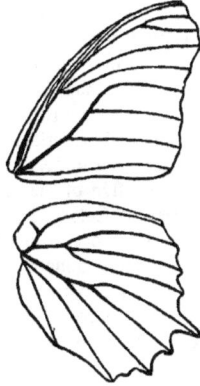

Fig. 113.—Neuration of the genus *Victorina*.

FOSSIL INSECTS

Investigations within comparatively recent times have led to the discovery of a host of fossil insects. A few localities in Europe and in North America are rich in such remains, and the number of species that have been described amounts to several thousands. Strangely enough, some of these fossil insects are very closely allied in form to species that are living at the present time, showing the extreme antiquity of many of our genera. One of the comparatively recent discoveries has been the fossil remains of a butterfly which Dr. Scudder, who has described it, declares to be very near to the African *Libythea labdaca*, which differs in certain minor anatomical respects from the American Libytheas which are figured in this work; and Dr. Scudder has therefore proposed a new generic name, *Dichora*, meaning "an inhabitant of two lands," which he applies to the African species because related to the extinct American butterfly. The strange discoveries, which have been made by palæontologists as to the huge character of many of the mammals, birds, and reptiles which at one time tenanted the globe, are paralleled by recent discoveries made in insect-bearing strata in France. M. Charles Brongniart of the Paris Museum is preparing an account of the collection which he has made at Commentry, and among the creatures which he proposes to figure is an insect which is regarded by Brongniart as one of the forerunners of our dragon-flies, which had an expanse of wing of two feet, a veritable giant in the insect world.

Of fossil butterflies there have thus far been discovered sixteen species. Of these, six belong to the subfamily of the *Nymphali-dæ*, and five of the six were found in the fossiliferous strata of Florissant, Colorado. Two species belong to the subfamily *Satyrinæ*, both occurring in deposits found in southern France, and representing genera more nearly allied to those now found in India and America than to the *Satyrinæ* existing at the present time in Europe. One of the fossils to which reference has already been made belongs to the subfamily of the *Libytheinæ*. The remainder represent the subfamilies of the *Pierinæ*, the *Papilioninæ*, and the family *Hesperiidæ*.

It is remarkable that the butterflies which have been found in a fossil state show a very close affinity to genera existing at the present time, for the most part, in the warmer regions of the earth. Though ages have elapsed since their remains were embedded in the mud which became transformed into stone, the processes of life have not wrought any marked structural changes in the centuries which have fled. This fixity of type is certainly remarkable in creatures so lowly in their organization.

Explanation of Plate XXIV

1. *Pyrrhanæa andria*, Scudder, ♀.
2. *Pyrrhanæa morrisoni*, Edwards. ♀.
3. *Pyrrhanæa portia*, Fabricius, ♂.
4. *Ageronia feronia*, Linnæus, ♂.
5. *Ageronia fornax*, Hübner, ♂, under side.
6. *Victorina steneles*, Linnæus, ♂.
7. *Cystineura amymone*, Ménétries, ♂.
8. *Synchloë crocale*, Edwards, ♂, under side.
9. *Synchloë crocale*, Edwards, ♂.
10. *Eurema lethe*, Fabricius, ♂.

PLATE XXIV.

SUBFAMILY SATYRINÆ (THE SATYRS)

> " Aught unsavory or unclean
> Hath my insect never seen;
> But violets and bilberry bells,
> Maple-sap and daffodils,
> Grass with green flag half-mast high,
> Succory to match the sky,
> Columbine with horn of honey,
> Scented fern and agrimony,
> Clover, catch-fly, adder's-tongue,
> And brier-roses dwelt among."
>
> EMERSON.

THE butterflies belonging to this subfamily are, for the most part, of medium size, and are generally obscure in color, being of some shade of brown or gray, though a few species within our territory are brightly colored. Gaily colored species belonging to this subfamily are more numerous in the tropics of both hemispheres. The wings are very generally ornamented, especially upon the under side, by eye-like spots, dark, pupiled in the center with a point of lighter color, and ringed around with one or more light circles. They are possessed of a weak flight, flitting and dancing about among herbage, and often hiding among the weeds and grasses. Most of them are forest-loving insects, though a few inhabit the cold and bleak summits of mountains and grassy patches near the margins of streams in the far North, while some are found on the treeless prairies of the West. In the warmer regions of the Gulf States a few species are found which have the habit of flitting about the grass of the roadsides and in open spaces about houses. The veins of the fore wings are generally greatly swollen at the base, enabling them thus to be quickly distinguished from all other butterflies of this family.

The eggs, so far as we have knowledge of them, are subspher-

197

ical, somewhat higher than broad, generany ribbed along the sides, particularly near the apex, and rounded at the base, which is generally broader than the apex.

The caterpillars at the time of emergence from the egg have the head considerably larger than the remainder of the body; but when they have reached maturity they are cylindrical, tapering a little from the middle to either end. They are bifurcated at the anal extremity, a character which enables them to be distinguished at a glance from the larvæ of all other American butterflies except those of the genus *Chlorippe*. They are mostly pale green or light brown in color, ornamented with stripes along the sides. They feed upon grasses and sedges, lying in concealment during the daytime, and emerging at dusk to take their nourishment.

The chrysalids are rather stout in form, but little angulated, and without any marked prominences or projections. They are green or brown in color. Most of them are pendant, but a few forms pupate at the roots of grasses or under stones lying upon the ground.

The butterflies of this subfamily have been arranged, so far as they are represented in the faunal region of which this book treats, in nine genera, which include about sixty species. It is quite possible that a number of species still remain to be discovered and described, though it is also true that some of the so-called species are likely to prove in the end little more than local races or varieties.

Genus DEBIS, Westwood

(The Eyed Nymphs)

" The wild bee and the butterfly
Are bright and happy things to see,
Living beneath a summer sky."

ELIZA COOK.

Butterfly.—Characterized by the stout but not greatly swollen costal vein of the fore wing, by the rather short costal vein of the hind wing, which terminates before quite reaching the outer angle, by the great length of the lower discocellular vein of the fore wing, and by the prolongation of the outer margin of the hind wing at the end of the third median nervule. The outer margin of the fore wing is either rounded or slightly excavated. The palpi are long and narrow, thickly clothed with hairs below; the antennæ are moderately long, gradually thickening toward the tip, without a well-marked club; the fore legs in both sexes greatly atrophied.

Egg.—Flattened spheroidal, broadly truncated at the base, the surface smooth.

Caterpillar.—Body long, slender, tapering from the middle; the head cleft, each half being produced upward as a conical horn; the anal segment provided with a pair of horns similar to those of the head, produced longitudinally backward.

Fig. 114.—Neuration of the genus *Debis*. (After Scudder.)

Chrysalis.—Strongly convex dorsally, concave ventrally, with a stout tubercular eminence on the thorax, without any other projecting tubercles or eminences; light green in color.

This genus is large, and is well represented in Asia and the Indo-Malayan region. I cannot see any good ground for generically separating the two species found in North America from their congeners of Asiatic countries, as has been done by some writers.

(1) **Debis portlandia**, Plate XVIII, Fig. 20, ♂; Plate III, Fig. 16, *larva;* Plate IV, Fig. 6, *chrysalis* (The Pearly Eye).

Butterfly.—The butterfly, the male of which is well depicted as to its upper side on the plate, does not differ greatly in the sexes. The hind wings on the under side are marked with a series of beautiful ocelli. In the North the insect is single-brooded; in the region of West Virginia and southward it is double-brooded. Expanse, 1.75–2.00 inches.

Early Stages.—The illustrations give a good idea of the mature larva and the chrysalis. The caterpillar, like most of the *Satyrinæ*, feeds upon grasses.

The range of this pretty insect is extensive, it being found from Maine to the Gulf of Mexico, and westward to the Rocky Mountains.

(2) **Debis creola**, Skinner, Plate XVIII, Fig. 18, ♂; Fig. 19, ♀ (The Creole).

Butterfly.—Easily distinguished from the preceding species by the elongated patches of dark raised scales upon the fore wings,

199

situated on the interspaces between the median nervules. The female has more yellow upon the upper side of the fore wings than *D. portlandia*. Expanse, 2.25 inches.

Early Stages.—Unknown.

Creola ranges from Florida to Mexico along the Gulf.

Genus SATYRODES, Scudder
(The Grass-nymphs)

Butterfly.—The head is moderately large; the eyes are not prominent, hairy; the antennæ are about half as long as the costa of the fore wing, not distinctly clubbed, gradually thickening toward the extremity. The palpi are slender, compressed, hairy below, with the last joint rather short and pointed. The fore and hind wings are evenly rounded on the outer margin. The costal vein of the fore wing is thickened, but not greatly swollen. The first and second subcostals are emitted well before the end of the cell, the third beyond it, and the fourth and fifth from a common stem, both terminating below the apex. The upper disco-cellular vein is wanting, and the upper radial, therefore, springs from the upper angle of the cell of the fore wing.

Fig. 115.—Neuration of the genus *Satyrodes*. (After Scudder.)

Egg.—Flattened spheroidal, broader than high, flat at the base and rounded above.

Caterpillar.—The head is full, the summit of either half produced upward and forward into a slender, conical horn. The body is nearly cylindrical, tapering backward, the last segment furnished with two pointed, backward projections, resembling the horns of the head.

Chrysalis.—Relatively longer and more slender than in the preceding genus, with the thoracic prominence more acute and the head more sharply pointed.

This genus was erected to receive the single species which, until the present time, is its sole representative.

(1) **Satyrodes canthus,** Boisduval and Leconte, Plate XXV, Fig. 1, ♂ ; Plate III, Fig. 9, *larva;* Plate IV, Fig. 9, *chrysalis* (The Common Grass-nymph).

200

Butterfly.—It always haunts meadows and hides among the tufts of tall grasses growing in moist places. It is rather common in New England and the Northern States generally. It is found in Canada and is reported from the cool upper mountain valleys in the Carolinas. It has a weak, jerking flight, and is easily taken when found. Expanse, 1.65–1.90 inch.

Early Stages.—These have been well described by various writers. The caterpillar feeds upon grasses.

Genus NEONYMPHA, Westwood
(The Spangled Nymphs)

" Oh! the bonny, bonny dell, whaur the primroses won,
Luikin' oot o' their leaves like wee sons o' the sun;
Whaur the wild roses hing like flickers o' flame,
And fa' at the touch wi' a dainty shame;
Whaur the bee swings ower the white-clovery sod,
And the butterfly flits like a stray thoucht o' God."

MacDonald.

Butterfly.—Eyes hairy. The costal and median veins of the fore wings are much swollen at the base. The palpi are thin, compressed, thickly clothed below with long hairs. The antennæ are comparatively short, gradually thickening toward the outer extremity, and without a well-defined club. Both the fore wing and the hind wing have the outer margin evenly rounded.

Egg.— Globular, flattened at the base, marked with irregular polygonal cells.

Caterpillar. — The head is large, rounded, the two halves produced conically and studded with little conical papillæ. The last segment of the body is bifurcate.

Chrysalis.— Relatively long, strongly produced at the vertex; elevated on the thorax into a blunt tubercular prominence; green in color.

Fig. 116. — Neuration of the genus *Neonympha*. (After Scudder.)

This genus, which has by some writers been sunk into the genus *Euptychia*, Hübner, is quite extensive. Nearly two hundred species are included in *Euptychia*, which is enormously developed in the tropical regions of the New World. Seven species of *Neonympha* are found within the region of which this book treats.

(1) **Neonympha gemma,** Hübner, Plate XXV, Fig. 2, ♂, *under side* (The Gemmed Brown).

Butterfly.—Upon the upper side the wings are pale mouse-gray, with a couple of twinned black spots on the outer margin of the hind wings. On the under side the wings are reddish-gray, marked with irregular ferruginous lines. Near the outer margin of the hind wings is a row of silvered spots, the spots corresponding in location to the dark marginal spots being expanded into a violet patch marked in the middle by a twinned black spot centered with silver. Expanse, 1.25–1.35 inch.

Early Stages.—These have been beautifully described and figured by Edwards in the third volume of "The Butterflies of North America."

The egg is somewhat globular, rather higher than wide, flattened at the base, and marked with numerous shallow reticulated depressions. The caterpillar of the spring brood is pale green, of the fall brood pale brown, marked respectively with numerous longitudinal stripes of darker green or brown. It has two long, elevated, horn-like projections upon the head, and on the anal segment two similar projections pointing straight backward. The chrysalis is small, green, or brown, strongly bifid at the head. The caterpillar feeds on grasses.

The insect ranges from West Virginia to Mexico.

(2) **Neonympha henshawi,** Edwards, Plate XXV, Fig. 8, ♂ (Henshaw's Brown).

Butterfly.— Much like *N. gemma*, but considerably larger and decidedly reddish upon the upper side of the wings. Expanse, 1.65 inch.

Early Stages.— Mr. Edwards has figured the egg, which is different in shape from that of the preceding species, being broader than high, subglobular, flattened broadly at the base, green in color, and almost devoid of sculpturings upon its surface. Of the other stages we know nothing.

Henshaw's Butterfly ranges through southern Colorado into Mexico.

(3) **Neonympha phocion,** Fabricius, Plate XXV, Fig. 7, ♂, *under side;* Plate III, Fig. 8, *larva;* Plate IV, Figs. 10 and 11 (The Georgian Satyr).

Butterfly.— The upper side is immaculate gray; beneath pale, with two ferruginous transverse lines. Between these lines is a ferruginous line on each wing, rudely describing a circle. In the circle on the fore wing are three or four eye-spots with a blue pupil and a yellow iris; in the circle on the hind wing are six eye-spots which are oblong and have the pupil oval. Expanse, 1.25 inch.

Early Stages.— These have been fully described, and are not unlike those of other species of the genus. The caterpillar feeds on grasses.

The insect ranges from New Jersey to the Gulf of Mexico as far west as Texas.

(4) **Neonympha eurytus**, Fabricius, Plate XXV, Fig. 4, ♂ ; Plate III, Figs. 3, 6, 10, 13, 14, *larva;* Plate IV, Fig. 28, *chrysalis* (The Little Wood-satyr).

Butterfly.—Easily distinguished from other species in our fauna by the presence of two more or less perfectly developed ocelli on the upper side of the fore wing and also of the hind wing. Expanse, 1.75 inch.

Early Stages.—This is a rather common butterfly, the larval stages of which have been fully described by various authors. The egg is even taller in proportion to its breadth than that of *N. gemma,* which it otherwise closely resembles in outline and sculpturing. The caterpillar is pale brown, conformed in general form to that of other species of the genus, but somewhat stouter. It feeds on grasses. The chrysalis is pale brown, mottled with darker brown.

The insect ranges through Canada and the United States to Nebraska, Kansas, and Texas.

(5) **Neonympha mitchelli**, French, Plate XXV, Fig. 6, ♂, *under side* (Mitchell's Satyr).

Butterfly.— Easily distinguished from the other species of the genus by the eye-spots on the under side of the wings, four on each of the primaries and six on each of the secondaries, arranged in a straight series on the outer third, well removed from the margin. These spots are black, ringed about with yellow and pupiled with blue.

Early Stages.— Unknown.

The species is local, and thus far is recorded only from northern New Jersey, near Lake Hopatcong, and the State of

Michigan. No doubt it occurs elsewhere, but has been over-looked by collectors.

(6) **Neonympha sosybius,** Fabricius, Plate XXV, Fig. 5, ♂, *under side* (The Carolinian Satyr).

Butterfly. — The upper surface is immaculate dark mouse-gray. On the under side the wings are paler, with three transverse undulatory lines, one defining the basal, the other the median area, and one just within the margin. Between the last two are rows of ocelli. The spots in these rows are obscure, except the first on the primaries and the second and last two on the secondaries, which are black, ringed about with yellow and pupiled with blue.

The female is like the male, but a trifle larger.

Early Stages. — These have been described by Edwards, French, and Scudder, and do not differ strikingly from those of other species.

The species ranges from the latitude of New Jersey southward, throughout the southern half of the Mississippi Valley to Mexico and Central America.

(7) **Neonympha rubricata,** Edwards, Plate XXV, Fig. 3, ♂ (The Red Satyr).

Butterfly. — Easily distinguished by its much redder color from all its congeners, among which it has its closest ally in *N. eurytus.* It has an eye-spot near the apex of the fore wing, and one near the anal angle of the hind wing. The basal area of the primaries beneath is bright reddish; the secondaries on this side are gray, crossed by two transverse lines as in the preceding species, and a double submarginal line. On the fore wings the double submarginal line is repeated, and in addition there is another line which runs upward from just before the inner angle to the costa, at about one third of its length from the apex. The eye-spots of the upper side reappear below, and in addition there is another near the outer angle of the secondaries, and a few silvery well-defined ocelli between the two on the secondaries.

Early Stages. — Unknown.

The Red Satyr is found in Texas, Arizona, Mexico, and Central America.

204

Explanation of Plate XXV

1. *Satyrodes canthus*, Boisd.-Lec., ♂.
2. *Neonympha gemma*, Hübner, ♂, under side.
3. *Neonympha rubricata*, Edwards, ♂.
4. *Neonympha eurytus*, Fabricius, ♂.
5. *Neonympha sosybius*, Fabricius, ♂, under side.
6. *Neonympha mitchelli*, French, ♂, under side.
7. *Neonympha phocion*, Fabricius, ♂, under side.
8. *Neonympha henshawi*, Edwards, ♂.
9. *Cœnonympha california*, Dbl.-Hew., var. *galactinus*, Boisd., ♂.
10. *Cœnonympha california*, Dbl.-Hew., var. *eryngii*, Henry Edwards, ♂.
11. *Cœnonympha ochracea*, Edwards, ♂.
12. *Cœnonympha ochracea*, Edwards, ♂, under side.
13. *Cœnonympha inornata*, Edwards, ♂, under side.
14. *Cœnonympha california*, Dbl.-Hew., ♀.

16. *Neominois dionysius*, Scudder, ♂.
17. *Erebia magdalena*, Strecker, ♂.
18. *Erebia sofia*, Strecker, = *ethela*, Edwards, ♀.
19. *Erebia discoidalis*, Kirby, ♂.
20. *Erebia tyndarus*, var. *callias*, Edwards, ♂.
21. *Cœnonympha ampelos*, Edwards, ♂, under side.
22. *Cœnonympha kodiak*, Edwards, ♀.
23. *Erebia disa*, var. *mancinus*, Dbl.-Hew., ♂.
24. *Cœnonympha haydeni*, Edwards, ♂.
25. *Cœnonympha elko*, Edwards, ♀, under side.
26. *Cœnonympha elko*, Edwards, ♂.
27. *Cœnonympha pamphiloides*, Reakirt, ♀.
28. *Erebia epipsodea*, Butler, ♂.
29. *Cœnonympha inornata*, Edwards, ♂.
30. *Cœnonympha ampelos*, Edwards, ♂.
31. *Cœnonympha pamphiloides*, Reakirt, ♂.

15. *Neominois ridingsii*, Edwards, ♂.

PLATE XXV.

Genus CŒNONYMPHA, Westwood
(The Ringlets)

" There is a difference between a grub and a butterfly; yet your butterfly was a grub."—SHAKESPEARE.

Butterfly. —Small butterflies. The costal, median, and submedian veins are all strongly swollen. The palpi are very heavily clothed with hairs, the last joint quite long and porrect. The antennæ are short, delicate, gradually but distinctly clubbed. The eyes are naked. Both wings on the outer margin are evenly rounded.

Egg.—The egg is conical, truncated, flat on the top, rounded at the base, with the sides marked with numerous low, narrow ribs, between which are slight cross-lines, especially toward the apex.

Caterpillar.—The head is globular; the body is cylindrical, tapering gradually backward, furnished in the last segment with two small horizontal cone-shaped projections.

FIG. 117.—Neuration of the genus *Cœnonympha.*

Chrysalis. — Ventrally straight, dorsally convex, strongly produced in a rounded, somewhat keeled eminence over the thorax; pointed at the end. Generally green or light drab in color, with dark markings on the sides of the wing-cases.

This genus is distributed throughout the temperate regions both of the Old and the New World, and includes in our fauna a number of forms, the most of which are peculiar to the Pacific coast.

(1) **Cœnonympha california,** Doubleday and Hewitson, Plate XXV, Fig. 14, ♀ ; form **galactinus,** Boisduval, Plate XXV. Fig. 9, ♂ ; form **eryngii,** Henry Edwards, Plate XXV, Fig. 10, ♂ (The California Ringlet).

Butterfly.—This little species is to be distinguished from its near allies by its white color. The form *galactinus* is the winter form; the form *california* the summer form. The former is characterized by the darker color of the hind wings on the under side and the more prominent development of the marginal ocelli. The form *eryngii* is simply a yellower form, with less dark shading on the under side.

Early Stages. —These have been most carefully and beautifully worked out by Edwards, and the reader, for a full knowledge of them, may consult the splendid plate in "The Butterflies of North America," vol. iii.

The species ranges from Vancouver's Island southward on the Pacific coast and eastward into Nevada.

(2) **Cœnonympha elko**, Edwards, Plate XXV, Fig. 25, ♀, *under side;* Fig. 26, ♂ (The Elko Ringlet).

Butterfly. —Yellow on both sides of the wings, the lower side paler than the upper, and the basal area lightly clouded with fuscous.

Early Stages. —Undescribed.

This species is found in Nevada and Washington.

(3) **Cœnonympha inornata**, Edwards, Plate XXV, Fig. 13, ♂, *under side;* Fig. 29, ♂ (The Plain Ringlet).

Butterfly. —The wings on the upper side are ochreous-brown, lighter on the disk. The costal margin of the fore wings and the outer margin of both fore and hind wings are gray. The ocellus at the apex of the fore wings on the under side is faintly visible on the upper side. On the under side the fore wings are colored as on the upper side as far as the termination of the discal area, which is marked by a narrow transverse band of pale yellow, followed by a conspicuous ocellus. The hind wings are gray, darkest toward the base, behind the irregular whitish transverse band which crosses the outer portion of the disk.

Early Stages. —Unknown.

The species occurs in Montana, Minnesota, British America, and Newfoundland. Newfoundland specimens, of which I possess a large series, are distinctly darker in color than those taken in the Northwest. Some recent writers are inclined to regard this as a variety of the European *C. typhon.* I am persuaded that they are mistaken.

(4) **Cœnonympha ochracea**, Edwards, Plate XXV, Fig. 11, ♂; Fig. 12, ♂, *under side* (The Ochre Ringlet).

Butterfly. —Glossy ochreous, yellow above, with no markings but those which show through from below. On the under side the wings are marked precisely as in the preceding species, except that there are two or three small rays on the secondaries near the base, one on the cell and one on either side of it, of the same tint as the discal transverse band, and in some specimens there is a series of incomplete marginal ocelli on the hind wings.

Early Stages.—Unknown.

Ochracea ranges from British Columbia to Arizona, as far east as Kansas.

(5) **Cœnonympha ampelos,** Edwards, Plate XXV, Fig. 21, ♂, *under side;* Fig. 30, ♀ (The Ringless Ringlet).

Butterfly.—Distinguished from its allies by the total absence of ocelli on both wings, above and below. Otherwise the species is very near *ochracea.*

Early Stages.—These have been described with minute accuracy by Edwards in the "Canadian Entomologist," vol. xix, p. 41.

Ampelos occurs from Nevada and Montana westward to Vancouver's Island.

(6) **Cœnonympha kodiak,** Edwards, Plate XXV, Fig. 22, ♀ (The Alaskan Ringlet).

Butterfly.—Much darker both on the upper and under sides than *C. california,* which in many other respects it resembles. The figure in the plate is that of the type. It is as yet rare in collections.

Early Stages.—Nothing is known of these. It is found in Alaska.

(7) **Cœnonympha pamphiloides,** Reakirt, Plate XXV, Fig. 27, ♀, *under side;* Fig. 31, ♂ (The Utah Ringlet).

Butterfly.—Rather larger than the other species of the genus found in North America. Easily distinguished by the marginal row of ocelli on the secondaries, which are always present, though often "blind," that is to say, without a distinct dark pupil. The author of the species named it from a supposed likeness to the European *C. pamphilus.* The resemblance is only superficial. *C. pamphilus* is a much smaller insect and much more plainly marked, judging from the large series of specimens I have received from various European localities. *Pamphilus* has no eye-spots on the hind wings. They are a conspicuous feature of *pamphiloides,* more so than in any other North American species except *C. haydeni.*

Early Stages.—Unknown.

Habitat, Utah and California.

(8) **Cœnonympha haydeni,** Plate XXV, Fig. 24, ♂, *under side* (Hayden's Ringlet).

Butterfly.—Dark immaculate mouse-gray on the upper side.

On the under side the wings are pale hoary gray, with the hind wings adorned by a marginal series of small ocelli, black, ringed about with yellow and pupiled with pale blue.

Early Stages.— Unknown.

Hayden's Ringlet is found in Montana, Idaho, Wyoming, and Colorado.

Genus EREBIA, Dalman

(The Alpines)

"Then we gather, as we travel,
Bits of moss and dirty gravel,
    And we chip off little specimens of stone;
And we carry home as prizes
Funny bugs of handy sizes,
    Just to give the day a scientific tone."
CHARLES EDWARD CARRYL.

Butterfly.— Medium-sized or small butterflies, dark in color, wings marked on the under side with eye-like spots; the antennæ short, with a gradually thickened club. The eyes are naked. The costal vein of the fore wing is generally strongly swollen at the base. The subcostal vein is five-branched; the first two nervules generally emitted before the end of the cell; the third nearer the fourth than the end of the cell; the fourth and fifth nervules spring from a common stem, the fourth terminating immediately on the apex. The lower radial is frequently projected inwardly into the cell from the point where it intersects the union of the middle and lower discocellular veins. The outer margins of both wings are evenly rounded.

FIG. 118.—Neuration of the genus *Erebia*, enlarged.

Egg.—Subconical, flattened at the base and at the top, the sides marked by numerous raised vertical ridges, which occasionally branch or intersect each other.

Caterpillar.—The head is globular, the body cylindrical, tapering gradually backward from the head, the last segment slightly bifurcate.

Chrysalis.— The chrysalis is formed about the roots of grass and on the surface of the ground, either lying loosely there or surrounded by a few strands of silk. The chrysalis is convex, both ventrally and dorsally, humped on the thorax, produced at the head; all the projections well rounded. The chrysalids are generally some shade of light brown or ashen-gray, with darker stripes and spots. This genus is arctic, and only found in the cooler regions of the North or upon elevated mountain summits. A few species range downward to lower levels in more temperate climates, but these are exceptional cases.

(1) **Erebia discoidalis,** Kirby, Plate XXV, Fig. 19, ♂ (The Red-streaked Alpine).

Butterfly.—Easily distinguished by the plain black wings, relieved by a reddish-brown shade on the disk of the primaries on the upper side.

Early Stages.—Hitherto undescribed.

This species is found in the far North. My specimens came from the shores of Hudson Bay.

(2) **Erebia disa,** var. **mancinus,** Doubleday and Hewitson, Plate XXV, Fig. 23, ♂ (The Alaskan Alpine).

Butterfly.—The wings are dark brown on the upper side. On the outer third below the apex are three or four black ocelli, broadly ringed with red and pupiled with white. The upper ocellus is generally bipupiled, that is to say, the black spot is twinned, and there are two small light spots in it. On the under side the fore wings are as on the upper side. The hind wings are broadly sown with gray scales, giving them a hoary appearance. The base is more or less gray, and there is a broad, regularly curved mesial band of dark gray, which in some specimens is very distinct, in others more or less obsolete. The female does not differ from the male, except that the ocelli on the fore wings are larger and more conspicuous.

Early Stages.—Unknown.

This species is found in Alaska and on the mountains of British Columbia.

(3) **Erebia callias,** Edwards, Plate XXV, Fig. 20, ♂ (The Colorado Alpine).

Butterfly.—Pale brown on the upper side, with a more or less indistinctly defined broad transverse band of reddish on the outer third of the fore wings. At the apical end of this band are two black ocelli, pupiled with white. The fore wings on the under side are reddish, with the costa and outer margin grayish. The ocelli on this side are as on the upper side. The hind wings are gray, dusted with brown scales and crossed by narrow, irregular, dark-brown subbasal, median, and submarginal lines.

Early Stages.— Unknown.

This species is not uncommon on the high mountains of Colorado and New Mexico. It is regarded as a variety of the European *E. tyndarus*, Esper, by many. All the specimens of *tyndarus* in my collection, and there are many, lack the ocelli on the fore wing, or they are very feebly indicated on the under side. Otherwise the two forms agree pretty closely.

(4) **Erebia epipsodea**, Plate XXV, Fig. 28, ♂ (The Common Alpine).

Butterfly.—The wings are dark brown on the upper side, with four or five black ocelli, pupiled with white and broadly surrounded by red near the outer margin of the fore wings, and with three or four similar ocelli located on the upper side of the hind wings. The spots on the upper side reappear on the under side, and in addition the hind wings are covered by a broad curved median blackish band.

Early Stages.—These have been carefully described by Edwards in "The Butterflies of North America," vol. iii, and by H. H. Lyman in the "Canadian Entomologist," vol. xxviii, p. 274. The caterpillar feeds on grasses.

The species ranges from New Mexico (at high elevations) northward to Alaska. It is common on the mountains of British Columbia.

(5) **Erebia sofia** Strecker (**ethela,** Edwards), Plate XXV, Fig. 18, ♀ (Sofia .

Butterfly.—Dark brown on the upper side, with an even submarginal band of red spots on the primaries, and five similar spots on the secondaries, the last two of the latter somewhat distant from each other and from the first three, which are nearer the outer angle. On the under side the primaries are reddish, with the submarginal band as on the upper side, but paler. On the secondaries, which are a little paler below than above, the spots of the upper side are repeated, but they are yellowish-white, standing forth conspicuously upon the darker ground-color.

Early Stages.—Hitherto undescribed.

Sofia has been found at Fort Churchill in British America, in the Yellowstone National Park, and in a few localities in Colorado. It is still rare in collections. The figure in the plate is that of the female type of Edwards' *ethela*, *ethela* being a synonym for *sofia*.

(6) **Erebia magdalena**, Strecker, Plate XXV, Fig. 17, ♂ (Magdalena).

Butterfly.— Uniformly dark blackish-brown on both sides of the wings, with no spots or markings.

Early Stages.—These have been partially described and figured by Edwards.

This species has thus far been found only in Colorado at an elevation of from ten to twelve thousand feet above sea-level.

There are two or three other species of this obscure genus, but they are rare boreal insects, of which little is as yet known.

Genus GEIROCHEILUS, Butler

Butterfly.—Medium-sized butterflies, dark in color, with light eye-like spots on the primaries and brown borders on the secondaries. The antennæ are short, with a gradually tapering club; the palpi are long, slender, compressed, well clothed with scales on the lower surface. The costa of the fore wings is strongly arched, the outer margin evenly rounded, the outer margin of the hind wings regularly scalloped. The costal vein of the primaries is somewhat thickly swollen at the base.

Early Stages.—Unknown.

(1) **Geirocheilus tritonia,** Edwards, Plate XVIII, Fig. 21, ♂ (Tritonia).

Butterfly.—The wings of the upper side are dark brown, with a submarginal row of white-centered ocelli below the apex of the primaries. The secondaries are marked with a submarginal band of red. On the under side the fore wings are as on the upper side.

FIG. 110.—Neuration of the genus *Geirocheilus.*

The hind wings have the submarginal band purplish-red, irrorated with whitish-

and dark-brown scales, on the inner edge relieved by a number of imperfectly developed ocelli, which are partially ringed about on the side of the base by pale yellow.

Early Stages.—Unknown.

Tritonia occurs in southern Arizona and northern Mexico.

Genus NEOMINOIS, Scudder

Butterfly.—Medium-sized, with the costa and inner margin of the fore wing straight, the outer margin of the same wing evenly rounded. The hind wings have the outer margin evenly rounded, and the costal margin quite strongly produced, or bent at an angle, just above the origin of the costal vein. The inner margin is straight. The costal vein of the fore wing is slightly swollen. The costal margin at the extremity of the second costal nervule is slightly bent inward; the upper discocellular vein is wanting; the lower radial vein is emitted from the lower discocellular a little below the point at which it unites with the middle discocellular. The middle discocellular of the hind wing appears as an inward continuation of the lower radial for some distance, when it bends upward suddenly to the origin of the upper radial. The head is small; the antennæ are short, with a thin, gradually developed club; the palpi are slender, compressed, well clothed with long hairs below.

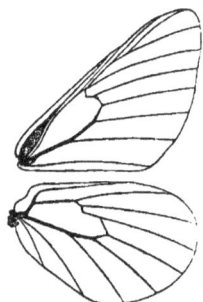

Fig. 120.—Neuration of the genus *Neominois*, enlarged.

Egg.—The egg is somewhat barrel-shaped, broader at the base than at the top, with the summit rounded. The sides are ornamented with fourteen or fifteen vertical raised ridges, which are quite broad, and sometimes fork or run into each other. On the sides these ridges seem to be regularly excised at their bases, and between them on the surface are many horizontal raised cross-lines, giving the depressed surface the appearance of being filled with shallow cells.

Caterpillar.—The mature caterpillar has the head globular, the body cylindrical, gradually tapering backward, and provided with two very short conical anal horns.

Chrysalis.—The chrysalis is formed under the surface of the earth; it is rounded, somewhat carinate, or keel-shaped, where the wing-cases unite on the ventral side. The head is rounded, the thorax strongly arched, the dorsal side of the abdomen very convex. On either side of the head are small clusters of fine processes shaped somewhat like an Indian club, the thickened part studded with little spur-like projections. These can only be seen under the microscope.

But two species of the genus are known within our faunal limits.

(1) **Neominois ridingsi,** Edwards, Plate XXV, Fig. 15, ♂ (Ridings' Satyr).

Butterfly.—The upper side is well depicted in the plate. The under side is paler than the upper side, and the basal and median areas of both wings are profusely mottled with narrow pale-brown striæ, the secondaries crossed by a darker mesial band, the outer margin of which is sharply indented. Expanse, 1.50 inch.

Early Stages.—These have been beautifully ascertained, described, and figured by Edwards in the third volume of "The Butterflies of North America." The egg, larva, and chrysalis agree with the generic description already given, which is based upon the researches of Edwards.

It is found in the Mountain States of the Pacific coast.

(2) **Neominois dionysius,** Scudder, Plate XXV, Fig. 16, ♂ (Scudder's Satyr).

Butterfly.—Distinguished from the preceding species by the larger and paler submarginal markings on the upper side of the wings and the pale color of the basal tract in both wings. On the under side the median band of the secondaries is narrower and more irregularly curved than in *ridingsi,* with the dentations of the outer margin more sharply produced. Expanse, 1.90 inch.

Early Stages.—Nothing has been written on the early stages, but no doubt they agree closely with those of the other species.

It is found in Utah, Colorado, and Arizona.

> " Hast thou heard the butterflies,
> What they say betwixt their wings ? "
>
> TENNYSON, *Adeline.*

Genus SATYRUS, Westwood

(The Wood-nymphs)

"Fluttering, like some vain, painted butterfly,
From glade to glade along the forest path."
ARNOLD, *Light of Asia.*

Butterfly. —Butterflies of medium size, their wings marked with eye-like spots, or ocelli. Upon the upper surface they are generally obscurely colored of some shade of gray or brown, occasionally marked with bands of yellow. On the under side the wings are generally beautifully striated and spotted, with the eye-like spots more prominent. The costal vein at the base is greatly swollen; the median and submedian veins less so. The first and second subcostal nervules arise very near the end of the cell, slightly before it. The outer margin of the fore wing is evenly rounded; the outer margin of the hind wing somewhat scalloped; the head small, the eyes of moderate size, full, naked; the antennæ gradually thickening to a broadly rounded club, which is slightly depressed; the palpi slender, compressed, profusely clothed beneath with long hairs. The fore legs are very small.

FIG. 121.— Neu-ration of the genus *Satyrus.* (After Scudder.)

Egg.—Short, barrel-shaped, greatly diminishing in size on the upper half; truncated at the summit; the sides furnished with a large number of vertical ribs, not very high, with numerous delicate cross-lines between them. At the summit the ribs are connected by a waved, raised elevation.

Caterpillar.—Head globular; body cylindrical, tapering from the middle forward and backward; provided with short and slender diverging anal horns.

Chrysalis.—Shaped very much as in the genus *Debis,* from which it is hardly distinguishable. Generally green in color.

This genus includes numerous species which are more or less subject to varietal modifications. In the following pages I have treated as species a number of forms which by some writers are reckoned as mere varieties. Whether the view of those who regard these forms in the light of varieties is correct is not per-

Explanation of Plate XXVI

1. *Satyrus alope*, Fabricius, ♂.
2. *Satyrus alope*, Fabricius, ♀.
3. *Satyrus nephele*, Kirby, ♂.
4. *Satyrus nephele*, Kirby, ♀, under side.
5. *Satyrus ariane*, Boisduval, ♂.
6. *Satyrus ariane*, Boisduval, ♀, under side.
7. *Satyrus œtus*, Boisduval, ♂.
8. *Satyrus œtus*, Boisduval, ♂, under side.
9. *Satyrus olympus*, Edwards, ♂.
10. *Satyrus olympus*, Edwards, ♀, under side.
11. *Satyrus charon*, Edwards, ♂.
12. *Satyrus charon*, Edwards, ♀.
13. *Satyrus meadi*, Edwards, ♀.
14. *Satyrus meadi*, Edwards, ♂, under side.
15. *Satyrus baroni*, Edwards, ♂.
16. *Satyrus baroni*, Edwards, ♂, under side.
17. *Satyrus gabbi*, Edwards, ♀, under side.
18. *Satyrus pegala*, Fabricius, ♀, under side.
19. *Satyrus paulus*, Edwards, ♂, under side.
20. *Satyrus sthenele*, Boisduval, ♂, under side.

PLATE XXVI.

fectly plain to me, and we cannot be sure until more extensive experiments in breeding have been carried out.

(1) **Satyrus pegala,** Fabricius, Plate XXVI, Fig. 18, ♀, *under side* (The Southern Wood-nymph).

Butterfly.—The largest species of the genus in our fauna, easily recognized by the broad yellow submarginal band on the primaries, marked with a single eye-spot in the male and two eye-spots in the female. The plate gives a correct idea of the under side of the wings. Expanse, 2.75 inches.

Early Stages.—These have only been partially ascertained. The caterpillar, like all others of the genus, feeds on grasses.

This insect is found in the Gulf States and as far north as New Jersey, and is probably only a large Southern form of the next species.

(2) **Satyrus alope,** Fabricius, Plate XXVI, Fig. 1, ♂ ; Fig. 2, ♀ ; Plate III, Fig. 18, *larva* (The Common Wood-nymph).

Butterfly.—Closely resembling the preceding species, but only two thirds of its size. The figures in our plate give a correct idea of its appearance. The number of the ocelli is not constant, and occasionally specimens occur in which they are almost wanting. Several varietal forms have been described : *S. maritima,* from Long Island and Martha's Vineyard, in which the wings are smaller, the band inclined to orange-yellow, and the upper side of the wings is darker than in the typical form; and *S. texana,* from the extreme South, in which the ground-color of the wings is paler brown, the yellow band ochreous, and the spots on the under side of the hind wings larger than in the other forms.

(*a*) **Satyrus alope,** form **nephele,** Kirby, Plate XXVI, Fig. 3, ♂ ; Fig. 4, ♀, *under side ;* Plate IV, Figs. 7, 8, *chrysalis* (The Clouded Wood-nymph).

This varietal form of *S. alope,* long held to be a species, but now known to be a dimorphic variety, is characterized by the partial or entire suppression of the yellow band on the primaries and the tendency of the eye-spots to become obsolete. It is the Northern form of the species, and is found in Canada, New England, and on the continent generally, from the Atlantic to the Pacific, north of the latitude of central New York and southward on the mountain masses of the Appalachian ranges.

(*b*) **Satyrus alope,** form **olympus,** Edwards, Plate XXVI, Fig. 9, ♂ ; Fig. 10, ♀, *under side* (Olympus).

This form of *S. alope* is common in the region west of the Mississippi. The males are a trifle darker and the females a shade paler than in the form *nephele*, which they closely approximate, and from which it would almost be impossible to separate them without a knowledge of the country whence they come.

(*c*) **Satyrus alope**, form **ariane**, Boisduval, Plate XXVI, Fig. 5, ♂ ; Fig. 6, ♀ , *under side* (Ariane).

In *ariane* we have a decidedly dwarfed form, in which the males and the females are quite dark. The ocelli, though small, are persistent, well defined, rarely showing a tendency to disappear completely. This form is found in British America, Oregon, and the northwestern portion of the United States.

(*d*) **Satyrus baroni**, Plate XXVI, Fig. 15, ♂ ; Fig. 16, ♂ , *under side* (Baron's Satyr).

This is another form, dark on the upper side and reddish below, in which the ocelli on the under side show a tendency to become obsolete, and in some specimens are wholly wanting.

There are other varietal forms, one of which, named *boöpis* by Behr, is commonly found on the Pacific coast in northern California, Oregon, and Washington, and the ocelli, while prominent on the upper side of the wings, are almost obsolete below.

Early Stages.—The early stages of *S. alope* (typical form) and its variety *nephele* have been well described by several authors. The caterpillar feeds on grasses. There is, however, a fine field for the entomologist to work out the causes of the rather remarkable variation to which the species is subject.

(3) **Satyrus gabbi**, Edwards, Plate XXVI, Fig. 17, ♀ , *under side* (Gabb's Satyr).

Butterfly.—The male is dark reddish-brown, the female pale fawn. The ocelli in both sexes are very well developed on both sides of the wings. The anal series on the secondaries consists of three spots, of which the one in the middle is always large. Expanse, 2.25 inches.

Early Stages.—Unknown.

Gabb's Satyr is found in Oregon and Utah.

(4) **Satyrus meadi**, Plate XXVI, Fig. 13, ♀ ; Fig. 14, ♂ , *under side* (Mead's Satyr).

Butterfly.—This well-marked species is comparatively small, and may easily be distinguished from all others by the bright red on the limbal area above and on the middle area of the primaries below. Expanse, 1.60–1.75 inch.

Early Stages.—These have been described and figured by Edwards in "The Butterflies of North America," vol. iii. The caterpillar is green, marked by paler stripes and lozenge-shaped spots of pale green on the side. The chrysalis is pale green. The egg is pale saffron. The caterpillars feed on grass.

Mead's Satyr ranges through Colorado, Montana, Utah, and Arizona.

(5) **Satyrus paulus**, Edwards, Plate XXVI, Fig. 19, ♂, *under side* (The Small Wood-nymph).

Butterfly.—A little smaller than *S. nephele*, dark brown above in both sexes, the fore wings always with two pupilate ocelli, one near the apex, the other near the inner angle, most conspicuously developed in the female. The secondaries have one or two spots of the same kind near the anal angle. On the under side the wings are pale reddish-brown, abundantly marked by transverse striæ. The primaries are marked with gray at the apex and on the outer margin, and have a submarginal and submedian transverse ferruginous line, between which the ocelli are located. The secondaries are crossed by a broad darker median band defined inwardly and outwardly by narrow dark lines. The outer third is pale gray, mottled with darker spots and lines, and traversed by a dark ferruginous submarginal line. Expanse, 1.75–2.00 inches.

Early Stages.—Unknown.

Paulus occurs in California and Nevada. It has been regarded as a variety of *sthenele* by some writers; but I am convinced of its distinctness, though there is considerable resemblance.

(6) **Satyrus charon**, Edwards, Plate XXVI, Fig. 11, ♂ ; Fig. 12, ♀ (The Dark Wood-nymph).

Butterfly.—The male is dark in color; the female is paler. There are two eye-spots on the fore wings in the usual location, indistinct on the upper, distinct on the lower side of the wings. The under sides of the wings are variable. In the type they are dark; in other specimens they are paler. They may or may not have ocelli on the secondaries. The form with obsolescent ocelli has been named *silvestris* by Edwards. Both the fore and hind wings are abundantly and evenly marked by little striæ, and crossed on either side of the median area by obscure, irregular, transverse dark lines, either one or both of which may be wanting in some specimens. Expanse, 1.50–1.75 inch.

Early Stages.—These have been described and beautifully

figured by Edwards in the third volume of his great work, to which the reader may refer. The caterpillar is green, cylindrical, tapering before and behind, marked with longitudinal pale-yellow lines. The chrysalis is green or black, striped with narrow white lines. The egg is somewhat firkin-shaped, flat at the top and base, vertically ribbed, and honey-yellow. The larva feeds on grasses.

Charon is found in the Northwest, ranging from British Columbia as far as New Mexico.

(7) **Satyrus œtus,** Boisduval, Plate XXVI, Fig. 7, ♂ ; Fig. 8, ♂ , *under side* (Boisduval's Satyr).

Butterfly.—Larger than *charon*, paler on the upper side, especially in the female sex, in which the outer third of the primaries is reddish-fawn. On the under side the secondaries of the male are without ocelli, or at most faint traces of ocelli appear. In the female the ocelli near the anal angle of the secondaries are usually well developed. Expanse, 1.60–2.00 inches.

Early Stages.—These await description.

The species is found in northern California.

(8) **Satyrus sthenele,** Boisduval, Plate XXVI, Fig. 20, ♂ , *under side* (The Least Wood-nymph).

Butterfly.—Quite small, superficially resembling *charon*. The female is paler and the ocelli are larger and more distinct than in *charon*. The distinguishing mark of this species is the irregular, dark, twice-strangulated band of the secondaries, bordered on both sides externally by whitish shades. This is shown in our figure. Expanse, 1.40–1.50 inch.

Early Stages.—Unknown.

The species is Californian.

Genus ŒNEIS, Hübner

(Chionobas, *Boisd.*)

(The Arctics)

" To reside
In thrilling region of thick-ribbed ice."
SHAKESPEARE.

Butterfly.—The antennæ are short; the eyes of moderate size; the front full, protuberant; the palpi slender; the fore wings somewhat produced at the tip, with the outer margins rounded and the hind margins very slightly, if at all, sinuated. The nervules of the fore wings are slightly dilated toward the base; the hind wings are elongated, oval, with the outer margins evenly rounded. The color of these butterflies is some shade of brown; the outer margin is generally lighter than the base of the wing, and is marked with black spots, sometimes pupiled with white. The wings are generally marbled and mottled on the under side, and sometimes crossed on the middle of the hind wings by a broad band of darker color. The fringes are brown, checkered with white.

Egg. — The egg is ovate-spherical, higher than broad, marked on the side from the apex to the base with raised sculptured ridges. These eggs are deposited, so far as we have been able to learn, on dried grass and the stems of plants in proximity to the growing plants upon which the young caterpillars are destined to feed.

Fig. 122.—Neuration of the genus *Œneis*, enlarged.

Caterpillar.—The head of the caterpillar when it emerges from the egg is somewhat larger than the rest of the body, but as it passes successive moults and attains maturity the relative thickness of the body increases, and the adult larva tapers a little from about the middle in either direction. The larvæ are pale green or brown, marked by darker stripes upon the back and on the sides, the markings on the sides being in most species more conspicuous than those on the back. The species all feed on grasses.

Chrysalis.—The chrysalids are stout, very slightly angulated, and are formed, so far as we know, unattached, under stones and at the roots of grasses. When pupating, the caterpillar often makes for itself a slight depression or cell in the soil, in which a few threads of silk have been deposited, though not enough to justify us in calling the structure a cocoon.

This genus is composed of butterflies which are mainly arctic in their habitat, or dwell upon the summits of lofty mountains, where the summer is but brief. Only a few species are found at comparatively low elevations, and these in British America, or the parts of the United States immediately contiguous to the Canadian line. The most widely known of all the species up to this time is the White Mountain Butterfly, *Œneis semidea*, Say, a colony of which has existed probably ever since the glacial period upon the loftiest summit of Mount Washington, in New Hampshire. A number of species are found in the region of the Rocky Mountains. One species, *Œneis jutta*, Hübner, occurs in Maine, Nova Scotia, and parts adjacent. There are in all about a score of species of this genus recognized by authors as occurring in our fauna. In spite of the fact that these insects are boreal or arctic in their habits, Mr. W. H. Edwards has with marvelous skill and patience succeeded in obtaining the eggs and rearing at his home in Coalburg, West Virginia, a number of species. We are indebted to him for more of our knowledge of the generic characteristics of these insects, in their early stages, than had been ascertained hitherto during a century of investigation. His work is one of the beautiful triumphs of that enduring zeal which is a supreme quality in the naturalist. In their early stages all of the species show a close likeness to one another.

(1) **Œneis gigas,** Butler, Plate XXVII, Fig. 1, ♂ ; Fig. 2, ♀ (The Greater Arctic).

Butterfly. —This, one of the largest species in the genus, occurs on Vancouver's Island. The butterfly hides among the dark mosses and upon the trunks of prostrate trees. The males are vigilant and inquisitive, and dart out suddenly when alarmed, or attracted by passing insects. The females have a slower and more leisurely flight and are more readily taken. Expanse, 2.00–2.25 inches.

Early Stages. —Edwards has figured the egg and the caterpillar in its first three stages, but the remaining life-history of the species awaits investigation.

(2) **Œneis iduna,** Edwards, Plate XXVII, Fig. 4, ♂ (The Iduna Butterfly).

Butterfly. —This insect, which even exceeds *Œ. gigas* in size, is found on the Coast Range in northern California. It is decidedly lighter on the outer third of the wings than the preceding species, the male being prevalently a pale yellowish-brown, with the basal and median areas of the fore wing dark brown. On the under side the wings are somewhat lighter than

1. *Œneis gigas*, Butler, ♂.
2. *Œneis gigas*, Butler, ♀.
3. *Œneis macouni*, Edwards, ♂.
4. *Œneis iduna*, Edwards, ♂.
5. *Œneis jutta*, Hübner, ♀.
6. *Œneis targete*, Hubner, ♂.

7. *Œneis brucei*, Edwards, ♂.
8. *Œneis varuna*, Edwards, ♂.
9. *Œneis ivallda*, Mead, ♂.
10. *Œneis chryxus*, Dbl.-Hew., ♂.
11. *Œneis semidea*, Say, ♂.
12. *Œneis uhleri*, Reakirt, ♀.

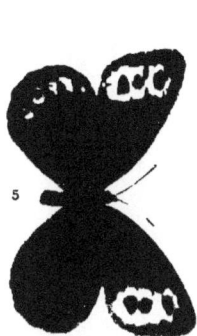

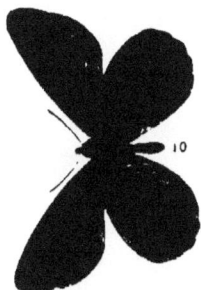

in the preceding species, and the transverse lines are more distinctly marked. Expanse, 2.00–2.30 inches.

Early Stages.—These have been most beautifully delineated by Edwards in the third volume of "The Butterflies of North America."

(3) **Œneis macouni,** Edwards, Plate XXVII, Fig. 3, ♂ (Macoun's Arctic).

Butterfly.—This species is closely allied to the two foregoing, but may be distinguished by the broad median band of dark brown traversing the under side of the hind wings, as well as by other peculiarities of marking. It lacks the bar of raised scales which is found in the male sex about the lower part of the cell of the fore wing in most of the species of the genus. It has been found thus far only on the north shore of Lake Superior and at the eastern base of the Rocky Mountains in the territory of Alberta. Expanse, 2.00–2.25 inches.

Early Stages.—For a knowledge of these in all their minute details the reader is again referred to the pages of the indefatigable Edwards.

(4) **Œneis chryxus,** Westwood, Plate XXVII, Fig. 10, ♂ (The Chryxus Butterfly).

Fig. 123.— Caterpillars of *Œneis macouni* (Riley).

Butterfly.—This species is widely distributed, being found in Colorado, British Columbia, and the vicinity of Hudson Bay. It is distinguished from other species by the darker brown color, which covers the basal and median areas of both the fore and hind wings, leaving a broad band of lighter brown on the outer margin. On the under side the wings are beautifully mottled with white and dark brown. *Œneis calais,* Scudder, is probably only a form of *chryxus,* which is somewhat lighter in color on the base of the wings. Expanse, 1.60–1.75 inch.

Early Stages.—The life-history is fully recorded in the pages of Edwards.

(5) **Œneis ivallda,** Mead, Plate XXVII, Fig. 9, ♂ (Mead's Arctic).

Butterfly.—This species is easily distinguished from all others by the peculiar pale ashen-brown of the upper side of the wings. It is not a common species, and is apparently restricted to the mountains of Nevada, principally about Lake Tahoe, though it probably occurs elsewhere. Expanse, 1.90–2.10 inches.

Early Stages.—Unknown.

(6) **Œneis varuna,** Edwards, Plate XXVII, Fig. 8, ♂ (The Varuna Butterfly).

Butterfly.—This species is much smaller than any of those which have thus far been mentioned. It is found in the prairie lands of Montana, North Dakota, and the parts of Canada adjacent. It is not uncommon about Calgary. It is light in color on the upper side of the wings, and on the under side it is mottled with brown, strongly marked with blackish blotches or shades. Expanse, 1.50–1.60 inch.

Early Stages.—These await description.

(7) **Œneis uhleri,** Reakirt, Plate XXVII, Fig. 12, ♂ (Uhler's Arctic).

Butterfly.—This species is found in Colorado. It is redder on the upper side than *varuna*, and the females are generally very richly ornamented with eye-spots on the outer borders of both the fore and hind wings. Expanse, 1.45–1.55 inch.

Early Stages.—These have been most thoroughly described and beautifully delineated by Edwards.

(8) **Œneis jutta,** Hübner, Plate XXVII, Fig. 5, ♀ (The Nova Scotian).

Butterfly.—This beautiful species, which is also found in Europe, is not uncommon in the State of Maine as far south as Bangor, and occurs also in Nova Scotia, and ranges thence westward to Ottawa and the Hudson Bay country. It is one of the more conspicuous species of the genus, the eye-like spots upon the wings having a very striking appearance. Expanse, 1.80–2.10 inches.

Early Stages.—For a thorough knowledge of these the reader may consult the pages of Scudder and Edwards.

(9) **Œneis semidea,** Say, Plate XXVII, Fig. 11, ♂ ; Plate III, Figs. 1, 2, 4, 7, 15, *larva;* Plate IV, Figs. 4, 5, *chrysalis* (The White Mountain Butterfly).

Butterfly.—This species has thin wings, and is much darker in color than any of the species which have thus far been mentioned. It is restricted in its habitat to the summit of Mount Washington, in New Hampshire, and only reappears on the high mountains of Colorado and in Labrador. Its life-history has been very carefully worked out. It is to be hoped that entomologists and tourists resorting to Mount Washington will not suffer it to disappear by reason of too wholesale a capture of the specimens, which hover about the barren rocks on which the race has existed since the great continental ice-sheet melted away and vanished from the face of New England. Expanse, 1.75 inch.

Early Stages.—The curious reader is again referred for a knowledge of these to the pages of Scudder and Edwards. They are similar to those of other species, and the generic description which has been given must suffice for all in this work.

(10) **Œneis brucei,** Edwards, Plate XXVII, Fig. 7, ♂ (Bruce's Arctic).

Butterfly.—Though somewhat closely related to the last species, Bruce's Arctic may at once be distinguished from it by the broad dark band on the under side of the secondaries and the great translucency of the wings, which permits a label to be read through them. It is found in Colorado and in British Columbia at an elevation of from twelve to thirteen thousand feet above sea-level. Expanse, 1.75 inch.

Early Stages.—All we know of these is contained in the pages of Edwards' great work.

(11) **Œneis taygete,** Hübner, Plate XXVII, Fig. 6, ♂ (The Labrador Arctic).

Butterfly.—Much like *Œ. brucei*, but the wings are not so translucent as in that species, and the broad mesial band on the under side of the hind wings is differently shaped, being more strongly directed outward just below the costa. The figure in the plate is from a specimen taken at Nain, in Labrador. Expanse, 1.75 inch.—

Early Stages.—Unknown.

There are eight or nine other species of *Œneis* in our fauna, but they are all arctic, and most of them very rare. Those we have described and figured will give a good idea of the genus.

IN THE FACE OF THE COLD

When the full moon hangs high overhead, the snow creaks underfoot, the north wind roars with furious blast, and the trees of the forests crack in the frost with a report like that of cannon, then, hanging in its little nest on the bare branches of the wind-tossed trees, the tiny caterpillar of the Viceroy keeps the spark of life where men freeze and die. Nothing in the realm of nature is more wonderful than the manner in which some of the most minute animal forms resist cold. The genera *Erebia* and *Œneis*, and many species of the genus *Brenthis*, are, as we have already learned, inhabitants of the arctic regions or of lofty Alpine summits, the climate of which is arctic. Their caterpillars often hibernate in a temperature of from forty to fifty, and even seventy, degrees below zero, Fahrenheit.

It has been alleged that caterpillars freeze in the winter and thaw out in the spring, at that time regaining their vitality. Thus far the writer is unable to ascertain that any experiments or observations have positively decided for or against this view. A number of recorded cases in which caterpillars are positively stated to have been frozen and to have afterward been found to be full of vitality when thawed are open to question.

The most ci.·umstantial account is that by Commander James Ross, R. N., F. R. S., quoted by Curtis in the Entomological Appendix to the "Narrative" of Sir John Ross's second voyage to the arctic regions. The specimens upon which the observations were made were the caterpillars of *Laria rossi*, a moth which is found abundantly in the arctic regions of North America. I quote from the account: "About thirty of the caterpillars were put into a box in the middle of September, and after being exposed to the severe winter temperature of the next three months, they were brought into a warm cabin, where, in less than two hours, every one of them returned to life, and continued for a whole day walking about; they were again exposed to the air at a temperature of about forty degrees below zero, and became immediately hard-frozen; in this state they remained a week, and on being brought again into the cabin, only twenty-three came to life; these were, at the end of four hours, put out once more into the air and again hard-frozen;

after another week they were again brought in, when only eleven were restored to life; a fourth time they were exposed to the winter temperature, and only two returned to life on being again brought into the cabin; these two survived the winter, and in May an imperfect *Laria* was produced from one, and six flies from the other."

The foregoing account seems to verify more thoroughly the stories that have been told than anything else I have been able to discover within the limits of entomological literature, but does not conclude argument. It would be interesting in these days, when methods of artificial freezing have been so highly perfected, to undertake a series of experiments to prove or disprove, as the case may be, the view which has been held since the time of the ancients. There is here a field for nice investigation on the part of some reader of this book. In making the experiment it probably would be well to select the larvæ of species which are known to hibernate during the winter and to be capable of withstanding a great degree of cold.

The effect of cold suddenly applied to the chrysalids of butterflies at the moment of pupation is often to produce remarkable changes in the markings. The spots upon the wings of butterflies emerging from chrysalids thus treated are frequently rendered more or less indistinct and blurred. The dark markings are intensified in color and enlarged; the pale markings are also in some cases ascertained to experience enlargement. Many of the strange and really beautiful aberrations known to collectors have no doubt been produced by the action of frost which has occurred at the season when the larva was pupating. The species believed by the writer to be most prolific in aberrations are species which pupate early in the spring from caterpillars which have hibernated or which pupate late in the autumn. Some are species found at considerable altitudes above sea-level, where late frosts and early frosts are apt to occur. A number of very beautiful experiments upon the effect of cold upon the color of butterflies have been made in recent years, and some very curious phenomena have been observed. The writer has in his collection a considerable number of strikingly aberrant specimens which emerged from chrysalids treated to a sudden artificial lowering of the temperature at the critical period of pupation.

225

SUBFAMILY LIBYTHEINÆ (THE SNOUT-BUTTERFLIES)

Butterfly.—The butterflies of this family are very readily distinguished from all others by their long projecting palpi, and by the fact that the males have four feet adapted to walking, while the females have six, in which respect they approach the Erycinidæ.

Only one genus is represented in our faunal region, the genus *Libythea*.

Genus LIBYTHEA, Fabricius
(The Snout-butterflies)

Butterfly.—Rather small in size, with the eyes moderately large; the antennæ with a distinct club at the end; the palpi with the last joint extremely long and heavily clothed with hair. The wings have the outer margin strongly excised between the first median nervule and the lower radial vein. Between the upper and lower radial veins the wing is strongly produced outwardly; the inner margin is bowed out toward the base before the inner angle. The costa of the hind wing is bent upward at the base and excised before the outer angle; the wing is produced at the ends of the subcostal vein, the third median nervule, and the extremity of the submedian vein. There is also a slight projection at the extremity of the first median nervule. Of these projections the one at the extremity of

FIG. 124.—Neuration of the genus *Libythea.*

226

the third median nervule is the most pronounced. The cell of the primaries and of the secondaries is lightly closed.

Egg.—The egg is ovoid, nearly twice as high as wide, with narrow vertical ridges on the sides, every other ridge much higher than its mate and increasing in height toward the vertex, where they abruptly terminate, their extremities ranging around the small depressed micropyle. Between these ridges are minute cross-lines.

Caterpillar.—The caterpillar has the head small, the anterior segments greatly swollen and overarching the head. The remainder of the body is cylindrical.

Chrysalis.—The chrysalis is of a somewhat singular shape, the abdomen conical, the head sharply pointed, a raised ridge running from the extremity of the head to the middle of the first abdominal segment on either side, and between these ridges is the slightly projecting thoracic tubercle. On the ventral side the outline is nearly straight.

The caterpillar feeds upon *Celtis occidentalis*. Three species are reckoned as belonging to our fauna. It is, however, doubtful whether these species are in reality such, and there is reason to believe that the three are merely varietal forms or races, no structural difference being apparent in any of them, and the only differences consisting in the ground-color of the wings.

(1) **Libythea bachmanni**, Kirtland, Plate XXVIII, Fig. 1, ♂ ; Fig. 2, ♂, *under side;* Plate V, Figs. 23, 24, *chrysalis* (The Snout-butterfly).

Butterfly.—Easily distinguished from the following species by the redder color of the light spots on the upper side of the wings. Expanse, 1.75 inch.

Early Stages.—The generic description must suffice for these. They have been frequently described.

The butterfly ranges from New England and Ontario southward and westward over the whole country as far as New Mexico and Arizona.

(2) **Libythea carinenta**, Cramer, Plate XXVIII, Fig. 3, ♂ (The Southern Snout-butterfly).

Butterfly.—Much like the preceding species, but readily distinguished from it by the paler yellowish-fulvous light markings of the upper side of the wings. Expanse, 1.75 inch.

Early Stages.—These have not been carefully described as yet.

L. carinenta ranges from New Mexico into South America.

FAMILY II. LEMONIIDÆ

SUBFAMILY ERYCININÆ (THE METAL-MARKS)

> " I wonder what it is that baby dreams.
> Do memories haunt him of some glad place
> Butterfly-haunted, halcyon with flowers,
> Where once, before he found this earth of ours,
> He walked with glory filling his sweet face ? "
> EDGAR FAWCETT.

Butterfly.—Small, the males having four ambulatory feet, the females six, in which respect they resemble the Libytheinæ, from which they may readily be distinguished by the small palpi. There is great variety in the shape and neuration of the wings. The genera of this subfamily have the precostal vein on the extreme inner margin of the wing; in some genera free at its end, and projecting so as to form a short frenulum, as in many genera of the moths. In addition the costal vein sends up a branch at the point from which the precostal is usually emitted. This apparent doubling of the precostal is found in no other group of butterflies, and is a strong diacritical mark by which they may

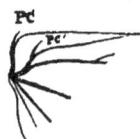

FIG. 125.— Neuration of base of hind wing of the genus *Lemonias: PC,* precostal vein ; *PC',* second precostal vein.

be recognized. They are said to carry their wings expanded when at rest, and frequently alight on the under surface of leaves, in this respect somewhat approaching in their habit the pyralid moths. Many of the species are most gorgeously colored; but those which are found within our region are for the most part not gaily marked. They may be distinguished from the Lycænidæ not only by the peculiar neuration and manner of carrying the wings, but by the relatively longer and more slender antennæ.

Early Stages.—Comparatively little is known of these, though in certain respects the larvæ and the chrysalis show a relationship

228

Explanation of Plate XXVIII

1. *Libythea bacbmanni*, Kirtland, ♂.
2. *Libythea bacbmanni*, Kirtland, ♂, under side.
3. *Libythea carinenta*, Cramer, ♂.
4. *Lemonias cythera*, Edwards, ♀, under side.
5. *Lemonias cythera*, Edwards, ♂.
6. *Lemonias virgulti*, Behr, ♂.
7. *Lemonias mormo*, Felder, ♂, under side.
8. *Lemonias nais*, Edwards, ♂.
9. *Lemonias nais*, Edwards, ♀.
10. *Lemonias durvi*, Edwards, ♀.
11. *Lemonias palmeri*, Edwards, ♂.
12. *Calephelis borealis*, Grote and Robinson, ♂, under side.
13. *Calephelis borealis*, Grote and Robinson, ♂.
14. *Calephelis nemesis*, Edwards, ♂.
15. *Calephelis australis*, Edwards, ♂.
16. *Calephelis carnius*, Linnæus, ♂.
17. *Lemonias cela*, Butler, ♂.
18. *Lemonias cela*, Butler, ♀.
19. *Lemonias cleis*, Edwards, ♂.
20. *Lemonias cleis*, Edwards, ♀.
21. *Eunisæa tarquinius*, Fabricius, ♂.
22. *Eumæus atala*, Poey, ♂, under side.
23. *Chrysophanus virginiensis*, Edwards, ♂.
24. *Chrysophanus virginiensis*, Edwards, ♀.
25. *Chrysophanus hypophlæus*, Boisduval, ♂.
26. *Chrysophanus editha*, Mead, ♂.
27. *Chrysophanus editha*, Mead, ♀.
28. *Chrysophanus epixanthe*, Boisd.-Lec., ♂.
29. *Chrysophanus xanthoides*, Boisduval, ♂.
30. *Chrysophanus xanthoides*, Boisduval, ♀.
31. *Chrysophanus thoë*, Boisd.-Lec., ♂.
32. *Chrysophanus thoë*, Boisd.-Lec., ♀.
33. *Chrysophanus helloides*, Boisduval, ♂.
34. *Chrysophanus helloides*, Boisduval, ♀.
35. *Chrysophanus gorgon*, Boisduval, ♂.
36. *Chrysophanus gorgon*, Boisduval, ♀.
37. *Chrysophanus mariposa*, Reakirt, ♂.
38. *Chrysophanus mariposa*, Reakirt, ♀.

PLATE XXVIII.

to the Lycænidæ, with which some writers have in fact grouped them, but erroneously, as the writer believes.

Almost all of the species are American, and the family attains its highest development in the tropical regions of South America.

Genus LEMONIAS, Westwood

Butterfly.—Small, brightly colored, the sexes often differing greatly in appearance from each other. The eyes are naked. The palpi are produced, porrect; the last joint is short, thin, pointed, and depressed. The antennæ are moderately long, provided with a gradually thickening, inconspicuous club. The upper discocellular vein is wanting in the fore wing. The middle and lower discocellulars are of equal length. The hind wing has the end of the cell obliquely terminated by the middle and lower discocellular veins. The apex of the fore wing is somewhat pointed, the outward margin straight. The outward margin of the hind wing is evenly rounded.

Fig. 126.— Neuration of the genus *Lemonias.*

Egg.—Flattened, turban-shaped, with a small, depressed, circular micropyle, the whole surface covered with minute hexagonal reticulations.

Caterpillar.—Short, flattened, tapering posteriorly; the segments arched; provided with tufts of hair ranged in longitudinal series, the hairs on the sides and at the anal extremity being long, bent outward and downward.

Chrysalis.—Short, suspended at the anal extremity, and held in position by a silk girdle, but not closely appressed to the surface upon which pupation has taken place; thickly covered with short, projecting hair.

The citadel of this genus is found about the head waters of the Amazon, where there are many species. Thence the genus spreads northward and southward, being represented in the limits of our fauna by only a few species, which are found on the extreme southern borders of the United States.

(1) **Lemonias mormo**, Felder, Plate XXVIII, Fig. 7, ♂, *under side* (The Mormon).

Butterfly.—The wings on the upper side are dark ashen-gray,

229

with the primaries from the base to the limbal area, and inwardly as far as the bottom of the cell and the first median nervule, red. The wings are profusely marked with white spots variously disposed. The under side is accurately depicted in our plate. Expanse, 1.10 inch.

Early Stages.—These have not been studied.

The Mormon is found in Utah, New Mexico, Arizona, and California.

(2) **Lemonias duryi,** Edwards, Plate XXVIII, Fig. 10, ♀ (Dury's Metal-mark).

Butterfly.—The only specimen as yet known is the type figured in our plate. I doubt whether it is entitled to specific rank, and am inclined to believe it to be a form of the succeeding species in which red has replaced the greater part of the gray on the upper side of both wings. Expanse, 1.25 inch.

Early Stages.—Unknown.

The specimen came from New Mexico.

(3) **Lemonias cythera,** Edwards, Plate XXVIII, Fig. 4, ♀, *under side;* Fig. 5, ♂ (Cythera).

Butterfly.—Distinguished from *L. mormo* by the red submarginal band on the secondaries on the upper side, the greater prevalence of red on the primaries, and by the tendency of the spots on the under side of the secondaries, just after the costa, to fuse and form an elongate pearly-white ray. The submarginal spots on the lower side of the fore wings are smaller than in *mormo.* The sexes do not differ except in size. Expanse, 1.00–1.30 inch.

Early Stages.—Unknown.

Cythera is found in Arizona and Mexico.

(4) **Lemonias virgulti,** Behr, Plate XXVIII, Fig. 6, ♂ (Behr's Metal-mark).

Butterfly.— Much like the preceding species on the upper side of the wings, but darker. The hind wings on the under side are much darker than in *L. cythera,* and the pearly-white spots relatively smaller, standing out very distinctly on this darker ground. Expanse, .90–1.10 inch.

Early Stages.— Undescribed.

Virgulti is common in southern California and Mexico.

(5) **Lemonias nais,** Edwards, Plate XXVIII, Fig. 8, ♂ ; Fig. 9, ♀ (Nais).

Butterfly.—The ground-color of the upper side is bright red, clouded with fuscous on the base of the hind wings and bordered with the same color. There is a small precostal white spot on the primaries near the apex. The wings are profusely marked with small black spots arranged in transverse series and bands. The fringes are checkered with white. On the under side the wings are pale reddish, mottled with buff on the secondaries. The black spots and markings of the upper side reappear on the under side and stand out boldly on the lighter ground-color. Expanse, 1.00–1.25 inch.

Early Stages.—These are beautifully delineated in "The Butterflies of North America," vol. ii. The egg is pale green, turban-shaped, covered with hexagonal reticulations. The caterpillar is rather stout and short, the first segment projecting over the head. The body is somewhat flattened and tapering behind, covered with tufts of hairs projecting outward and downward on all sides, only the two rows of short tufts on the back sending their hairs upward. The color is mouse-gray, striped longitudinally on the back with yellowish-white, the tufts more or less ringed about at their base with circles of the same color. The chrysalis is blackish-brown, attached at the anal end, held in place by a girdle, but not closely appressed to the surface on which pupation has taken place, and thickly studded with small projecting hairs. The larva lives on the wild plum.

Nais occurs from Colorado to Mexico east of the Rocky Mountains.

(6) **Lemonias palmeri,** Edwards, Plate XXVIII, Fig. 11, ♂ (Palmer's Metal-mark).

Butterfly.—Smaller than any of the preceding species. The ground-color of the wings is mouse-gray, spotted with white; on the under side the wings are whitish-gray, laved with pale red at the base of the fore wings. The white spots of the upper side reappear on the under side. Expanse, .75–.95 inch.

Early Stages.—These are, so far as they have been worked out by Edwards, quite similar in many respects to those of the preceding species.

The range of the species is from Utah southward to Mexico.

(7) **Lemonias zela,** Butler, Plate XXVIII, Fig. 17, ♂ ; Fig. 18, ♀ (Zela).

Butterfly.—The upper side of both sexes is delineated in the

231

plate. On the under side the wings are pale red, marked with a few black spots, representing on the under side the markings of the upper side. Of these, the spots of the median and submarginal bands are most conspicuous. Expanse, 1.00–1.35 inch.

(*a*) **Lemonias zela,** Butler, var. **cleis,** Edwards, Plate XXVIII, Fig. 19, ♂ ; Fig. 20, ♀ (Cleis).

The pale variety, *cleis*, is sufficiently well represented in our plate to need no description. On the under side it is like *L. zela*.

The species occurs in Arizona and Mexico.

Genus CALEPHELIS, Grote and Robinson

Butterfly.—Very small, brown or reddish in color, with metallic spots upon the wings. Head small; eyes naked; antennæ

relatively long, slender, with a bluntly rounded club. Palpi very short; the third joint small, pointed. The accompanying cut shows the neuration.

Early Stages.—Entirely unknown.

(1) **Calephelis cænius,** Linnæus, Plate XXVIII, Fig. 16, ♂ (The Little Metal-mark).

Fig. 127.— Neuration of the genus *Calephelis.*

Butterfly.—Very small, reddish-brown on the upper side, brighter red on the under side. On both the upper and under sides the wings are profusely spotted with small steely-blue metallic markings, arranged in more or less regular transverse series, especially on the outer margin. Expanse, .75 inch.

Early Stages.—The life-history is unknown.

Cænius is common in Florida, and ranges thence northward to Virginia and westward to Texas.

(2) **Calephelis borealis,** Grote and Robinson, Plate XXVIII, Fig. 12, ♂, *under side;* Fig. 13, ♂ (The Northern Metal-mark).

Butterfly.—Fully twice as large as the preceding species. The wings on the upper side are sooty-brown, spotted with black, and marked by a marginal and submarginal series of small metallic spots. On the under side the wings are light red, spotted with a multitude of small black spots arranged in regular series.

The two rows of metallic spots near the margins are repeated more distinctly on this side. Expanse, 1.15 inch.

Early Stages.—Unknown.

This rare insect has been taken from New York to Virginia, and as far west as Michigan and Illinois. The only specimen I have ever seen in life I took at the White Sulphur Springs in West Virginia. It settled on the under side of a twig of black birch, with expanded wings, just over my head, and by a lucky stroke of the net I swept it in.

(3) **Calephelis australis**, Edwards, Plate XXVIII, Fig. 14, ♂ (The Southern Metal-mark).

Butterfly.—The wings in the male sex are more pointed at the apex than in the preceding species, and in both sexes are smaller in expanse. The color of the upper side of the wings is dusky, on the under side pale yellowish-red. On both sides the wings are obscurely marked with dark spots arranged in transverse series. The marginal and submarginal metallic bands of spots are as in the preceding species. Expanse, 1.00 inch.

Early Stages.— Unknown.

Australis ranges from Texas and Arizona into Mexico.

(4) **Calephelis nemesis**, Edwards, Plate XXVIII, Fig. 15, ♂ (The Dusky Metal-mark).

Butterfly.—Very small,— as small as *cænius*,— but with the fore wings at the apex decidedly pointed in the male sex. The wings are dusky-brown above, lighter obscure reddish below. Both the primaries and the secondaries on the upper side are crossed by a dark median band, broader on the primaries at the costa. The metallic markings are quite small and indistinct. Expanse, .85 inch.

Early Stages.— Unknown.

Nemesis occurs in Arizona and southern California.

UNCLE JOTHAM'S BOARDER

"I 've kep' summer boarders for years, and allowed
  I knowed all the sorts that there be;
But there come an old feller this season along,
  That turned out a beater for me.
Whatever that feller was arter, I vow
  I hain't got the slightest idee.

Uncle Jotham's Boarder

"He had an old bait-net of thin, rotten stuff
    That a minner could bite his way through;
But he never went fishin'— at least, in the way
    That fishermen gen'ally do;
But he carried that bait-net wherever he went;
    The handle was j'inted in two.

"And the bottles and boxes that chap fetched along!
    Why, a doctor would never want more;
If they held pills and physic, he 'd got full enough
    To fit out a medicine-store.
And he 'd got heaps of pins, dreffle lengthy and slim,
    Allers droppin' about on the floor.

"Well, true as I live, that old feller just spent
    His hull days in loafin' about
And pickin' up hoppers and roaches and flies —
    Not to use for his bait to ketch trout,
But to kill and stick pins in and squint at and all.
    He was crazy 's a coot, th' ain't no doubt.

"He 'd see a poor miller a-flyin' along,—
    The commonest, every-day kind,—
And he 'd waddle on arter it, fat as he was,
    And foller up softly behind,
Till he 'd flop that-air bait-net right over its head,
    And I 'd laugh till nigh out of my mind.

"Why, he 'd lay on the ground for an hour at a stretch
    And scratch in the dirt like a hen;
He 'd scrape all the bark off the bushes and trees,
    And turn the stones over; and then
He 'd peek under logs, or he 'd pry into holes.
    I 'm glad there ain't no more sech men.

"My wife see a box in his bedroom, one day,
    Jest swarmin' with live caterpillars;
He fed 'em on leaves off of all kinds of trees —
    The ellums and birches and willers;
And he 'd got piles of boxes, chock-full to the top
    With crickets and bees and moth-millers.

"I asked him, one time, what his business might be.
    Of course, I fust made some apology.
He tried to explain, but such awful big words!
    Sorto' forren, outlandish, and collegey.
'S near 's I can tell, 'stead of enterin' a trade,
    He was tryin' to jest enter *mology*.

234

"And Hannah, my wife, says she 's heerd o' sech things;
  She guesses his brain warn't so meller.
There 's a thing they call Nat'ral Histerry, she says,
  And, whatever the folks there may tell her,
Till it 's settled she 's wrong she 'll jest hold that-air man
  Was a Nat'ral Histerrical feller."

ANNIE TRUMBULL SLOSSON.

MIMICRY

Protective mimicry as it occurs in animals may be the simulation in form or color, or both, of natural objects, or it may be the simulation of the form and color of another animal, which for some reason enjoys immunity from the attacks of species which ordinarily prey upon its kind. Of course this mimicry is unconscious and is the result of a slow process of development which has, no doubt, gone on for ages.

Remarkable instances of mimicry, in which things are simulated, are found in the insect world. The "walking-sticks," as they are called, creatures which resemble the twigs of trees; the "leaf-insects," in which the foliage of plants is apparently reproduced in animate forms; the "leaf-butterfly" of India, in which the form and the color and even the venation of leaves are reproduced, are illustrations of mimicry which are familiar to all who have given any attention to the subject.

Repulsive objects are frequently mimicked. A spider has been lately described from the Indo-Malayan region, which, as it rests upon the leaves, exactly resembles a patch of bird-lime. The resemblance is so exact as to deceive the most sagacious, and the discovery of the creature was due to the fact that the naturalist who happened to see it observed, to his surprise, that what he was positive was a mass of ordure was actually in motion. A similar case of mimicry is observable among some of the small acontiid moths of North America. One of these is pure white, with the tips of the fore wings dark greenish-brown. It sits on the upper side of leaves, with its fore wings folded over, or rolled about the hind wings, and in this attitude it so nearly approximates in appearance the ordure of a sparrow as to have often deceived me when collecting.

FAMILY III. LYCÆNIDÆ

(THE BLUES, THE COPPERS, THE HAIR-STREAKS)

SUBFAMILY LYCÆNINÆ

"Mark, while he moves amid the sunny beam,
O'er his soft wings the varying lusters gleam.
Launched into air, on purple plumes he soars,
Gay nature's face with wanton glance explores;
Proud of his varying beauties, wings his way,
And spoils the fairest flowers, himself more fair than they."
Quoted as from Haworth by Scudder.

Butterfly.—Small, in both sexes having all feet adapted to
walking. There is exceeding diversity of form in the various genera composing this family. Many of the genera are characterized
by the brilliant blue on the upper side of the wings; in other
genera shades of coppery-red predominate. The hair-streaks
frequently have the hind wings adorned with one or more slender, elongated tails. In Africa and in Asia there are numerous
genera which strongly mimic protected insects belonging to the
Acræinæ.

Egg.—The eggs are for the most part flattened or turban-
shaped, curiously and beautifully adorned with ridges, minute
eminences, and reticulations. Some of them under the microscope strongly resemble the shells of "sea-biscuits" with the
rays removed (see p. 4, Fig. 7).

Caterpillar.—The caterpillars are for the most part slug-
shaped, flattened. They are vegetable feeders, save the larvæ of
two or three genera, which are aphidivorous, feeding upon mealy
bugs or plant-lice.

Chrysalis.—The chrysalids are short, compressed, attached at
the anal extremity, with a girdle or cincture about the middle,
closely fastened to the surface upon which pupation takes place.

236

Explanation of Plate XXIX

1. *Chrysophanus arota*, Boisduval, ♂.
2. *Chrysophanus arota*, Boisduval, ♀.
3. *Chrysophanus sirius*, Edwards, ♂.
4. *Chrysophanus sirius*, Edwards, ♀.
5. *Chrysophanus rubidus*, Behr, ♂.
6. *Chrysophanus rubidus*, Behr, ♀.
7. *Chrysophanus snowi*, Edwards, ♂.
8. *Chrysophanus snowi*, Edwards, ♀.
9. *Thecla halesus*, Cramer, ♂.
10. *Thecla m-album*, Boisd.-Lec., ♂.
11. *Thecla crysalus*, Edwards, ♂.
12. *Thecla grunus*, Boisduval, ♂.
13. *Thecla autolycus*, Edwards, ♀.
14. *Thecla alcestis*, Edwards, ♀.
15. *Thecla acadica*, Edwards, ♂.
16. *Thecla acadica*, Edwards, ♀.
17. *Thecla itys*, Edwards, ♀.
18. *Thecla cecrops*, Hübner, ♀, *under side.*
19. *Thecla wittfeldi*, Edwards, ♀.
20. *Thecla wittfeldi*, Edwards, ♂, *under side.*
21. *Thecla spinetorum*, Boisduval, ♀.
22. *Thecla favonius*, Smith and Abbot, ♂.
23. *Thecla laeta*, Edwards, ♂.
24. *Thecla laeta*, Edwards, ♂, *under side*
25. *Thecla adenostomatis*, Henry Edwards, ♂.
26. *Thecla calanus*, Hübner, ♂.
27. *Thecla edwardsi*, Saunders, ♀.
28. *Thecla liparops*, Boisd.-Lec., ♀.
29. *Thecla damon*, Cramer, var. *discoidalis*, Skinner, ♂.
30. *Thecla tacita*, Henry Edwards, ♂.
31. *Thecla melinus*, Hübner, form *humuli*, Harris, ♂.
32. *Thecla damon*, Cramer, ♀, *under side.*
33. *Thecla saepium*, Boisduval, ♂.
34. *Thecla saepium*, Boisduval, ♂, *under side.*
35. *Thecla ines*, Edwards, ♂.
36. *Thecla chalcis*, Behr, ♂.
37. *Thecla chalcis*, Behr, ♀, *under side.*
38. *Thecla acis*, Drury, ♂, *under side.*
39. *Thecla simaethis*, Drury, ♂, *under side.*

\

PLATE XXIX

Genus EUMÆUS, Hübner

Butterfly.— Medium size or small; dark in color, with the under side and the borders of the upper sides beautifully adorned with spots having a metallic luster. The palpi are divergent, longer in the female than in the male. The antennæ are stout, rather short, with a gradually thickened club. The eyes are naked. The veins on the fore wing are stout. The accompanying cut gives a clear idea of the neuration.

Early Stages.—Nothing is known of these.

Three species are reckoned as belonging to the genus, two of them being found sparingly in the extreme southern limits of our fauna.

FIG. 128.—Neuration of the genus *Eumæus.*

(1) **Eumæus atala**, Poey, Plate XXVIII, Fig. 22, ♂, *under side* (Atala).

Butterfly.— Easily distinguished by the figure in the plate from all other species except its congener *E. minyas*, Hübner, which can be readily separated from it by its larger size. Expanse, 1.65–1.75 inch.

Early Stages.—These await description.

Atala is found in Florida and Cuba. *Minyas* occurs in southwestern Texas, and thence southward to Brazil.

Genus THECLA, Fabricius
(The Hair-streaks)

" These be the pretty genii of the flow'rs,
Daintily fed with honey and pure dew."

Hood.

Butterfly.— Small or medium-sized; on the upper side often colored brilliantly with iridescent blue or green, sometimes dark brown or reddish; on the under side marked with lines and spots variously disposed, sometimes obscure in color, very frequently most brilliantly colored.

Various subdivisions based upon the neuration of the wings have been made in the genus in recent years, and these subdivisions are entitled to be accepted by those who are engaged in a

237

comparative study of the species belonging to this great group. Inasmuch, however, as most American writers have heretofore classified all of these insects under the genus *Thecla*, the author has decided not to deviate from familiar usage, and will therefore not attempt to effect a subdivision according to the views of recent writers, which he nevertheless approves as scientifically accurate.

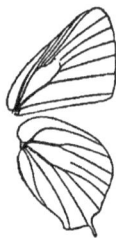

Egg. — Considerable diversity exists in the form of the eggs of the various species included under this genus as treated in this book, but all of them may be said to be turban-shaped, more or less depressed at the upper extremity, with their surfaces beautifully adorned with minute projections arranged in geometric patterns.

FIG. 129.—Neuration of *Thecla edwardsi.* (After Scudder.) Typical neuration of the genus.

Caterpillar. — The caterpillars are slug-shaped, their heads minute, the body abruptly tapering at the anal extremity. They feed upon the tender leaves of the ends of branches, some of them upon the leaves of flowers of various species.

Chrysalis. — What has been said concerning the chrysalids of the family applies likewise to the chrysalids of this and the succeeding genera. They lie closely appressed to the surface upon which they are formed, and are held in place by an attachment at the anal extremity, as well as by a slight girdle of silk about the middle. In color they are generally some shade of brown.

(1) **Thecla grunus,** Boisduval, Plate XXIX, Fig. 12, ♂ (Boisduval's Hair-streak).

Butterfly. — The wings are brown on the upper side, lighter on the disk; in some specimens, more frequently of the female sex, bright orange-tawny. On the under side the wings are pale tawny, with transverse marginal and submarginal series of small dark spots on both wings. Two or three of the marginal spots near the anal angle are black, each crowned with a metallic-green crescent. Expanse, 1.10–1.20 inch.

Early Stages. — These have, in part, been described by Dyar, "Canadian Entomologist," vol. xxv, p. 94. The caterpillar is short, flattened, the segments arched, the body tapering backward, bluish-green, covered with little dark warty prominences bearing tufts of hairs, obscurely striped longitudinally with broken,

238

pale lines, and having a diamond-shaped shield back of the head. The chrysalis is thick and conformed to the generic type of structure. The color is pale green, striped and dotted with pale yellow on the abdomen. The caterpillar feeds in the Yosemite Valley upon the young leaves of the live-oak (*Quercus chryso-lepis*).

The insect is found in California and Nevada.

(2) **Thecla crysalus,** Edwards, Plate XXIX, Fig. 11, ♂ (The Colorado Hair-streak).

Butterfly.—The wings on the upper side are royal purple, broadly margined with black. On the fore wings a broad oblique black band runs from the middle of the costa to the middle of the outer margin. At the inner angles of both wings are conspicuous orange spots. On the under side the wings are fawn, marked with white lines edged with brown. The orange spots reappear on this side, but at the anal angle of the hind wings are transformed to red eye-spots, pupiled with black and margined with metallic green. The hind wings are tailed. Expanse, 1.50 inch.

The variety **citima,** Henry Edwards, differs in being without the orange spots and having the ground-color of the under side ashen-gray. Specimens connecting the typical with the varietal form are in my possession.

Early Stages.—Unknown.

Found in southern Colorado, Utah, Arizona, and southern California.

(3) **Thecla halesus,** Cramer, Plate XXIX, Fig. 9, ♂ (The Great Purple Hair-streak).

Butterfly.—The hind wings have a long tail, and are lobed at the anal angle. The wings are fuscous, iridescent bluish-green at the base. The body is bluish-green above. On the under side the thorax is black, spotted with white, the abdomen bright orange-red. The wings on the under side are evenly warm sepia, spotted with crimson at their bases, glossed with a ray of metallic green on the fore wings in the male sex, and in both sexes splendidly adorned at the anal angle by series of metallic-green and iridescent blue and red spots. Expanse, 1.35–1.50 inch.

Early Stages.—All we know of them is derived from the drawings of Abbot, published by Boisduval and Leconte, and this is but little. The caterpillar is said by Abbot to feed on various oaks.

It is very common in Central America and Mexico; is not scarce in the hot parts of the Gulf States; and is even reported as having been captured in southern Illinois. It also occurs in Arizona and southern California.

(4) **Thecla m-album,** Boisduval and Leconte, Plate XXIX, Fig. 10, ♂ (The White-M Hair-streak).

Butterfly.—Smaller than the preceding species; on the upper side somewhat like it; but the iridescent color at the base of the wings is blue, and not so green as in *balesus*. On the under side the wings are quite differently marked. The fore wing is crossed by a submarginal and a median line of white, shaded with brown, the median line most distinct. This line is continued upon the hind wings, and near the anal angle is zigzagged, so as to present the appearance of an inverted M. Near the outer angle of the M-spot is a rounded crimson patch. The anal angle is deep black, glossed with iridescent blue. Expanse, 1.35–1.45 inch.

Early Stages.—All we know of this pretty species is based upon the account and drawings of Abbot made in the last century. We need better information. According to Abbot, the caterpillar feeds on *astragalus* and different oaks.

This species has been taken as far north as Jersey City and Wisconsin, and ranges southward as far as Venezuela. Its citadel is found in the live-oak hummocks of the Gulf States and the oak forests on the highlands cf Mexico and more southern countries.

(5) **Thecla martialis,** Herrich-Schäffer, Plate XXX, Fig. 18, ♀, *under side* (The Martial Hair-streak).

Butterfly.—The insect figured in the plate, which may easily be recognized by its under side, has been determined by Dr. Skinner to be the above species. My specimens coming from the Edwards collection are labeled *Thecla acis,* ♀. They were taken at Key West. A comparison with the under side of *T. acis* (see Plate XXIX, Fig. 38) will reveal the great difference. Expanse, 1.00 inch.

Early Stages.—Unknown.

Habitat, southern Florida and Cuba.

(6) **Thecla favonius,** Abbot and Smith, Plate XXIX, Fig. 22, ♂ (The Southern Hair-streak).

Butterfly.—The wings are dusky-brown above, with a small pale oval sex-mark in the male near the upper edge of the cell in the primaries. On either side of the second median nervule, near the outer margin of both wings, are bright orange-red patches, most conspicuous in the female. The hind wings near the anal angle are blackish, margined with a fine white line. On the under side the wings are marked much as in *m-album*, but in the region of the median nervules, midway between their origin and termination, is a rather broad transverse carmine streak, edged inwardly with dark lines. This is largest and most conspicuous in the female sex. Expanse, 1.00–1.15 inch.

Early Stages. —These have been described, in part, by Abbot and Smith and Packard. The caterpillar feeds on oaks.

Favonius is found in the Gulf States, and as far north as South Carolina.

(7) **Thecla wittfeldi,** Edwards, Plate XXIX, Fig. 19, ♀ ; Fig. 20, ♂, *under side* (Wittfeld's Hair-streak).

Butterfly. —The figures in the plate give a correct idea of both the upper and under sides of this insect. It is much darker in ground-color than any of its congeners. Expanse, 1.25–1.35 inch.

Early Stages. —Unknown.

The types which are in my possession came from the Indian River district in Florida.

(8) **Thecla autolycus,** Edwards, Plate XXIX, Fig. 13, ♀ (The Texas Hair-streak).

Butterfly. —On the upper side resembling *favonius*, but with the orange-red spots on the wings much broader, ranging from the lower radial vein to the submedian in the fore wings. The carmine spots on the under side of the wings are not arranged across the median nervules, as in *favonius*, but are in the vicinity of the anal angle, crowning the black crescents near the inner end of the outer margin. Expanse, 1.15–1.30 inch.

Early Stages. —Unknown.

This species is found in Texas, and is also said to have been found in Missouri and Kansas.

(9) **Thecla alcestis,** Edwards, Plate XXIX, Fig. 14, ♀ (Alcestis).

Butterfly. —Uniformly slaty-gray on the upper side of the wings, with the usual oval sex-mark on the fore wing of the male, and a few bluish scales near the anal angle. The ground-color of the wings on the under side is as above, but somewhat paler. A white bar closes the cell of both wings. Both wings

241

are crossed by white lines, much as in *m-album*. The anal angle
is marked with black, followed outwardly by a broad patch of
iridescent greenish-blue scales. Between the end of the submarginal vein and the first median nervule is a black spot surmounted with carmine, edged inwardly with black; three or four
carmine crescents similarly edged, but rapidly diminishing in size,
extend as a transverse submarginal band toward the costa. Expanse, 1.25 inch.

Early Stages.—Unknown.

Alcestis is found in Texas and Arizona.

(10) **Thecla melinus,** Hübner, Plate XXIX, Fig. 31, ♂ ;
Plate XXXII, Fig. 20, ♂ ; Plate V, Fig. 39, *chrysalis* (The Common Hair-streak).

Butterfly.—Much confusion has arisen from the fact that this
insect has received a number of names and has also been confounded with others. Fig. 31 in Plate XXIX represents the insect labeled *humuli*, Harris, in the Edwards collection; Fig. 20 in Plate XXXII represents
the insect labeled *melinus*, Hübner. There is a very
large series of both in the collection, but a minute
comparison fails to reveal any specific difference.
Humuli of Harris is the same as *melinus* of Hübner;
and recent authors, I think, are right in sinking the
name given by Harris as a synonym. This common
little butterfly may easily be recognized by its plain
slaty upper surface, adorned by a large black spot,
crowned with crimson between the origin of the two
tails of the secondaries. Expanse, 1.10-1.20 inch.

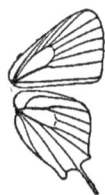

Fig. 130.—
Neuration of
*Thecla meli-
nus.* (After
Scudder.) Typ-
ical of subge-
nus *Uranotes.*

Early Stages.—These are in part well known. The
caterpillar feeds on the hop-vine. *Melinus* is found all over temperate North America, and ranges southward into Mexico and
Central America at suitable elevations.

(11) **Thecla acadica,** Edwards, Plate XXIX, Fig. 15, ♂ ;
Plate V, Fig. 35, *chrysalis* (The Acadian Hair-streak).

Butterfly.—The male is pale slaty-gray above, with some ill-
defined orange spots near the anal angle, the usual oval sex-
mark on the fore wing. The female is like the male above; but
the orange spots at the anal angle of the hind wings are broader,
and in some specimens similar spots appear on the fore wings
near the inner angle. On the under side in both sexes the

wings are pale wood-brown, adorned by a black bar at the end of the cells, submarginal and median bands of small black spots surrounded with white, and on the secondaries by a submarginal series of red crescents diminishing in size from the anal angle toward the outer angle. Near the anal angle are two black spots separated by a broad patch of bluish-green scales. Expanse, 1.15–1.25 inch.

Early Stages.—For a knowledge of what is known of these the reader may consult the pages of Scudder and Edwards. The caterpillar feeds upon willows.

It is found all over the Northern States, ranging from Quebec to Vancouver's Island. It seems to be very common on Mount Hood, from which I have a large series of specimens.

(12) **Thecla itys,** Edwards, Plate XXIX, Fig. 17, ♀ (Itys).

Butterfly.—The only specimen of this species known to me is figured in the plate. It is the type. Of its early stages nothing is known. It was taken in Arizona. Expanse, 1.25 inch.

(13) **Thecla edwardsi,** Plate XXIX, Fig. 27, ♀ *under side;* Plate V, Fig. 29, *chrysalis* (Edwards' Hair-streak).

Butterfly.—Dark plumbeous-brown on the upper side, with a pale sex-mark on the fore wing of the male. On the under side the wings are paler and a trifle warmer brown, with their outer halves marked with numerous fine white broken lines arranged in pairs, with the space between them darker than the ground-color of the wing. The usual black spots, green scales, and red crescents are found near the anal angle on the under side.

Early Stages.—For all that is known of these the reader will do well to consult the pages of Scudder. The caterpillar feeds on oaks.

The species ranges from Quebec westward to Colorado and Nebraska, being found commonly in New England.

(14) **Thecla calanus,** Hübner, Plate XXIX, Fig. 26, ♂ ; Plate V, Figs. 25, 27, *chrysalis* (The Banded Hair-streak).

Butterfly.—On the upper side resembling the preceding species very closely, but a trifle darker, and warmer brown. On the under side the wings are marked by fine white lines on the outer half, which are not broken, as in *edwardsi,* but form continuous bands. Expanse, 1.15 inch.

Early Stages.—The caterpillar feeds on oaks. The life-history is described with minute exactness by Scudder in "The Butterflies of New England," vol. ii, p. 888.

This insect has a wide range, being found from the province of Quebec to Texas and Colorado. It is common in western Pennsylvania.

(15) **Thecla liparops**, Boisduval and Leconte, Plate XXIX, Fig. 28, ♀, *under side;* Plate V, Fig. 28, *chrysalis* (The Striped Hair-streak).

Butterfly.—Dark brown on the upper side, grayish below. The lines are arranged much as in *T. edwardsi*, but are farther apart, often very narrow, scarcely defining the dark bands between them. The spots at the anal angle are obscure and blackish. Expanse, 1.15 inch.

Early Stages.—Much like those of the allied species. Scudder, in "The Butterflies of New England," gives a full account of them. The caterpillar feeds on a variety of plants—oaks, willows, the wild plum, and other rosaceous plants, as well as on the *Ericaceæ*.

It ranges through the northern Atlantic States and Quebec to Colorado and Montana, but is local in its habits, and nowhere common.

(16) **Thecla chalcis**, Behr, Plate XXIX, Fig. 36, ♂ ; Fig. 37, ♀, *under side* (The Bronzed Hair-streak).

Butterfly.—On the upper side uniformly brown. On the under side dark, with a narrow submarginal and an irregular median transverse band, and a pale short bar closing the cell on both wings; a black spot at the anal angle of the secondaries, preceded by a few bluish-green scales. Expanse, 1.00–1.10 inch.

Early Stages.—Unknown.

Habitat, California and Utah.

(17) **Thecla sæpium**, Boisduval, Plate XXIX, Fig. 33, ♂ ; Fig. 34, ♀ (The Hedge-row Hair-streak).

Butterfly.—Almost identically like the preceding species, except that the wings on the upper side are a trifle redder, on the under side paler; the lines on the under side of the wings are narrowly defined externally by white, and the anal spots are better developed and defined on the hind wings. Expanse, 1.20 inch.

Early Stages.—Unknown.

This species is found throughout the Pacific States, and I am inclined to believe it identical with *chalcis*. If this should be proved to be true the latter name will sink as a synonym.

(18) **Thecla adenostomatis,** Henry Edwards, Plate XXIX, Fig. 25, ♂ (The Gray Hair-streak).

Butterfly.—Mouse-gray on the upper side, with a few white lines on the outer margin near the anal angle; hoary-gray on the under side, darker on the median and basal areas. The limbal area is defined inwardly by a fine white line, is paler than the rest of the wing, and on the secondaries is marked by a full, regularly curved submarginal series of small dark lunules. Expanse, 1.30 inch.

Early Stages.—Undescribed.

Habitat, California.

(19) **Thecla spinetorum,** Boisduval, Plate XXIX, Fig. 21, ♀ (The Thicket Hair-streak).

Butterfly.—Dark blackish on the upper side, with both wings at the base shot with bluish-green. On the under side the wings are pale reddish-brown, marked much as in the following species, but the lines and spots are broader, more distinct, and conspicuous. Expanse, 1.15 inch.

Early Stages.—This species is reported, so far, from Colorado, California, and Washington.

(20) **Thecla nelsoni,** Boisduval, Plate XXX, Fig. 8, ♀, *under side;* Fig. 13, ♀ (Nelson's Hair-streak).

Butterfly.—Bright fulvous on the upper side, with the costa, the outer margins, the base, and the veins of both fore and hind wings fuscous. On the under side the wings are paler red, with an incomplete narrow white line shaded with deep red just beyond the median area, and not reaching the inner margin. This line is repeated on the hind wing as an irregularly curved median line. Between it and the outer margin on this wing are a few dark lunules near the anal angle. Expanse, 1.00 inch.

Early Stages.—I cannot discover any account of these.

The species has been found in California and Colorado.

(21) **Thecla blenina,** Hewitson, Plate XXX, Fig. 9, ♂, *under side* (Hewitson's Hair-streak).

Butterfly.—Brown on the upper side, in some specimens bright fulvous bordered with brown. On the under side the wings are pale red, shot with pea-green on the secondaries and at the base of the primaries. The markings of the under side are much as in the preceding species, but the line on the hind wing dividing the discal from the limbal area is broader and

245

very white, and the spots between it and the margin more conspicuous. Expanse, 1.12 inch.

Early Stages.—Unknown.

It is reported from Arizona and southern California. It has been named *siva* by Edwards, and the figure is from his type so labeled.

(22) **Thecla damon,** Cramer, Plate XXIX, Fig. 32, ♂, *under side;* var. **discoidalis,** Skinner, Plate XXIX, Fig. 29, ♂ ; Plate V, Figs. 30, 31, *chrysalis* (The Olive Hair-streak).

Butterfly.—On the upper side bright fulvous, with the costa, the outer margins, and the veins of both wings blackish, darkest at the apex. On the under side the wings are greenish, crossed on the fore wing by a straight, incomplete white line, and on the hind wing by a similar irregular line. Both of these lines are margined internally by brown. There are a couple of short white lines on the hind wing near the base, and the usual crescentic spots and markings on the outer border and at the anal angle. Expanse, .90–1.00 inch.

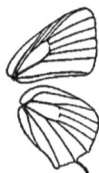

Early Stages.—These have been described by several authors. The caterpillar feeds on the red cedar (*Juniperus virginiana*, Linnæus). It is double-brooded in the North and triple-brooded in the South.

FIG. 131.— Neuration of *Thecla damon,* enlarged. Type of subgenus *Mitura,* Scudder.

Damon ranges from Ontario to Texas over the entire eastern half of the United States.

(23) **Thecla simæthis,** Drury, Plate XXIX, Fig. 39, ♂, *under side* (Simæthis).

Butterfly.—Resembling the preceding species, but the white band on the secondaries is straight, and the outer margins are heavily marked with brown. Expanse, .85–1.00 inch.

Early Stages.—Unknown.

This species occurs in Texas, Mexico, and southward.

(24) **Thecla acis,** Drury, Plate XXIX, Fig. 38, ♀, *under side* (Drury's Hair-streak).

Butterfly.—The upper side of the wings is dark brown. The under side is shown in the plate. Expanse, .90 inch.

Early Stages.—Unknown.

This very pretty species is found in the extreme southern portions of Florida and the Antilles.

(25) **Thecla cecrops,** Hübner, Plate XXX, Fig. 7, ♂ ; Plate XXIX, Fig. 18, ♀, *under side* (Cecrops).

Explanation of Plate XXX

1. *Thecla dumetorum*, Boisduval, ♂.
2. *Thecla dumetorum*, Boisduval, ♂, under side.
3. *Thecla affinis*, Edwards, ♀, under side.
4. *Thecla behri*, Edwards, ♂.
5. *Thecla behri*, Edwards, ♂, under side.
6. *Thecla clytie*, Edwards, ♀.
7. *Thecla cecrops*, Hübner, ♂.
8. *Thecla nelsoni*, Boisduval, ♀, under side.
9. *Thecla blenina*, Hewitson, ♂, under side. (The figure is that of the type of *T. siva*, Edwards.)
10. *Thecla titus*, Fabricius, ♂.
11. *Thecla niphon*, Hübner, ♀.
12. *Thecla irus*, Godart, ♂.
13. *Thecla nelsoni*, Boisduval, ♀.
14. *Thecla titus*, Fabricius, ♂, under side.
15. *Thecla augustus*, Kirby, ♀.
16. *Lycaena fuliginosa*, Edwards, ♂, under side.
17. *Thecla cryphon*, Boisduval, ♀, under side.
18. *Thecla martialis*, ♀, under side.
19. *Lycaena pseudargiolus*, Boisd.-Lec., var. *marginata*, Edwards, ♂, under side.
20. *Lycaena pseudargiolus*, Boisd.-Lec., var. *lucia*, Kirby, ♂, under side.
21. *Thecla henrici*, Grote and Robinson, ♀.
22. *Thecla niphon*, Hübner, ♀, under side.
23. *Lycaena couperi*, Grote, ♂.
24. *Lycaena fulla*, Edwards, ♂.
25. *Lycaena fulla*, Edwards, ♀.
26. *Lycaena clara*, Henry Edwards, ♀.
27. *Lycaena marina*, Reakirt, ♀, under side.
28. *Lycaena dædalus*, Behr, ♀, under side.
29. *Lycaena icarioides*, Boisduval, ♂, under side.
30. *Lycaena enoptes*, Boisduval, ♀, under side.
31. *Lycaena glaucon*, Edwards, ♀, under side.
32. *Lycaena pseudargiolus*, Boisd.-Lec., ♂, under side.
33. *Lycaena isola*, Reakirt, ♂, under side. (The figure is that of the type of *L. alce*, Edwards.)
34. *Lycaena couperi*, Grote, ♂, under side.
35. *Lycaena antiacis*, Boisduval, ♂, under side.
36. *Lycaena antiacis*, Boisduval, ♂.
37. *Lycaena pheres*, Boisduval, ♂.
38. *Lycaena isola*, Reakirt, ♀.
39. *Lycaena glaucon*, Edwards, ♂.
40. *Lycaena aster*, Edwards, ♂.
41. *Lycaena antiacis*, Boisduval, ♀.
42. *Lycaena pheres*, Boisduval, ♀, under side.
43. *Lycaena xerxes*, Boisduval, ♂, under side.
44. *Lycaena sagittigera*, Felder, ♀, under side.
45. *Lycaena ammon*, Lucas, ♀, under side.
46. *Lycaena aster*, Edwards, ♀.
47. *Lycaena aster*, Edwards, ♂, under side.
48. *Lycaena scudderi*, Edwards, ♂.
49. *Lycaena scudderi*, Edwards, ♀.
50. *Lycaena lygdamas*, Doubleday, ♀, under side.
51. *Lycaena enoptes*, Boisduval, ♂.

PLATE XXX.

Butterfly.—Dark brown, glossed at the base of the wings and on the inner margin of the secondaries with blue. The under side is well delineated in the plate. Expanse, 1.00 inch.

Early Stages.—These await description.

Cecrops is common in the Southern States, and has been taken as far north as West Virginia, Kentucky, and southern Indiana.

(26) **Thecla clytie,** Edwards, Plate XXX, Fig. 6, ♀ (Clytie).

Butterfly.—Blue above, with the apical two thirds of the fore wings black. The wings on the under side are white, with the usual marginal and transverse markings quite small and faint. Expanse, .90 inch.

Early Stages.—Unknown.

Habitat, Texas and Arizona.

(27) **Thecla ines,** Edwards, Plate XXIX, Fig. 35, ♂ (Ines).

Butterfly.—Much like the preceding species, but smaller, with the secondaries marked with blackish on the costa. On the under side the wings are slaty-gray, with numerous fine lines and a broad median dark shade on the hind wings, running from the costa to the middle of the wing. Expanse, .75 inch.

Early Stages.—Unknown.

Ines is found in Arizona.

(28) **Thecla behri,** Edwards, Plate XXX, Fig. 4, ♂ ; Fig. 5, ♂ , *under side* (Behr's Hair-streak).

Butterfly.—Both sides are well displayed in the plate, and therefore need no particular description. Expanse, 1.10 inch.

Early Stages.—Unknown.

This species is found in northern California and Oregon, and eastward to Colorado.

(29) **Thecla augustus,** Kirby, Plate XXX, Fig. 15, ♀ (The Brown Elfin).

Butterfly.—Brown on the upper side; paler on the under side. The fore wings are marked by a straight incomplete median band, and the hind wings by an irregularly curved median band or line. Back of these lines toward the base both wings are darker brown. Expanse, .90 inch.

Early Stages.—These are not well known. Henry Edwards describes the caterpillar as "carmine-red, covered with very short hair, each segment involute above, with deep double foveæ." The chrysalis is described by the same observer as being " pitchybrown, covered with very short bristly hair, the wing-cases paler." The food-plant is unknown.

This species is boreal in its haunts, and is found in New England and northward and westward into the British possessions.

(30) **Thecla irus,** Godart, Plate XXX, Fig. 12, ♂ ; Plate V, Figs. 32–34, *chrysalis* (The Hoary Elfin).

Butterfly.—Grayish-brown on the upper side. The wings on the under side are of the same color, paler on the outer margins, and darker toward the base. The species is subject to considerable variation. The variety *arsace,* Boisduval, has the hind wings marked with reddish near the anal angle, and the outer margin below marked with hoary-purple. The usual small crescentic spots appear on the outer margin of the hind wings, or they may be absent. Expanse, 1.10 inch.

Early Stages.—An epitome of all that is known is to be found in "The Butterflies of New England." The caterpillar feeds on young plums just after the leaves of the blossom have dropped away.

The species is rather rare, but has been found from the Atlantic to the Pacific in the latitude of New England.

(31) **Thecla henrici,** Grote and Robinson, Plate XXX, Fig. 21, ♀ (Henry's Hair-streak).

Butterfly.—Much like the preceding species on the upper side, but with the outer half of the wings broadly reddish-brown. The secondaries on the under side are broadly blackish-brown on the basal half, with the outer margin paler. The division between the dark and light shades is irregular and very sharply defined, often indicated by a more or less perfect irregularly curved median white line. Expanse, 1.00–1.10 inch.

Early Stages.—These have been described by Edwards in the "American Naturalist," vol. xvi, p. 123. The habits of the larva are identical with those of the preceding species.

It occurs from Maine to West Virginia, but is rare.

(32) **Thecla eryphon,** Boisduval, Plate XXX, Fig. 17, ♀, *under side* (Eryphon).

Butterfly.—Closely resembling the following species both on the upper and under side of the wings, but easily distinguished by the fact that, on the under side of the fore wings, the inner of the two dark bands on the outer third of the wing is not sharply angulated below the third median nervule, as in *T. niphon,* but is more even, and in general parallel with the submarginal line. Expanse, 1.15 inch.

Early Stages.—These have not been described.

Eryphon replaces the Eastern *T. niphon* on the Pacific coast.

(33) **Thecla niphon**, Hübner, Plate XXX, Fig. 11, ♀ ; Fig. 22, ♀ , *under side;* Plate V, Figs. 38, 40, *chrysalis* (The Banded Elfin)

Butterfly.—Reddish-brown on the upper side. The under side is accurately depicted in the plate. Expanse, 1.10 inch.

Early Stages.—These have been elaborately described by Scudder in his great work. The caterpillars feed upon pine.

The Banded Elfin is found from Nova Scotia to Colorado, in the Northern States, where its food-plant occurs, but is never abundant.

Fig. 132.—Neuration of *Thecla niphon*, enlarged. Typical of subgenus *Incisalia*, Minot.

(34) **Thecla affinis**, Edwards, Plate XXX, Fig. 3, ♀ , *under side* (The Green-winged Hairstreak).

Butterfly.—On the upper side closely resembling the following species. On the under side the wings are uniformly bright green. Expanse, 1.00 inch.

Early Stages.—These await description.

The types came from Utah. I also have specimens from California.

(35) **Thecla dumetorum**, Boisduval, Plate XXX, Fig. 1, ♂ ; Fig. 2, ♂ , *under side* (The Green White-spotted Hair-streak).

Butterfly.—Dark fawn-color above, sometimes tinged externally with reddish. On the under side both wings are green, the primaries having a short straight band of white spots on the outer third, and the secondaries a small white spot on the costa beyond the middle, and two or three conspicuous white spots near the anal angle. Expanse, 1.10 inch.

Early Stages.—The eggs are laid on the unopened flower-heads of *Hosackia argophylla.* This is all we know of the life-history.

The species ranges from Oregon and California eastward as far as Colorado.

(36) **Thecla læta**, Edwards, Plate XXIX, Fig. 23, ♂ ; Fig. 24, ♂ , *under side* (The Early Hair-streak).

Butterfly.—The wings brown, glossed with bright blue above;

249

on the under side pale fawn, with a band of pale-red spots on both wings about the middle, and a few similar spots on the outer and inner margins of the hind wings. Expanse, .75 inch.

Early Stages.—Only the egg, described and figured by Scudder, is known.

It ranges from Quebec to southern New Jersey, and westward to West Virginia, and has been taken on Mount Graham, in Arizona. It appears in early spring. It is still rare in collections.

(37) **Thecla titus,** Fabricius, Plate XXX, Fig. 10, ♂ ; Fig. 14, ♂ , *under side;* Plate V, Fig. 37, *chrysalis* (The Coral Hair-streak).

FIG. 133.—Neuration of *Thecla titus,* enlarged. Typical of subgenus *Strymon,* Hübner.

Butterfly.—Uniformly gray-brown on the upper side. Some specimens of the female have a few red spots at the anal angle of the hind wing. On the under side the wings are colored as on the upper side; but the hind wings have a conspicuous submarginal band of coral-red spots on their outer third. Expanse, 1.30 inch.

Early Stages.—These have been well described by several authors. The fullest account is given by Scudder. The caterpillar feeds on the leaves of the wild cherry and the wild plum.

The insect occurs from the Atlantic to the Pacific, from Maine to Georgia. It is not very common.

There are some ten or more other species of this genus found in our fauna, but the species figured in our plates will suffice to give a good idea of the genus.

Genus FENISECA, Grote

(The Harvesters)

" Upon his painted wings, the butterfly
Roam'd, a gay blossom of the sunny sky."
WILLIS G. CLARK.

Butterfly.—Small, bright orange-yellow, on the upper side spotted with black, on the under side more or less mottled and shaded with gray and brown, the markings of the upper side reappearing. The cut shows the neuration, which need not be minutely described.

Egg.—Subglobular, much wider than high, its surface smooth, marked with a multitude of very fine and indistinct raised ridges, giving it the appearance of being covered by very delicate polygonal cells.

Caterpillar.—In its mature stage the caterpillar is short, slug-shaped, covered with a multitude of bristling hairs, upon which it gathers the white exudations or scales of the mealy bugs upon which it feeds.

Chrysalis.— Small, brown in color; when viewed dorsally showing a remarkable and striking likeness to the face of a monkey, a singular phenomenon which also appears even more strikingly in chrysalids of the allied genus *Spalgis*, which is found in Africa and Asia.

FIG. 134.—Neuration of the genus *Feniseca*, enlarged.

But one species of the genus is known.

(1) **Feniseca tarquinius,** Fabricius, Plate XXVIII, Fig. 21, ♂ ; Plate V, Figs. 45, 46, *chrysalis* (The Harvester).

Butterfly.—The upper side of the wings is well depicted in the plate. There is considerable variation, however, in the size of the black markings upon the upper surface, and I have specimens in which they almost entirely disappear. On the under side the wings are paler; the spots of the upper side reappear, and, in addition, the hind wings are mottled profusely with small pale-brown spots. Expanse, 1.30 inch.

Early Stages.—What has been said of these in the description of the genus will suffice for the species.

This curious little insect, which finds its nearest allies in Asia and Africa, ranges all over the Atlantic States from Nova Scotia to the Carolinas, and throughout the valley of the Mississippi.

Genus CHRYSOPHANUS, Doubleday
(The Coppers)

"Atoms of color thou hast called to life
(We name them butterflies) float lazily
On clover swings, their drop of honey made
By thee, dear queen, already for their need."
MARY BUTTS.

Butterfly.—Small butterflies, with the upper side of the wings some shade of coppery-red or orange, frequently glossed with purple. On the under side the wings are marked with a multitude of small spots and lines. The neuration of the wing is delineated in the figure herewith given, and needs no further description.

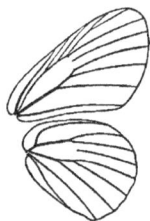

Egg.—The eggs are hemispherical, flattened on the base, the upper surface deeply pitted with polygonal or somewhat circular depressions.

Caterpillar.— The caterpillars, so far as known, are decidedly slug-shaped, thickest in the middle, tapering forward and backward, and having a very small head.

FIG. 135. — Neuration of *Chrysophanus thoë*, enlarged. Typical of the genus.

Chrysalis. — The chrysalids are small, rounded at either end, and held in place by a girdle of silk a little forward of the middle.

This genus is found in the temperate regions of both the New and the Old World, and also in South Africa.

(1) **Chrysophanus arota**, Boisduval, Plate XXIX, Fig. 1, ♂ ; Fig. 2, ♀ (Arota).

Butterfly.—The plate gives a good idea of the upper side of the wings in both sexes. On the under side the fore wings are pale gray in the male and pale red in the female, with the outer margin lavender. The spots of the upper side reappear on the disk. The hind wings on the under side are purplish-gray on the inner two thirds and paler gray on the outer third, with many black spots on the disk, margined with white. Expanse, 1.10–1.25 inch.

Early Stages.—These have been partially described by Dyar in the "Canadian Entomologist," vol. xxiii, p. 204. The caterpillar feeds on the wild gooseberry (*Ribes*).

Arota is a Californian species.

(2) **Chrysophanus virginiensis**, Edwards, Plate XXVIII, Fig. 23, ♂ ; Fig. 24, ♀ (The Nevada Copper).

Butterfly.—Allied to the preceding species, but easily distinguished by the submarginal white bands of crescent-shaped spots on the under side. These are particularly distinct on the hind wings. Expanse, 1.25–1.30 inch.

Early Stages.—Unknown.

Virginiensis, so named because the first specimens came from Virginia City, ranges in California, Nevada, and Colorado.

(3) **Chrysophanus xanthoides,** Boisduval, Plate XXVIII, Fig. 29, ♂ ; Fig. 30, ♀ (The Great Copper).

Butterfly.—The student will easily recognize it by its larger size, it being the largest species of the genus in North America, and by its creamy-white under surface, spotted with distinct small black spots, in large part reproducing the spots of the upper side. Expanse, 1.50–1.65 inch.

(4) **Chrysophanus editha,** Mead, Plate XXVIII, Fig. 26, ♂ ; Fig. 27, ♀ (Editha).

Butterfly.—This is a much smaller species than the last, which it somewhat resembles on the upper side. On the under side it is wholly unlike *xanthoides,* the wings being pale pearly-gray, pale ochreous on the outer margins, the spots of the fore wings black and of the hind wings ochreous, narrowly margined with white or fine black lines. Expanse, 1.10–1.25 inch.

Early Stages.—Entirely unknown.

This species is found in Nevada.

(5) **Chrysophanus gorgon,** Boisduval, Plate XXVIII, Fig. 35, ♂ ; Fig. 36, ♀ (Gorgon).

Butterfly.—Somewhat like the preceding species, but with the fore wings of the male redder on the upper side, and of the female more broadly mottled with pale red, the spots in some specimens inclining to buff. The under side of the wings is white, marked with the usual series of black spots. The secondaries have a marginal series of elongated pale-red spots, tipped at either end with black. Expanse, 1.25–1.30 inch.

Early Stages.—We as yet know nothing of these.

Gorgon is found in California and Nevada.

(6) **Chrysophanus thoë,** Boisduval, Plate XXVIII, Fig. 31, ♂ ; Fig. 32, ♀ ; Plate V, Fig. 50, *chrysalis* (The Bronze Copper).

Butterfly.—The plate makes a description of the upper side of the wings unnecessary. On the under side the fore wing in both sexes is bright tawny-red, pale gray at the apex; the hind wings are bluish-gray, with a broad band of carmine on the outer margin. Both wings are profusely adorned with small black spots. Expanse, 1.30–1.40 inch.

Early Stages.—These are only partially known. The caterpillar feeds on *Rumex.*

It is not uncommon in northern Indiana, Illinois, and Pennsylvania, and ranges from Maine to Kansas and Colorado.

(7) **Chrysophanus mariposa**, Reakirt, Plate XXVIII, Fig. 37, ♂ ; Fig. 38, ♀ (Reakirt's Copper).

Butterfly.—Small, with a broad dusky band on the hind wing of the male and on the fore wing of the female. The male is purplish-red above, the female bright red, with the usual spots. On the under side the ground-color of the fore wings is pale red, of the hind wings clear ashen-gray, with the characteristic markings of the genus. Expanse, 1.10 inch.

Early Stages.—Undescribed.

The insect ranges from British Columbia into northern California, Montana, and Colorado.

(8) **Chrysophanus helloides**, Boisduval, Plate XXVIII, Fig. 33, ♂ ; Fig. 34, ♀ (The Purplish Copper).

Butterfly.—The male has the fore wings broadly shot with iridescent purple. The female is well delineated in the plate. On the under side the fore wings are pale red, the hind wings reddish-gray, with a marginal row of brick-red crescents. The usual black spots are found on both wings. Expanse, 1.15–1.30 inch.

Early Stages.—We know next to nothing of these.

The Purplish Copper is found in the Northwestern States from northern Illinois and Iowa to Vancouver's Island.

(9) **Chrysophanus epixanthe**, Boisduval and Leconte, Plate XXVIII, Fig. 28, ♂ (The Least Copper).

Butterfly.—The smallest species of the genus in North America. On the upper side the wings of the male are dark fuscous, shot with purple, and having a few red spots near the anal angle of the secondaries. The female on the upper side is pale gray, and more profusely marked with black spots. On the under side the wings are light gray, bluish at the base, and marked with the usual spots. Expanse, .85–.95 inch.

Early Stages.—Little is known of these.

This is a Northern species, ranging from Newfoundland, where it is common, to British Columbia, never south of the latitude of New England.

(10) **Chrysophanus hypophlæas**, Boisduval, Plate XXVIII, Fig. 25, ♂ ; Plate V, Fig. 49, *chrysalis* (The American Copper).

Butterfly.—This is one of the commonest butterflies in the United States. The figure in the plate will serve to recall it to the mind of every reader. It is abundant everywhere except in

254

the Gulf States, and ranges as far north as Manitoba and the Hudson Bay region. Expanse, 1.00 inch.

Early Stages.—These have often been described. The caterpillar, which is small and slug-shaped, feeds upon the common sorrel (*Rumex acetosella*).

(11) **Chrysophanus snowi,** Edwards, Plate XXIX, Fig. 7, ♂ ; Fig. 8, ♀ (Snow's Copper).

Butterfly.—This is a medium-sized species, easily recognized by the even, rather wide black border on both wings on the upper side, and the dirty-gray color of the hind wings on the under side. Expanse, 1.15–1.25 inch.

Early Stages.—Unknown.

Snow's Copper, which is named in honor of the amiable Chancellor of the University of Kansas, occurs in Colorado at high elevations, and is reported from Alberta and British Columbia.

(12) **Chrysophanus rubidus,** Behr, Plate XXIX, Fig. 5, ♂ ; Fig. 6, ♀ (The Ruddy Copper).

Butterfly.—This is a rather large species. The male on the upper side is prevalently pale, lustrous red, with a narrow black marginal band and uniformly conspicuous white fringes. The upper side of the female is accurately depicted in the plate. On the under side the wings are shining white, the secondaries immaculate. Expanse, 1.30–1.50 inch.

Early Stages.—These are altogether unknown.

This exceedingly beautiful species is found in Oregon, Nevada, and Montana.

(13) **Chrysophanus sirius,** Edwards, Plate XXIX, Fig. 3, ♂ ; Fig. 4, ♀ (Sirius).

Butterfly.—The male closely resembles the preceding species on the upper side, but is brighter red, especially along the nervules of the fore wings. The female on the upper side is dusky. On the under side the wings are whitish or pale gray, but the hind wings are not without spots, as in the preceding species, and carry the characteristic markings of the genus. Expanse, 1.20–1.30 inch.

Early Stages.—Unknown.

The species has been found from Fort McCleod, in British America, as far south as Arizona, among the North American Cordilleras.

THE UTILITY OF ENTOMOLOGY

All the forces of nature are interdependent. Many plants would not bear seeds or fruit were it not for the activity of insects, which cause the pollen to be deposited upon the pistil and the seed-vessel to be fertilized. Attempts were made many years ago to grow clover in Australia, but the clover did not make seed. All the seed required for planting had to be imported at much expense from Europe. It was finally ascertained that the reason why the clover failed to make seed was because throughout Australia there were no bumblebees. Bumblebees were introduced, and now clover grows luxuriantly in Australia, making seed abundantly; and Australian meats, carried in the cold-storage rooms of great ocean steamers, are used to feed the people of Manila, Hong-Kong, Yokohama, and even London.

A few years ago the orange-groves in southern California became infested with a scale-insect, which threatened to ruin them and to bring orange-growing in that part of the land to an unprofitable end. The matter received the careful attention of the chief entomologist of the United States Department of Agriculture, the lamented Professor C. V. Riley. In the course of the studies which he and his associates prosecuted, it was ascertained that the same scale-insect which was ruining the orange-groves of California is found in the orange-groves of Queensland, but that in Queensland this insect did comparatively small injury to the trees. Investigation disclosed the fact that in Queensland the scale-insect was kept down by the ravages of a parasitic insect which preyed upon it. This parasite, by order of the chief entomologist, was immediately imported, in considerable numbers, into southern California, and let loose among the orange-groves. The result has been most beneficial.

These are two illustrations, from among hundreds which might be cited, of the very practical value of entomological knowledge.

The annual loss suffered by agricultural communities through ignorance of entomological facts is very great. Every plant has its insect enemy, or, more correctly, its insect lover, which feeds upon it, delights in its luxuriance, but makes short work, it may be of leaves, it may be of flowers, it may be of fruit. It has

EXPLANATION OF PLATE XXXI

1. *Lycæna pseudargiolus*, Boisd.-Lec.,
   var. *lucia*, Kirby, ♂.
2. *Lycæna pseudargiolus*, Boisd.-Lec.,
   var. *marginata*, Edwards, ♂.
3. *Lycæna pseudargiolus*, Boisd.-Lec.,
   var. *marginata*, Edwards, ♀.
4. *Lycæna pseudargiolus*, Boisd.-Lec.,
   var. *nigra*, Edwards, ♂.
5. *Lycæna pseudargiolus*, Boisd.-Lec.,
   var. *violacea*, Edwards, ♂.
6. *Lycæna pseudargiolus*, Boisd.-Lec.,
   ♂.
7. *Lycæna pseudargiolus*, Boisd.-Lec.,
   ♀.
8. *Lycæna pseudargiolus*, Boisd.-Lec.,
   var. *neglecta*, Edwards, ♂.
9. *Lycæna pseudargiolus*, Boisd.-Lec.,
   var. *neglecta*, Edwards, ♀.
10. *Lycæna pseudargiolus*, Boisd.-Lec.,
   var. *piasus*, Boisduval, ♂.
11. *Lycæna dædalus*, Behr, ♂.
12. *Lycæna dædalus*, Behr, ♀.
13. *Lycæna heteronea*, Boisduval, ♂
14. *Lycæna heteronea*, Boisduval, ♀.
15. *Lycæna sæpiolus*, Boisduval, ♂.
16. *Lycæna sæpiolus*, Boisduval, ♀.
17. *Lycæna lygdamas*, Doubleday, ♂.
18. *Lycæna lygdamas*, Doubleday, ♀.
19. *Lycæna sagittigera*, Felder, ♂.
20. *Lycæna sagittigera*, Felder, ♀.
21. *Lycæna sonorensis*, Felder, ♂.
22. *Lycæna sonorensis*, Felder, ♀.
23. *Lycæna sbasta*, Edwards, ♂.
24. *Lycæna sbasta*, Edwards, ♀.
25. *Lycæna melissa*, Edwards, ♂.
26. *Lycæna melissa*, Edwards, ♀.
27. *Lycæna acmon*, Dbl.-Hew., ♂.
28. *Lycæna acmon*, Dbl.-Hew., ♀.
29. *Lycæna comyntas*, Godart, ♂.
30. *Lycæna comyntas*, Godart, ♀.
31. *Lycæna ammon*, Lucas, ♀.
32. *Lycæna marina*, Reakirt, ♀.

been estimated that every known species of plant has five or six species of insects which habitually feed upon it. Where the plant is one that is valuable to man and is grown for his use, the horticulturist or the farmer finds himself confronted, presently, by the ravages of these creatures, and unless he has correct information as to the best manner in which to combat them, he is likely to suffer losses of a serious character. We all have read of the havoc wrought by the Kansas locust, or grasshopper, and of the ruin brought about by insects of the same class in Asia and in Africa. We all have heard of the Hessian fly, of the weevil, and of the army-worm. The legislature of Massachusetts has in recent years been spending hundreds of thousands of dollars in the attempt to exterminate the gipsy-moth. The caterpillar of the cabbage-butterfly ruins every year material enough to supply sauer-kraut to half of the people. The codling-moth, the little pinkish caterpillar of which worms its way through apples, is estimated to destroy five millions of dollars' worth of apples every year within the limits of the United States. And what shall we say of the potato-bug, that prettily striped beetle, which, starting from the far West, has taken possession of the potato-fields of the continent, and for the extermination of which there is annually spent, by the agricultural communities of the United States, several millions of dollars in labor and in poisons?

A few facts like these serve to show that the study of entomology is not a study which deserves to be placed in the category of useless pursuits. Viewed merely from a utilitarian standpoint, this study is one of the most important, far outranking, in its actual value to communities, the study of many branches of zoölogical science which some people affect to regard as of a higher order.

The legislature of Pennsylvania acted wisely in passing a law which demands that in every high school established within the State there shall be at least one teacher capable of giving instruction in botany and in entomology. The importance of entomology, while not perceived by the masses as yet, has been recognized by almost all the legislatures of the States; and not only the general government of the United States, but the governments of the individual commonwealths, are at the present time employing a number of carefully trained men, whose business is to

257

ascertain the facts and instruct the people as to the best ma ner in which to ward off the attacks of the insect swarms, which are respecters neither of size nor beauty in the vegetable world, attacking alike the majestic oak and the lowliest mosses.

Genus LYCÆNA, Fabricius

(The Blues)

" Bright butterflies
Fluttered their vans, azure and green and gold."
Sir Edwin Arnold.

Butterfly.—Generally small, for the most part blue on the upper side of the wings, white or gray on the under side, variously marked with spots and lines.

What has been said in reference to the subdivision of the genus *Thecla* may be repeated in regard to the genus which we are considering. It has been in recent years subdivided by writers who have given close attention to the matter, and these subdivisions are entirely defensible from a scientific standpoint. Nevertheless, owing to the close resemblance which prevails throughout the group, in this book, which is intended for popular use, the author has deemed it best not to separate the species, as to do so presupposes a minute anatomical knowledge, which the general reader is not likely to possess.

Egg.—The eggs are for the most part flattened, turban-shaped (see p. 4, Fig. 7).

Caterpillar.—Slug-shaped, as in the preceding genera, feeding upon the petals and bracts of flowers, or upon delicate terminal leaves.

Chrysalis.—Closely resembling the chrysalids of the preceding genera.

This genus is very widely distributed in the temperate regions of both hemispheres. Many of the species are inhabitants of the cold North or high mountain summits, while others are found in the tropics.

(1) **Lycæna fuliginosa**, Edwards, Plate XXX, Fig. 16, ♂, *under side* (The Sooty Gossamer-wing).

Butterfly.—Dark gray on the upper side in both sexes. On

the under side the figure in the plate gives a correct representation of the color and markings. Expanse, 1.10 inch.

Early Stages.—Unknown.

The species occurs in northern California, Utah, Nevada, Oregon, and Washington.

(2) **Lycæna heteronea**, Boisduval, Plate XXXI, Fig. 13, ♂ ; Fig. 14, ♀ ; Plate XXXII, Fig. 19, ♀, *under side* (The Varied Blue).

Butterfly.—On the upper side the male is blue, the female brown. On the under side the wings are white, with faint pale-brown spots on the hind wings and distinct black spots on the fore wings, more numerous than in *L. lycea*, which it closely resembles on the under side. It is the largest species of the genus, and the female reminds us by its markings on the upper side of the females of *Chrysophanus*. Expanse, 1.25–1.40 inch.

Early Stages.—These await description.

Heteronea ranges from Colorado to California, at suitable elevations among the mountains.

(3) **Lycæna clara,** Henry Edwards, Plate XXX, Fig. 26, ♀ (The Bright Blue).

Butterfly.—The figure in the plate is that of the type of the female, the only specimen in my collection. Expanse, 1.15 inch.

Early Stages.—These are entirely unknown.

The type came from southern California.

(4) **Lycæna lycea,** Edwards, Plate XXXII, Fig. 18, ♂, *under side* (Lycea).

Butterfly.—The perfect insect is very nearly as large as *L. heteronea*. The male is lilac-blue on the upper side, with the margins dusky. The black spots of the under side do not show through on the upper side, as in *L. heteronea*. The female is dusky, with the wings shot with blue at their bases, more especially on the fore wing. There are no black spots on the upper side of the wings in this sex, as in *L. heteronea*. On the under side the wings are whitish. The spots on this side are well delineated in our figure in Plate XXXII. Expanse, 1.30 inch.

Early Stages.—These await description.

The butterfly is found in the region of the Rocky Mountains, from New Mexico to Montana.

(5) **Lycæna fulla,** Edwards, Plate XXX, Fig. 24, ♂ ; Fig. 25, ♀ (Fulla).

Butterfly.— Smaller than the preceding species. The upper

side of the male is not lilac-blue, but ultramarine. The female is almost indistinguishable on the upper side from the female of *L. lycea.* On the under side the wings are pale stone-gray, with a black spot at the end of the cell of the primaries and a large white spot at the end of the cell of the secondaries. The other spots, which are always ringed about with white, are located much as in *L. icarioides* (see Plate XXX, Fig. 29). Expanse, 1.15–1.20 inch.

Early Stages.—Unknown.

Fulla occurs in northern California, Oregon, Washington, and British Columbia.

(6) **Lycæna icarioides,** Boisduval (**mintha,** Edwards), Plate XXX, Fig. 29, ♂, *under side* (Boisduval's Blue).

Butterfly.—The insect on the upper side closely resembles the preceding species in both sexes. On the under side it may be at once distinguished from the following species by the absence on the margin of the hind wings of the fine black terminal line, and by having only one, not two rows of submarginal black spots. There are other marked and striking differences, and the merging of *L. dædalus,* Behr, with this species, which has been advocated by some recent writers, is no doubt due to their lack of sufficient and accurately identified material. Expanse, 1.35 inch.

Early Stages.—Unknown.

This species, which is not common, is found in southern California.

(7) **Lycæna dædalus,** Behr, Plate XXXI, Fig. 11, ♂ ; Fig. 12, ♀ ; Plate XXX, Fig. 28, ♀, *under side* (Behr's Blue).

Butterfly.—The wings of the male on the upper side are deep lustrous blue, with darker borders and white fringes. The wings of the female are brown, margined with reddish. The name *archaja* was applied to this sex by Dr. Behr, before it was known to be the female of his *L. dædalus.* Expanse, 1.12 inch.

Early Stages.—These have not yet been studied.

Dædalus is common in southern California.

(8) **Lycæna sæpiolus,** Boisduval, Plate XXXI, Fig. 15, ♂ ; Fig. 16, ♀ (The Greenish Blue).

Butterfly.—The male on the upper side has the wings blue, shot in certain lights with brilliant green. The female on the same side is dusky, with greenish-blue scales at the bases of the wings, and often with reddish markings on the outer margin of the hind wings. On the under side the wings are gray or pale wood-brown, with greenish-blue at their base and a profusion of small black spots margined with white. Now and then the black spots are lost, the white margins spreading inwardly and usurping the place of the black. Expanse, .95–1.10 inch.

Early Stages.—These await further study.

The species ranges from British Columbia to Colorado.

(9) **Lycæna pheres**, Boisduval, Plate XXX, Fig. 37, ♂ ; Fig. 42, ♀, *under side* (Pheres).

Butterfly.—The male is pale shining blue above, with dusky borders. The female is dusky, with a little blue at the base of the wings on the same side. Below, the spots on the fore wings are strongly defined; on the hind wings they are white on a pale stone-gray ground. Expanse, 1.20 inch.

Early Stages.—We know no more of these than we do of those of the preceding species.

Pheres has nearly the same range as *sæpiolus*.

(10) **Lycæna xerxes**, Boisduval, Plate XXX, Fig. 43, ♂, *under side* (Xerxes).

Butterfly.—The wings in both sexes are dusky above, shot with blue, more widely in the male than in the female. On the under side the wings are dark stone-color, with all the spots on both wings white, very rarely slightly pupiled with blackish. Expanse, 1.25 inch.

Early Stages.—Unknown.

The species is found in central California.

(11) **Lycæna antiacis**, Boisduval, Plate XXX, Fig. 35, ♂, *under side;* Fig. 36, ♂ ; Fig. 41, ♀ (The Eyed Blue).

Butterfly.—On the upper side the male is pale lilac-blue, the female dusky, heavily marked with blue at the base of the wings. On the under side the wings are deep, warm stone-gray. There is a single quite regular band of large-sized black spots on the fore wing beyond the middle, and a triply festooned curved band of similar spots on the hind wing. These spots are all margined with white. Expanse, 1.15–1.25 inch.

Early Stages.—These await description.

The insect is found in California.

(12) **Lycæna couperi**, Grote, Plate XXX, Fig. 34, ♂, *under side* (Couper's Blue).

Butterfly.—The wings of the male above are pale shining blue,

261

with a narrow black border; of the female darker blue, broadly margined externally with dusky. On the under side the wings are dark brownish-gray, with the spots arranged much as in *L. antiacis*, but with those of the hind wings generally white, and without a dark pupil. The series on the fore wing is usually distinctly pupiled with black. Expanse, 1.25 inch.

Early Stages.—Unknown.

The species is found in Newfoundland, Labrador, Anticosti, and westward and northward. It is a boreal form.

(13) **Lycæna lygdamas**, Doubleday, Plate XXXI, Fig. 17, ♂ ; Fig. 18, ♀ ; Plate XXX, Fig. 50, ♀, *under side* (The Silvery Blue).

Butterfly.—The male has the upper side of the wings pale silvery-blue, narrowly edged with black; the wings of the female on the upper side are darker blue, dusky on the borders, with a dark spot at the end of the cell of the primaries. On the under side the wings are pale chocolate-brown, with a submarginal band of black spots, margined with white, on both wings, as well as a spot at the end of the cells, and one or two on the costa of the secondaries. Expanse, .85–1.10 inch.

Early Stages.—These are yet to be ascertained.

The insect is reported from Michigan to Georgia.

(14) **Lycæna sagittigera**, Felder. Plate XXXI, Fig. 19, ♂ ; Fig. 20, ♀ ; Plate XXX, Fig. 44, ♀, *under side* (The Arrow-head Blue).

Butterfly.—The wings in both sexes are variable pale blue, dusky on the margins, with white fringes checkered with dusky at the ends of the veins. On the under side the wings are dark gray, profusely spotted, the most characteristic markings being a white ray in the cell of the hind wings, a broad submarginal band of white arrow-shaped markings on both wings, with a black spot at the tip of each sagittate maculation and a dark triangular shade between the barbs. These markings are not shown as they should be in Plate XXX, Fig. 44. They are only faintly indicated. Expanse, 1.25–1.30 inch.

Early Stages.—These await description.

This butterfly ranges from Oregon to Mexico, and eastward as far as Colorado on the mountains.

(15) **Lycæna speciosa**, Henry Edwards, Plate XXXII, Fig. 1, ♂ ; Fig. 2, ♀, *under side* (The Small Blue).

Butterfly.—Quite small; the male pale blue above, edged with dusky; the female dusky, with the inner two thirds shot with blue. The maculation of the under side is as represented in the plate. Expanse, .80 inch.

Early Stages.—Unknown.

Habitat, southern California.

(16) **Lycæna sonorensis**, Felder, Plate XXXI, Fig. 21, ♂ ; Fig. 22, ♀ (The Sonora Blue).

Butterfly.—Easily distinguished from all other species of the genus by the red spots in the region of the median nervules on the upper side. Expanse, .87 inch.

Early Stages.—Unknown.

This lovely little insect is found rather abundantly in southern California and northern Mexico.

(17) **Lycæna podarce**, Felder, Plate XXXII, Fig. 15, ♂ ; Fig. 16, ♀ (The Gray Blue).

Butterfly.—The male is grayish-blue above, with dusky margins, lighter on the disk of both the fore and hind wings. There are a few dark marginal crescents on the hind wings. On the under side the wings are very pale, profusely spotted, the spot at the end of the cell of the secondaries being large and whitish, without a pupil, the rest being black ringed about with white. The female is dark brown above, the fore wings having a black spot ringed about with yellowish at the end of the cell. Expanse, 1.05 inch.

Early Stages.—These have never been described.

The species is thus far known from California, Nevada, and Colorado. It is alpine in its habits.

(18) **Lycæna aquilo**, Boisduval, Plate XXXII, Fig. 9, ♂ ; Fig. 10, ♂ , *under side* (The Labrador Blue).

Butterfly.—The male is dusky bluish-gray on the upper side; the female somewhat darker. It is easily distinguished from other species by the dark-brown shades on the under side of the secondaries. Expanse, .80 inch.

Early Stages.—Unknown.

It is found in Labrador and arctic America.

(19) **Lycæna rustica**, Edwards, Plate XXXII, Fig. 17, ♂ , *under side* (The Rustic Blue).

Butterfly.—Much like the preceding species, but a third larger, and brighter blue on the upper side of the wings of the male. On the under side the disposition of the spots and markings is precisely as in *L. aquilo*, but on the secondaries the dark spots and shades are all replaced by white on a pale-gray ground. Expanse, .90–1.00 inch.

Early Stages.—We are in complete ignorance as to these.

The butterfly is found in British America and on the Western Cordilleras.

(20) **Lycæna enoptes,** Boisduval, Plate XXX, Fig. 30, ♀, *under side;* Fig. 51, ♂ (The Dotted Blue).

Butterfly.—The wings on the upper side are purplish-blue,—pale in the male, darker in the female,—bordered with dusky, more heavily in the female than in the male. The fringes are white, checkered with dusky at the ends of the veins. The female sometimes has the hind wings marked on the upper side with red marginal spots on the inner half of the border. On the under side the wings are pale bluish-gray, marked with a profusion of small black spots, those on the outer margin arranged in two parallel lines, between which, on the hind wings, are red spots. Expanse, 1.00 inch.

Early Stages.—Awaiting description.

Enoptes ranges from Washington to Arizona.

(21) **Lycæna glaucon,** Edwards, Plate XXX, Fig. 31, ♀, *under side;* Fig. 39, ♂ (The Colorado Blue).

Butterfly.—Purplish-blue, closely resembling the preceding species, but having the black margin of the wings broader than in *L. enoptes*, with the dark crescents of the marginal series on the under side showing through as darker spots in the margins of the hind wings. The female has a band of orange spots on the margins of the secondaries. The two marginal rows of spots on the lower side of the wings are arranged and colored as in the preceding species. Expanse, 1.00 inch.

Early Stages.—Of these we must again confess ignorance.

Glaucon ranges from Washington into California, and eastward to Colorado, where it is quite common in the mountain valleys.

(22) **Lycæna battoides,** Behr, Plate XXXII, Fig. 11, ♂ (Behr's Blue).

Butterfly.—On the upper side paler blue than the preceding species, with the hind margin tinged with reddish, shining through from below, and small crescentic dark spots. On the under side

264

the wings are smoky-gray, with all the black spots, which are arranged as in the preceding species, greatly enlarged and quadrate, and a broad submarginal border of orange on the hind wings. The female is like the male, but with more orange on the upper side of the hind wings.

Early Stages.—But little is, as yet, known of these.

The insect ranges from California and Arizona to Colorado.

(23) **Lycæna shasta,** Edwards, Plate XXXI, Fig. 23, ♂ ; Fig. 24, ♀ (The Shasta Blue).

Butterfly.—The figures in the plate give a fairly good idea of the upper side of this species in both sexes, though the male is not quite so dark a blue as represented. On the under side the wings have the usual black spots, on a dirty-gray ground, and, in addition, on the hind wings there are a number of small marginal spots surmounted by metallic-colored bluish-green scales, somewhat like those found in some species of the genus *Thecla.* Expanse, 1.00 inch.

Early Stages.— So far as I know, these have never been described.

My specimens are all from Montana and Nevada. It is also reported from northern California, Oregon, and Kansas, though I question the latter locality.

(24) **Lycæna melissa,** Edwards, Plate XXXI, Fig. 25, ♂ ; Fig. 26, ♀ (The Orange-margined Blue).

Butterfly.—The male on the upper side is pale blue, with a narrow black marginal line and white fringes. The female is brown or lilac-gray, with a series of orange-red crescents on the margins of both wings. On the under side the wings are stone-gray, with the usual spots, and on the secondaries the orange-colored marginal spots are oblong, tipped inwardly with black and outwardly by a series of metallic-green maculations. Expanse, .90–1.15 inch.

Early Stages.—We know very little about these.

It is found from Kansas to Arizona, and northward to Montana.

(25) **Lycæna scudderi,** Edwards, Plate XXX, Fig. 48, ♂ ; Fig. 49, ♀ ; Plate V, Fig. 41, *chrysalis* (Scudder's Blue).

Butterfly. — The commonest Eastern representative of the group to which the preceding four or five and the following three species belong. On the upper side the male cannot be distinguished from *L. melissa;* the female is darker and has only a few orange crescents on the outer margin of the hind wing. On the under side the wings are shining white, the spots are much reduced in size, the large orange spots found in *L. melissa* are replaced by quite small yellowish or ochreous spots, and the patches of metallic scales defining them externally are very minute. Expanse, 1.00–1.10 inch.

Early Stages. —These are accurately described by Dr. Scudder in his great work, " The Butterflies of New England," and by others. The caterpillar feeds upon the lupine, and probably other leguminous plants.

It is widely distributed through the basin of the St. Lawrence, the region of the Great Lakes, and northward as far as British Columbia, being also found on the Catskill Mountains. I have found it very common at times about Saratoga, New York.

(26) **Lycæna acmon,** Doubleday and Hewitson, Plate XXXI, Fig. 27, ♂ ; Fig. 28, ♀ (Acmon).

Butterfly. —The plate gives a good representation of the male and the female of this pretty species, which may at a glance be distinguished from all its allies by the broad orange-red band on the hind wings, marked by small black spots. On the under side it is marked much as *L. melissa.* Expanse, .90–1.10 inch.

Early Stages. —Unknown.

It is found from Arizona to Washington and Montana.

(27) **Lycæna aster,** Plate XXX, Fig. 40, ♂ ; Fig. 46, ♀ ; Fig. 47, ♂ , *under side* (The Aster Blue).

Butterfly. —On the under side this species is very like *enoptes* and other allied species. The male looks like a dwarfed specimen of *L. scudderi.* The female is dull bluish-gray above, with black spots on the outer margins of the wing, most distinct on the secondaries, and, instead of a band of orange spots before them, a diffuse band of blue spots, paler than the surrounding parts of the wing. Expanse, .95–1.00 inch.

Early Stages. —These furnish a field for investigation.

The insect is reported thus far only from Newfoundland, from which locality I obtained, through the purchase of the Mead collection, a large and interesting series.

(28) **Lycæna annetta,** Mead, Plate XXXII, Fig. 13, ♂ ; Fig. 14, ♀ (Annetta).

Butterfly. —The male closely resembles the male of *L. melissa*

EXPLANATION OF PLATE XXXII

1. *Lycæna speciosa*, Henry Edwards, ♂.
2. *Lycæna speciosa*, Henry Edwards, ♀, under side.
3. *Lycæna banno*, Stoll, ♂, under side.
4. *Lycæna isophthalma*, Herrich-Schäffer, ♂.
5. *Lycæna exilis*, Boisduval, ♂.
6. *Lycæna theonus*, Lucas, ♀.
7. *Lycæna amyntula*, Boisduval, ♂.
8. *Lycæna amyntula*, Boisduval, ♀.
9. *Lycæna aquilo*, Boisduval, ♂.
10. *Lycæna aquilo*, Boisduval, ♂, under side.
11. *Lycæna battoides*, Behr, ♂.
12. *Lycæna comyntas*, Godart, ♂, under side.
13. *Lycæna annetta*, Mead, ♂.
14. *Lycæna annetta*, Mead, ♀.
15. *Lycæna podarce*, Felder, ♂.
16. *Lycæna podarce*, Felder, ♀.
17. *Lycæna rustica*, Edwards, ♂, under side.
18. *Lycæna lycea*, Edwards, ♂, under side.
19. *Lycæna heteronea*, Boisduval, ♀, under side.
20. *Thecla melinus*, Hübner, ♂.
21. *Nathalis iole*, Boisduval, ♂.
22. *Nathalis iole*, Boisduval, ♀.
23. *Euchloë creusa*, Dbl.-Hew., ♂.
24. *Euchloë ausonides*, Boisduval, ♂.
25. *Euchloë ausonides*, Boisduval, ♀.
26. *Euchloë cethura*, Felder, ♂.
27. *Euchloë cethura*, Felder, ♀.
28. *Euchloë sara*, Boisduval, ♂.
29. *Euchloë sara*, Boisduval, ♀.
30. *Euchloë lanceolata*, Boisduval, ♂.
31. *Euchloë sara*, Boisduval, var. *reakirti*, Edwards, ♂.
32. *Euchloë sara*, Boisduval, var. *reakirti*, Edwards, ♀.
33. *Euchloë pima*, Edwards, ♂.
34. *Euchloë sara*, Boisduval, var. *julia*, Edwards, ♂.
35. *Euchloë sara*, Boisduval, var. *stella*, Edwards, ♂.
36. *Euchloë sara*, Boisduval, var. *stella*, Edwards, ♀.
37. *Euchloë genutia*, Fabricius, ♂.
38. *Euchloë genutia*, Fabricius, ♀.
39. *Euchloë olympia*, Edwards, var. *rosa*, Edwards, ♂, under side.

PLATE XXXII.

on the upper side. The female is paler than the male, which is unusual in this genus, and has a " washed-out " appearance. On the under side the markings are very like those found in *L. scudderi*. Expanse, 1.15 inch.

Early Stages. —Entirely unknown.

The types which I possess came from Utah.

(29) **Lycæna pseudargiolus,** Boisduval and Leconte, Plate XXXI, Fig. 6, ♂ ; Fig. 7, ♀ ; Plate XXX, Fig. *32, ♂, under side;* Plate V, Figs. *36, 43,* 44, *chrysalis* (The Common Blue).

Butterfly. —This common but most interesting insect has been made the subject of most exhaustive and elaborate study by Mr. W. H. Edwards, and the result has been to show that it is highly subject to variation. It illustrates the phenomena of polymorphism most beautifully. The foregoing references to the plate cite the figures of the typical summer form. In addition to this form the following forms have received names :

(*a*) Winter form **lucia,** Kirby, Plate XXXI, Fig. 1, ♂ ; Plate XXX, Fig. 20, ♂, *under side.* This appears in New England in the early spring from overwintering chrysalids, and is characterized by the brown patch on the middle of the hind wing on the under side.

(*b*) Winter form **marginata,** Edwards, Plate XXXI, Fig. 2, ♂ ; Fig. 3, ♀ ; Plate XXX, Fig. 19, ♂, *under side.* This appears at the same time as the preceding form. The specimens figured in the plate were taken in Manitoba. This form is characterized by the dark margins of the wings on the under side.

(*c*) Winter form **violacea,** Edwards, Plate XXXI, Fig. 5, ♂ . This is the common winter form. The spots below are distinct, but never fused or melted together, as in the two preceding forms.

FIG. 136.— Neuration of *Lycæna pseudargiolus,* enlarged. Typical of subgenus *Cyaniris,* Dalman.

(*d*) Form **nigra,** Edwards, Plate XXXI, Fig. 4, ♂ . The wings on the under side are as in *violacea,* but are black above. It is found in West Virginia and occurs also in Colorado.

(*e*) Summer form **neglecta,** Edwards, Plate XXXI, Fig. 8, ♂ ; Fig. 9, ♀ . This is smaller than the typical form *pseudargiolus,* also has the dark spots on the under side of the wings more distinct, and the hind wings, especially in the female, paler.

(*f*) Southern form **piasus,** Plate XXXI, Fig. 10, ♂. This form, which is uniformly darker blue on the upper side than the others, is found in Arizona.

There are still other forms which have been named and described.

Early Stages.—These have been traced through all stages with minutest care. The egg is delineated in this book on p. 4, Fig. 7. The caterpillar is slug-shaped, and feeds on the tender leaves and petals of a great variety of plants.

The range of the species is immense. It extends from Alaska to Florida, and from Anticosti to Arizona.

(30) **Lycæna amyntula,** Boisduval, Plate XXXII, Fig. 7, ♂ ; Fig. 8, ♀ (The Western Tailed Blue).

Butterfly.—Closely resembling *L. comyntas,* of which it may be only a slightly modified Western form. Until the test of breeding has been applied we cannot be sure of this. The figures in the plate give a very good representation of the upper side of the wings of this species.

Early Stages.—But little has been found out concerning these.

It ranges from the eastern foot-hills of the Rocky Mountains to the Pacific in British America and the northern tier of Western States.

Fig. 137.—
Neuration of
Lycæna comyntas, enlarged.
Typical of the
subgenus *Everes,* Hübner.

(31) **Lycæna comyntas,** Godart, Plate XXXI, Fig. 29, ♂ ; Fig. 30, ♀ ; Plate XXXII, Fig. 12, ♂, *under side;* Plate V, Figs. 42, 47, 48, *chrysalis* (The Eastern Tailed Blue).

Butterfly.—The blue of the upper side of the male in the plate is too dark ; but the female and the under side of the wings are accurately delineated. The species is generally tailed, but specimens without tails occur. Expanse, 1.00–1.10 inch.

Early Stages.—These are well known and have been fully described. The caterpillar feeds on leguminous plants.

This delicate little species ranges from the valley of the Saskatchewan to Costa Rica, and from the Atlantic to the foot-hills of the Western Cordilleras. It is common in the Middle and Western States, flitting about roadsides and weedy forest paths.

(32) **Lycæna isola,** Reakirt, Plate XXX, Fig. 33, ♀, *under side;* Fig. 38, ♀ (Reakirt's Blue).

Butterfly.—The male on the upper side is pale lilac-blue, with the outer borders and the ends of the veins narrowly dusky. The female is brownish-gray on the upper side, with the wings at their base glossed with blue. In both sexes there is a rather conspicuous black spot on the margin of the hind wings between the first and second median nervules. The under side is accurately depicted in our plate, to which the student may refer. Expanse, 1.00 inch.

Early Stages.—Unknown.

The species occurs in Texas, Arizona, and Mexico.

(33) **Lycæna hanno,** Stoll, Plate XXXII, Fig. *3*, *♂*, *under side* (The Florida Blue).

Butterfly.—Larger than the preceding species, on the upper side resembling *L. isola;* but the blue of the male is not lilac, but bright purplish, and the female is much darker. On the under side a striking distinction is found in the absence on the fore wing of the postmedian band of large dark spots so conspicuous in *L. isola.* Expanse, .85 inch.

Early Stages.—We have no information as to these.

The insect occurs in Florida and throughout the Antilles and Central America.

(34) **Lycæna isophthalma,** Herrich-Schäffer, Plate XXXII, Fig. 4, *♂* (The Dwarf Blue).

Butterfly.—Light brown on the upper side in both sexes, with the outer margin of the hind wings set with a row of dark spots, which on the under side are defined by circlets of metallic scales. The under side is pale brown, profusely marked by light spots and short bands. Expanse, .75 inch.

Early Stages.—Up to this time we have learned very little concerning them.

The species occurs in the Gulf States and the Antilles.

(35) **Lycæna exilis,** Boisduval, Plate XXXII, Fig. 5, *♂* (The Pygmy Blue).

Butterfly.—On the upper side this, which is the smallest of North American butterflies, very closely resembles the foregoing species, but may be instantly distinguished by the white spot at the inner angle of the fore wing and the white fringes of the same wing near the apex. The hind wings on the under side are set with a marginal series of dark spots ringed about with metallic scales. Expanse, .65 inch.

Early Stages.—Unknown.

The Pygmy is found in the Gulf States and throughout tropical America.

(36) **Lycæna ammon,** Lucas, Plate XXXI, Fig. 31, ♀ ; Plate XXX, Fig. 45, ♀, *under side* (The Indian River Blue).

Butterfly.—The male is brilliant lilac-blue on the upper side; the female shining violet-blue, with very dark and wide black borders on the fore wings and one or two conspicuous black eye-spots near the anal angle of the hind wings, each surmounted by a carmine crescent. The figure in Plate XXX gives a correct representation of the under side. Expanse, .95–1.10 inch.

Early Stages.—Unknown.

⸺his beautiful little insect is not uncommon in southern Florida, and also occurs in the Antilles and tropical America.

(37) **Lycæna marina,** Reakirt, Plate XXXI, Fig. 32, ♀ ; Plate XXX, Fig. 27, ♀, *under side* (The Marine Blue).

Butterfly.—The male, on the upper side, is pale dusky-lilac, the dark bands of the lower side appearing faintly on the upper side. The female is dark brown on the upper side, with the wings at the base shot with bright lilac-blue; the dark bands on the disk in this sex are prominent, especially on the fore wings. The under side of the wings is accurately depicted in Plate XXX and therefore requires no description. Expanse, 1.10 inch.

Early Stages.—Unknown.

Marina is found in Texas, Arizona, southern California, and southward.

(38) **Lycæna theonus,** Lucas, Plate XXXII, Fig. 6, ♀ (The West Indian Blue).

Butterfly.—The male is shining lavender-blue, this color glossing the dark outer borders of the wings; the female is white, with the outer costal borders heavily blackish, the primaries shot with shining sky-blue toward the base. On the under side the wings are crossed by dark bands of spots, arranged much as in *L. marina,* but darker. The hind wings, near the anal angle, have conspicuous eye-spots both above and below. Expanse, .80 inch.

Early Stages.—Unknown.

This lovely insect is found in the Gulf States and all over the hot lands of the New World.

SIZE

Size, like wealth, is only relative. The farmer who owns a hundred acres appears rich to the laborer whom he employs to cut his wheat; but many a millionaire spends in one month as much as would purchase two such farms. The earth seems great to us, and the sun still greater; but doubtless there are suns the diameter of which is equal to the distance from the earth to the sun, in which both earth and sun would be swallowed up as mere drops in an ocean of fire. In the animal kingdom there are vast disparities in size, and these disparities are revealed in the lower as well as in the higher classes. In the class of mammals we find tiny mice and great elephants; in the insect world we find beetles which are microscopic in size, and, not distantly related to them, beetles as large as a clenched fist. The disparity between a field-mouse and a sulphur-bottomed whale is no greater than the disparity in size which exists between the smallest and the largest of the beetle tribe. And so it is with the lepidoptera. It would take several thousands of the Pygmy Blue, *Lycæna exilis*, to equal in weight one of the great bird-wing butterflies of the Australian tropics. The greatest disparity in size in the order of the lepidoptera is not, however, shown in the butterflies, but among the moths. There are moths the wings of which do not cover more than three sixteenths of an inch in expanse, and there are moths with great bulky bodies and wings spreading from eight to nine inches. It would require ten thousand of the former to equal in weight one of the latter, and the disproportion in size is as great as that which exists between a shrew and a hippopotamus, or between a minnow and a basking-shark.

It is said that, taking the sulphur-bottomed whale as the representative of the most colossal development of flesh and blood now existing on land or in the sea, and then with the microscope reaching down into the realm of protozoan life, the common blow-fly will be ascertained to occupy the middle point on the descending scale. Man is, therefore, not only mentally, but even physically, a great creature, though he stands sometimes amazed at what he regards as the huge proportions of other creatures belonging to the vertebrates.

FAMILY IV

PAPILIONIDÆ (THE SWALLOWTAILS AND ALLIES)

THE butterflies of this family in both sexes are provided with six ambulatory feet. The caterpillars are elongate, and in the genera *Papilio* and *Ornithoptera* have osmateria, or protrusive scent-organs, used for purposes of defense.

The chrysalids in all the genera are more or less elongate, attached at the anal extremity, and held in place by a girdle of silk, but they never lie appressed to the surface upon which pupation takes place, as is true in the *Erycinidæ* and *Lycænidæ*.

SUBFAMILY PIERINÆ (THE SULPHURS AND WHITES)

> " Fly, white butterflies, out to sea,
> Frail pale wings for the winds to try;
> Small white wings that we scarce can see
>   Fly.
> Here and there may a chance-caught eye
> Note, in a score of you, twain or three
> Brighter or darker of tinge or dye;
> Some fly light as a laugh of glee,
> Some fly soft as a long, low sigh:
> All to the haven where each would be,—
>   Fly."   SWINBURNE.

Butterfly.—For the most part medium-sized or small butterflies, white or yellow in color, with dark marginal markings. In many genera the subcostal vein of the fore wing has five, or even in some cases six nervules, and the upper radial is lacking in this wing.

Early Stages.—The eggs are spindle-shaped, marked with vertical ridges and cross-lines. The caterpillars are cylindrical, relatively long, generally green in color, longitudinally striped with darker or paler lines. The chrysalids are generally more or less pointed at the head, with the wing-cases in many of the genera greatly developed on the ventral side, forming a deep, keel-shaped projection upon this surface.

272

Explanation of Plate XXXIII

1. *Catopsilia agarithe*, Boisduval, ♂.
2. *Catopsilia eubule*, Linnæus, ♂.
3. *Catopsilia eubule*, Linnæus, ♂, under side.
4. *Catopsilia philea*, Linnæus, ♂.
5. *Colias eurytheme*, Boisduval, ♂, under side.
6. *Pyrameis huntera*, Fabricius, ♂, under side.

This subfamily is very large, and is enormously developed in the tropics of both hemispheres. Some of the genera are very widely distributed in temperate regions, especially the genera *Pieris* and *Colias.*

Genus DISMORPHIA, Hübner

"I saw him run after a gilded butterfly; and when he caught it, he let it go again; and after it again; and over and over he comes, and up again; catched it again." SHAKESPEARE, *Coriolanus.*

Butterfly.—The butterflies are medium sized, varying much in the form of wing, in some species greatly resembling other *Pierinæ* in outline, but more frequently resembling the Ithomiid and Heliconiid butterflies, which they mimic. Some of them represent transitional forms between the type commonly represented in the genus *Pieris* and the forms found in the two above-mentioned protected groups. The eyes are not prominent. The palpi are quite small. The basal joint is long, the middle joint oval, and the third joint small, oval, or slightly club-shaped. The antennæ are long, thin, terminating in a gradually enlarged spindle-shaped club; the fore wings being sometimes oval, more frequently elongated, twice, or even

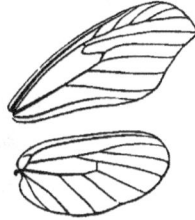

FIG. 138.—Neuration of the genus *Dismorphia.*

three times, as long as broad, especially in the male sex; the apex pointed, falcate, or rounded. The cell is long and narrow. The first subcostal vein varies as to location, rising either before or after the end of the cell, and, in numerous cases, coalescing with the costal vein, as is shown in the cut.

Early Stages.— Of the early stages of these interesting insects we have no satisfactory knowledge.

The species of the genus belong exclusively to the tropical regions of the New World. There are about a hundred species which have already been named and described, and undoubtedly there are many more which remain to be discovered. These insects can always be distinguished from the protected genera which they mimic by the possession of six well-developed ambulatory feet in both sexes, the protected genera being possessed of only four feet adapted to walking.

(1) **Dismorphia melite,** Linnæus, Plate XXXVII, Fig. 17, ♂ ; Fig. 18, ♀ (The Mimc).

Butterfly.—The figures in the plate make a description of the upper side unnecessary. On the under side the wings of the male are shining white, except the costa, which is evenly dull ochreous from the base to the apex. The hind wings are ochreous, mottled with pale brown. The female, on the under side, has the fore wings very pale yellow, with the black spots of the upper side reproduced; the hind wings are deeper yellow, mottled with pale-brown spots and crossed by a moderately broad transverse pale-brown band of the same color.

Early Stages.—Unknown.

The species is credited to our fauna on the authority of Reakirt. It is abundant in Mexico. It mimics certain forms of *Ithomiinæ.*

Genus NEOPHASIA, Behr

" It was an hour of universal joy.
The lark was up and at the gate of heaven,
Singing, as sure to enter when he came;
The butterfly was basking in my path,
His radiant wings unfolded."

ROGERS.

Butterfly.—Medium sized, white in color, more nearly related in the structure of its wings to the European genus *Aporia* than to any other of the American pieridine genera. The upper radial

FIG. 130.—Neuration of the genus *Neophasia.*

is lacking, and the subcostal is provided with five branches, the first emitted well before the end of the cell; the second likewise emitted before the end of the cell and terminating at the apex; the third, fourth, and fifth rising from a common stalk at the outer upper angle of the cell.

Early Stages.—The egg is flask-shaped, fluted on the sides, recalling the shape of the "pearltop" lamp-chimney. The caterpillar, in its mature form, is about an inch long. The body is cylindrical, terminating in two short anal tails. The color is dark green, with a broad white band on each side, and a narrow band of white on the back. The feet are black, and the prolegs greenish-yellow. The chrysalis is dark green, striped

274

with white, resembling the chrysalids of the genus *Colias*, but somewhat more slender. The caterpillar feeds upon conifers. But one species is known.

(1) **Neophasia menapia**, Felder, Plate XXXIV, Fig. 7, ♂ (The Pine White).

Butterfly.—The insect on the under side sometimes has the outer margin of the secondaries marked with spots of bright pink-ish-red, resembling in this style of coloration certain species of the genus *Delias* of the Indo-Malayan fauna.

Early Stages.—These have been thoroughly described by Ed-wards in his third volume. The caterpillar infests the pine-trees and firs of the northern Pacific States. The larva lets itself down by a silken thread, often a hundred feet in length, and pupates on the ferns and shrubbery at the foot of the trees. It sometimes works great damage to the pine woods.

Genus TACHYRIS, Wallace

" The virtuoso thus, at noon,
Broiling beneath a July sun,
The gilded butterfly pursues
O'er hedge and ditch, through gaps and mews;
And, after many a vain essay
To captivate the tempting prey,
Gives him at length the lucky pat,
And has him safe beneath his hat;
Then lifts it gently from the ground;
But, ah! 't is lost as soon as found.
Culprit his liberty regains,
Flits out of sight, and mocks his pains."

COWPER.

This genus, which includes about seventy species, may be distinguished from all other genera belonging to the *Pierinæ* by the two stiff brush-like clusters of hairs which are found in the male sex attached to the abdominal clasps. All of the species belong-ing to the genus are found in the Old World, with exception of the species described in this book, which has a wide range throughout the tropical and subtropical regions of the New World. The peculiarities of neuration are well shown in the accompanying cut, in which the hind wing has been somewhat unduly magnified in proportion to the fore wing.

Early Stages.—The life-history of our species has not been thoroughly studied, but we have ascertained enough of the early

275

stages of various species found in the tropics of the Old World to know that there is a very close relationship between this genus and that which follows in our classification.

(1) **Tachyris ilaire,** Godart, Plate XXXV, Fig. 4, ♂ ; Fig. 5, ♀ (The Florida White).

Butterfly.—The hind wings of the male on the under side, which is not shown in the plate, are very pale saffron. The under side of the wings in the female is pearly-white, marked with bright orange-yellow at the base of the primaries. A melanic form of the female sometimes occurs in which the wings are almost wholly dull blackish on both sides.

Early Stages.—We know, as yet, but little of these.

The insect is universally abundant in the tropics of America, and occurs in southern Florida.

Fig. 140.—Neuration of the genus *Tachyris.* Hind wing relatively enlarged.

Genus PIERIS, Schrank

(The Whites)

"And there, like a dream in a swoon, I swear
I saw Pan lying,—his limbs in the dew
And the shade, and his face in the dazzle and glare
Of the glad sunshine; while everywhere,
Over, across, and around him blew
Filmy dragon-flies hither and there,
And little white butterflies, two and two,
In eddies of odorous air."

JAMES WHITCOMB RILEY.

Butterfly.—Medium-sized butterflies, white in color, marked in many species on both the upper and under sides with dark brown. The antennæ are distinctly clubbed, moderate in length. The palpi are short, delicate, compressed, with the terminal joint quite short and pointed. The subcostal vein of the primaries has four branches, the first subcostal arising before the end of the cell, the second at its upper outer angle, and the third and fourth from a common stem emitted at the same point. The outer margin of the primaries is straight, the outer margin of the secondaries more or less evenly rounded.

Egg.—The egg is spindle-shaped, with vertical raised ridges.

Caterpillar.—Elongate, the head hemispherical, very slightly, if at all, larger in diameter than the body. The caterpillars feed upon cruciferous plants.

Chrysalis.—Attached by the anal extremity, and held in place by a silk girdle; slightly concave on the ventral side; convex on the dorsal side, with a distinct or pointed hump-like projection on the thorax. At the point where the thoracic and abdominal segments unite in some species there is in addition a distinct keel-shaped eminence, and at the head the chrysalis is furnished with a short conical projection.

Fig. 141.—Neuration of the genus *Pieris*.

(1) **Pieris monuste,** Linnæus, Plate XXXV, Fig. 1, ♂ ; Fig. 2, ♀ (The Great Southern White).

Butterfly.—The upper side of the wings, depicted in the plate, requires no comment. On the under side the black marginal markings of the primaries reappear as pale-brown markings. The hind wing is pale yellow or grayish-saffron, crossed by an ill-defined pale-brown transverse band of spots, and has the veins marked with pale brown, and interspersed between them pale-brown rays on the interspaces.

Early Stages.—What we know of these is derived principally from Abbot through Boisduval, and there is opportunity here for investigation.

The species has a wide range through tropical America, and is not uncommon in the Gulf States.

(2) **Pieris beckeri,** Edwards, Plate XXXIV, Fig. 8, ♂ ; Fig. 9, ♀ (Becker's White).

Butterfly.—This species, through the green markings of the under side of the hind wings, concentrated in broad blotches on the disk, recalls somewhat the species of the genus *Euchloë,* and by these markings it may easily be discriminated from all other allied species.

Early Stages.—These have been in part described by Edwards in the second volume of "The Butterflies of North America."

The species ranges from Oregon to central California, and eastward to Colorado.

(3) **Pieris occidentalis,** Reakirt, Plate XXXIV, Fig. 13, ♂ (The Western White).

Butterfly.—Not unlike the preceding species on the upper side, but easily distinguished by the markings of the under side of the wings, which are not concentrated in blotches, but extend as broad longitudinal rays on either side of the veins from the base to the outer margin.

Early Stages.—These require further investigation. We do not, as yet, know much about them.

The species has a wide range in the mountain States of the West, where it replaces the Eastern *P. protodice.*

(4) **Pieris protodice,** Boisduval and Leconte, Plate XXXIV, Fig. 10, ♂ ; Fig. 11, ♀ ; Plate II, Fig. 7, *larva;* Plate V, Figs. 66, 67, *chrysalis* (see also p. 12, Fig. 26) (The Common White).

Butterfly.—Allied to the foregoing species, especially to *P. occidentalis;* but it may always be quickly distinguished by the pure, immaculate white color of the hind wings of the male on the under side, and by the fact that in the female the hind wings are more lightly marked along the veins by gray-green.

Winter form **vernalis,** Edwards, Plate XXXIV, Fig. 18, ♂. What has been said of the typical or summer form does not hold true of this winter form, which emerges from chrysalids which have withstood the cold from autumn until spring. The butterflies emerging from these are generally dwarfed in size, and in the males have the dark spots on the upper side of the wings almost obsolete or greatly reduced, and the dark markings along the veins on the under side well developed, as in *P. occidentalis.* The females, on the contrary, show little reduction in the size and intensity of any of the spots, but rather a deepening of color, except in occasional instances.

Early Stages.—The life-history of this insect has often been described. The caterpillar feeds upon cruciferous plants, like many of its congeners.

It ranges from the Atlantic to the Rocky Mountains, from Canada to the Gulf States.

(5) **Pieris sisymbri,** Boisduval, Plate XXXIV, Fig. 12, ♂ (The California White).

Butterfly.—Smaller in size than the preceding species, with the veins of the fore wing black, contrasting sharply with the white ground-color. All the spots are smaller and more regular, especially those on the outer margin of the fore wing, giving the edge an evenly checkered appearance. On the under side the hind wings have the veins somewhat widely bordered with gray, interrupted about the middle of the wing by the divergence of the lines on either side of the veins in such a way as to produce the effect of a series of arrow-points with their barbs directed toward the base. The female is like the male, with the markings a little heavier. A yellow varietal form is sometimes found.

Early Stages.—The life-history is given and illustrated by Edwards in his second volume. The caterpillar, which is green, banded with black, feeds upon the *Cruciferæ*.

(6) **Pieris napi**, Esper, Plate II, Figs. 8, 9, *larva;* Plate V, Figs. 57, 63, 64, *chrysalis* (The Mustard White).

Butterfly.—This is a Protean species, of which there exist many forms, the result of climatic and local influences. Even the larva and chrysalis show in different regions slight microscopic differences, for the influences which affect the imago are operative also in the early stages of development. The typical form which is found in Europe is rarely found in North America, though I have specimens from the northern parts of the Pacific coast region which are absolutely indistinguishable from European specimens in color and markings. I give a few of the well-marked forms or varieties found in North America to which names have been given.

(*a*) Winter form **oleracea-hiemalis**, Harris, Plate XXXIV, Fig. 16, ♂ (see also p. 5, Fig. 9, and p. 13, Fig. 27). The wings are white above in both sexes. Below the fore wings are tipped with pale yellow, and the entire hind wing is yellow. The veins at the apex of the fore wings and on the hind wings are margined with dusky.

(*b*) Aberrant form **virginiensis**, Edwards, Plate XXXIV, Fig. 14, ♂. The wings are white above, slightly tipped at the apex of the fore wings with blackish. Below the wings are white, faintly, but broadly, margined with pale dusky.

(*c*) Form **pallida**, Scudder, Plate XXXIV, Fig. 15, ♀. In this form the wings are white above and below, with a small black spot on the fore wing of the female above, and hardly any trace of dark shading along the veins on the under side.

(*d*) Alpine or arctic form **bryoniæ**, Ochsenheimer, Plate XXXIV, Fig. 17, ♀. In this form, which is found in Alaska,

279

Siberia, and the Alps of Europe, the veins above and below are strongly bordered with blackish, and the ground-color of the hind wings and the apex of the fore wings on the under side are distinctly bright yellow.

(*c*) Newfoundland variety **acadica**, Edwards, Plate XXXIV, Fig. 19, ♀. This form is larger than the others, and in markings intermediate between *pallida* and *bryoniæ*. The under side in both sexes and the upper side in the female are distinctly yellowish.

Early Stages.—These are well known and have often been described, but some of the varietal forms need further study.

The species ranges from the Atlantic to the Pacific, and from Alaska to the northern limits of the Gulf States.

(7) **Pieris rapæ**, Linnæus, Plate XXXV, Fig. 3, ♀ ; Plate II, Figs. 11, 12, *larva;* Plate V, Figs. 58, 65, *chrysalis* (The Cabbage-butterfly).

Butterfly.—This common species, which is a recent importation from Europe, scarcely needs any description. It is familiar to every one. The story of its introduction and the way in which it has spread over the continent has been well told by Dr. Scudder in the second volume of "The Butterflies of New England," p. 1175. The insect reached Quebec about 1860. How it came no man knows; perhaps in a lot of cabbages imported from abroad; maybe a fertile female was brought over as a stowaway. At all events, it came. Estimates show that a single female of this species might be the progenitor in a few generations of millions. In 1863 the butterfly was already common about Quebec, and was spreading rapidly. By the year 1881 it had spread over the eastern half of the continent, the advancing line of colonization reaching from Hudson Bay to southern Texas. In 1886 it reached Denver, as in 1884 it had reached the head waters of the Missouri, and it now possesses the cabbage-fields from the Atlantic to the Pacific, to the incalculable damage of all who provide the raw material for sauer-kraut. The injury annually done by the caterpillar is estimated to amount to hundreds of thousands of dollars.

INSTINCT

Two city fathers were standing in the market-place beside a pile of cabbages. A naturalist, who was their friend, came by. As he approached, a cabbage-butterfly, fluttering about the place,

Explanation of Plate XXXIV

1. *Euchloë celbura*, Felder, var. *morri-soni*, Edwards, ♂.
2. *Euchloë creusa*, Dbl.-Hew., ♂, under side
3. *Euchloë ausonides*, Boisduval, ♂, under side
4. *Euchloë sara*, Boisduval, var. *flora*, Wright, ♂.
5. *Euchloë sara*, Boisduval, var. *flora*, Wright, ♀.
6. *Euchloë sara*, Boisduval, var. *julia*, Edwards, ♀, under side
7. *Neophasia menapia*, Felder ♂.
8. *Pieris beckeri*, Edwards, ♂.
9. *Pieris beckeri*, Edwards, ♀.
10. *Pieris protodice*, Boisd.-Lec., ♂.
11. *Pieris protodice*, Boisd.-Lec. ♀.
12. *Pieris sisymbri*, Boisduval, ♀.
13. *Pieris occidentalis*, Reakirt, ♂
14. *Pieris virginiensis*, Edwards, ♂
15. *Pieris napi*, Esper, var. *pallida*, Scudder, ♀.
16. *Pieris napi*, Esper, var. *oleracea-hiemalis*, Harris, ♂.
17. *Pieris napi*, Esper, var. *hiemae*, Ochsenheimer, ♀.
18. *Pieris protodice*, Boisd.-Lec., var. *vernalis*, Edwards, ♂.
19. *Pieris napi*, Esper, var. *acadica*, Edwards, ♀.
20. *Kricogonia lyside*, Godart, ♂
21. *Kricogonia lyside*, Godart, ♀.

PLATE XXXIV.

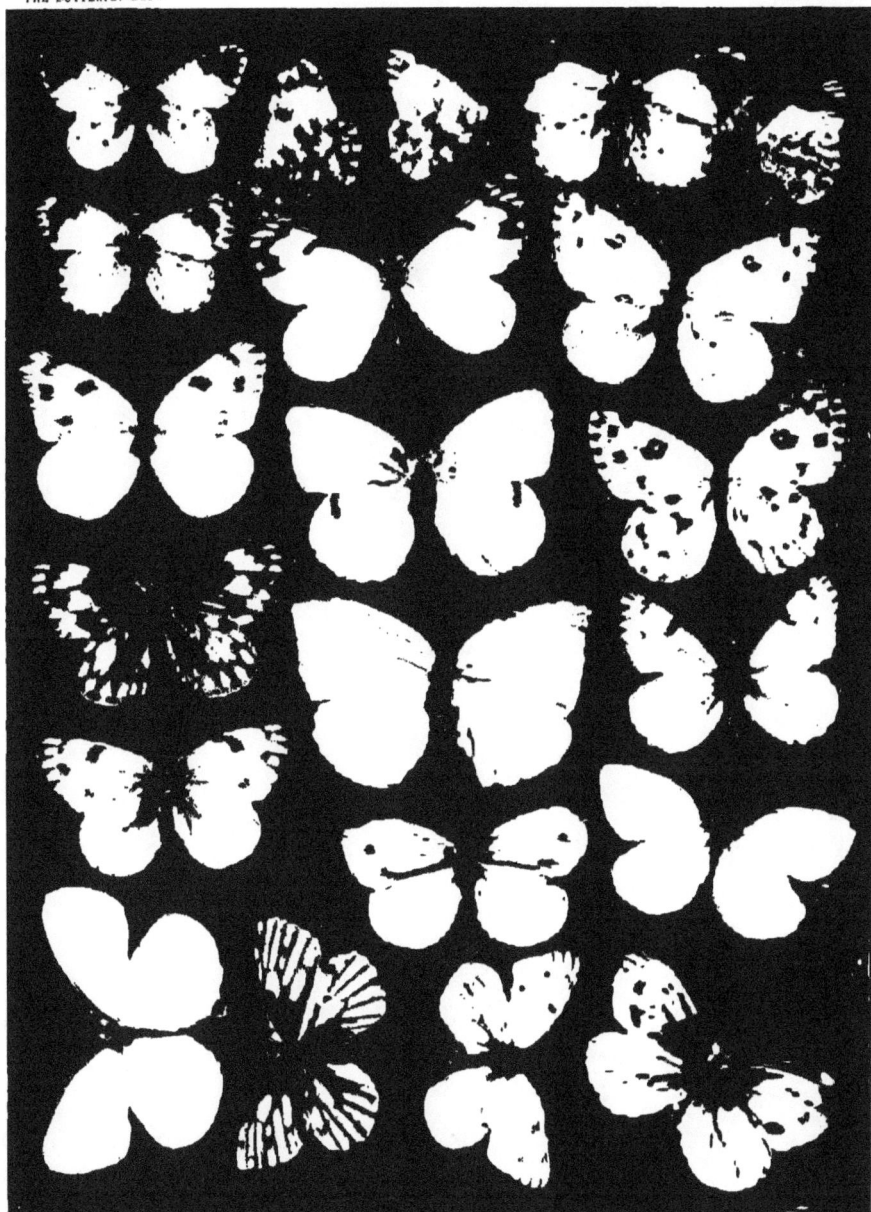

lit on the straw hat of one of the dignitaries. The naturalist, accosting him, said: "Friend, do you know what rests upon your head?" "No," said he. "A butterfly." "Well," said he, "that brings good luck." "Yes," replied the naturalist; "and the insect reveals to me the wonderful instinct with which nature has provided it." "How is that?" quoth the city father. "It is a cabbage-butterfly that rests upon your head."

Genus NATHALIS, Boisduval

" The butterflies, gay triflers
Who in the sunlight sport."
HEINE.

Butterfly.—The butterfly is very small, yellow, margined with black. The upper radial vein in the fore wing is wanting. The subcostal has four nervules, the third and fourth rising from a common stalk emitted from the upper outer corner of the cell, the first and second from before the end of the cell. The precostal vein on the hind wing is reduced to a small swelling beyond the base. The palpi are slender; the third joint long and curved; the second joint oval; the third fine, spindle-shaped, and pointed. The antennæ are rather short, with a somewhat thick and abruptly developed club.

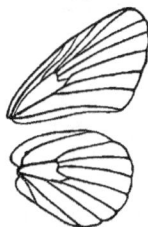

FIG. 142.—Neuration of the genus *Nathalis*, enlarged.

Early Stages.—Very little is known of these. Three species belong to this genus, which is confined to the subtropical regions of the New World, one species only invading the region of which this volume treats.

(1) **Nathalis iole**, Boisduval, Plate XXXII, Fig. 21, ♂ ; Fig. 22, ♀ (The Dwarf Yellow).

Butterfly.—This little species, which cannot be mistaken, and which requires no description, as the plate conveys more information concerning it than could be given in mere words, ranges from southern Illinois and Missouri to Arizona and southern California. Its life-history has not yet been described. Expanse, 1.00–1.25 inch.

The identification of this species with *N. felicia*, Poey, which is found in Cuba, is doubtfully correct. The two species are very closely allied, but, nevertheless, distinct from each other.

Genus EUCHLOË, Hübner

(Anthocharis *of authors*)

(The Orange-tips)

" When daffodils begin to peer,
With, heigh! the doxy over the dale,
Why, then comes in the sweet o' the year;
For the red blood reigns in the winter's pale."

SHAKESPEARE.

Butterfly.—Small butterflies, white in color, with the apical region of the primaries dark brown, marked with spots and bands of yellowish-orange or crimson. On the under side the wings are generally more or less profusely mottled with green spots and striæ.

Egg.—Spindle-shaped (see p. 4, Fig. 6), laterally marked with raised vertical ridges, between which are finer cross-lines.

Caterpillar.—The caterpillar, in its mature stage, is relatively long, with the head small.

FIG. 143.—Neuration of the genus *Euchloë.*

Chrysalis.—With the head relatively enormously projecting; wing-cases compressed, and uniting to form a conspicuous keel-shaped projection, the highest point of which lies at the juncture of the two ends of the silk girdle where they are attached to the supporting surface.

There are numerous species of this genus, and all are exceedingly pretty.

(1) **Euchloë sara**, Boisduval, Plate XXXII, Fig. 28, ♂ ; Fig. 29, ♀ (Sara).

Butterfly.—The wings on the upper side in both sexes are shown in the figures above cited. On the under side the hind wings are marked with dark irregular patches of greenish-brown scales loosely scattered over the surface, and having a "mossy" appearance.

There are several forms which are regarded by recent writers as varieties and may probably be such. Of these we give the following:

(*a*) Variety **reakirti**, Edwards, Plate XXXII, Fig. 31, ♂ ; Fig. 32, ♀ (Reakirt's Orange-tip) = **flora**, Wright, Plate XXXIV, Fig. 4, ♂ ; Fig. 5, ♀ . This form hardly differs at all from the form

282

sara, except in being smaller, and having the margins of the hind wings marked with dark spots at the ends of the veins.

(*b*) Variety **stella,** Edwards, Plate XXXII, Fig. 35, ♂ ; Fig. 36, ♀ (Stella). The females of this form are prevalently yellowish on the upper side of the wings; otherwise they are marked exactly as the preceding variety.

(*c*) Variety **julia,** Edwards, Plate XXXII, Fig. 34, ♂ ; Plate XXXIV, Fig. 6, ♀, *under side* (Julia). The only distinction in this form is the fact that the black bar dividing the red apical patch from the white on the remainder of the wing is broken, or tends to diminution at its middle.

Early Stages.—Unknown.

The species, in all its forms, belongs to the mountain States of the Pacific coast. *Flora*, Wright, is regarded by Beutenmüller, who has given us the latest revision of the genus, as identical with *sara*. It comes nearer the variety *reakirti* than any other form, as will be seen by an examination of the plates which give figures of the types. Expanse, 1.25–1.75 inch.

(2) **Euchloë ausonides,** Boisduval, Plate XXXII, Fig. 24, ♂ ; Fig. 25, ♀ ; Plate XXXIV, Fig. 3, ♂, *under side* (Ausonides).

Butterfly.—On the under side the fore wings are greenish; the hind wings are marked with three irregular green bands, the outer one forking into six or seven branches toward the outer and inner margins. Expanse, 1.65–1.90 inch.

Early Stages.—The larva and chrysalis are described by Edwards in "The Butterflies of North America," vol. ii. The caterpillar is pale whitish-green, with dark-green longitudinal stripes on the side and back. It feeds on cruciferous plants.

Ausonides ranges from Arizona to Alaska, and eastward to Colorado.

(3) **Euchloë creusa,** Doubleday and Hewitson, Plate XXXII, Fig. 23, ♂ : Plate XXXIV, Fig. 2, ♀, *under side* (Creusa).

Butterfly.—Similar to the preceding species, but smaller, the white more lustrous on the under side, and the green markings on the under side of the wings heavier. Expanse, 1.20–1.40 inch.

Early Stages.—We know very little of these.

The species is reported from California, Colorado, and Alberta. I possess a singular varietal form or aberration from Arizona, in which the black spot on the upper side of the primaries fills the outer half of the cell.

(4) **Euchloë rosa**, Edwards, Plate XXXII, Fig. 39, ♂, *under side* (Rosa).

Butterfly.—Pure white, without any red at the tip of the primaries. The transapical black band is broken in the middle, and a black bar closes the cell. The under side is well represented in the plate. Expanse, 1.35–1.40 inch.

Early Stages.—Entirely unknown.

The species is found in Texas.

(5) **Euchloë cethura**, Felder, Plate XXXII, Fig. 26, ♂ ; Fig. 27, ♀ ; form **morrisoni**, Edwards, Plate XXXIV, Fig. 1, ♂ (Cethura).

Butterfly.—This delicate little insect, for the identification of which the plates will abundantly serve, has been regarded as existing in two varietal forms, one of which has been named after the indefatigable collector Morrison, whose death is still lamented by the elder generation of American entomologists. The varietal form is characterized by the heavier green markings of the under side of the wings. Expanse, 1.25–1.40 inch.

(6) **Euchloë pima**, Edwards, Plate XXXII, Fig. 33, ♂ (The Pima Orange-tip).

Butterfly.—This beautiful and well-marked species, the most brilliant of the genus, is yellow on the upper side in both sexes. The red of the upper side appears on the lower side. The hind wings are heavily marked with solid green bands. Expanse, 1.50 inch.

Early Stages.— Unknown.

The only specimens thus far known have come from Arizona.

(7) **Euchloë genutia**, Fabricius, Plate XXXII, Fig. 37, ♂ ; Fig. 38, ♀ ; Plate II, Fig. 5, *larva;* Plate V, Fig. 59, *chrysalis;* Fig. 6, p. 4, *egg* (The Falcate Orange-tip).

Butterfly.—This species is readily recognized by the decidedly falcate tip of the fore wings. The first brood appears in early spring. It is single-brooded in the Northern States, but is double-brooded in the western portions of North Carolina, where I have taken it quite abundantly late in the autumn. Expanse, 1.30–1.50 inch.

Early Stages.—The life-history is well known. The caterpillar feeds on *Sisymbrium, Arabis, Cardamine,* and other cruciferous plants.

It ranges from New England to Texas, but is not found, so

far as is known, in the regions of the Rocky Mountains and on the Pacific coast.

(8) **Euchloé lanceolata,** Boisduval, Plate XXXII, Fig. 30, ♂ (Boisduval's Marble).

Butterfly.—The figure gives a correct idea of the upper surface of the male. The female on the upper side is marked with light-black spots on the outer margin near the apex. On the under side in both sexes the apex of the primaries and the entire surface of the secondaries, except a small spot on the costa, are profusely sprinkled with small brown scales. The veins of the hind wing are brown. Expanse, 1.65–1.95 inch.

Early Stages.—The caterpillar, which feeds upon *Turritis,* is green, shaded on the sides with pale blue, striped laterally with white, and covered with transverse rows of minute black points, each bearing a short black bristle. We know nothing of the other stages.

The species ranges from northern California to Alaska.

Genus CATOPSILIA, Hübner

(The Great Sulphurs)

" A golden butterfly, upon whose wings
  There must be surely character'd strange things,
      .   .   .   .

  Onward it flew, . . . then high it soar'd,
  And downward suddenly began to dip,
  As if, athirst with so much toil, 't would sip
  The crystal spout-head; so it did, with touch
  Most delicate, as though afraid to smutch
  Even with mealy gold the waters clear."

                          KEATS, *Endymion.*

Butterfly.—Large butterflies, brilliant lemon-yellow or orange-yellow, marked with a few darker spots and with a narrow band of brown, especially in the female sex, on the outer margin of the primaries. They are very quick and vigorous in flight, more so than is the case in any of the preceding genera.

Egg.—The eggs are spindle-shaped, flat at the base, and acutely pointed, with a few longitudinal ribs and a multitude of delicate cross-lines.

Caterpillar.—The caterpillar is relatively long, with the head

285

small; the segments somewhat moniliform, resembling beads strung together, the surface covered with a multitude of minute papillæ ranged in transverse rows.

Chrysalis.—The chrysalis is strongly concave on the dorsal side, with the head greatly produced as a long, pointed, conical projection; the wing-cases are compressed and form a very wide, keel-shaped projection on the ventral side. This peculiar formation of the wing-cases reaches its greatest development in this genus.

The butterflies of this genus are mainly tropical. Four or five species, however, are found in the warmer parts of the United States, and one of them ranges north as far as northern New Jersey, and has been occasionally taken even in northern Illinois.

(1) **Catopsilia eubule,** Linnæus, Plate XXXIII, Fig. 2, ♂; Fig. 3, ♂, *under side;* Plate II, Figs. 2, 4, *larva;* Plate V, Figs. 60–62, *chrysalis* (The Cloudless Sulphur).

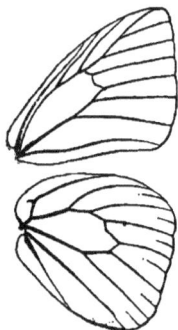

Fig. 144.—Neuration of the genus *Catopsilia.*

Butterfly.—This splendid and vigorous butterfly is found from New England and Wisconsin to Patagonia, being very abundant in the tropics, where it congregates in great swarms upon moist places by the side of streams. It haunts in great numbers the orange-groves of the South, and is very fond of flowers. It is rare on the northern limits of its range, though quite common on the coast of New Jersey. Expanse, 2.50 inches. The caterpillar feeds on leguminous plants, but especially upon the different species of *Cassia.*

(2) **Catopsilia philea,** Linnæus, Plate XXXIII, Fig. 4, ♂ (The Red-barred Sulphur).

Butterfly.—This is another noble species of this fine genus, which includes some of the showiest insects of the subfamily. It may be readily recognized by the bar of deep orange crossing the cell of the primaries, and by the orange tint on the outer margin of the hind wings. Expanse, 3.00–3.50 inches.

Early Stages.—But little is as yet known of these. The larva feeds on the same kinds of plants as the larva of *C. eubule.* It occurs in Texas, and is said to have also been found in Illinois as a straggler. It is abundant in Mexico, Central America, and southward.

(3) **Catopsilia agarithe,** Boisduval, Plate XXXIII, Fig. 1, ♂ (The Large Orange Sulphur).

Butterfly.—About the same size as *C. eubule,* but deep orange on both sides of the wings. The wings of the female are bordered somewhat heavily with brown, and are duller in color than those of the male. Expanse, 2.50–2.75 inches.

Early Stages.—The caterpillar, which resembles that of *eubule,* feeds upon various species of *Cassia.* The chrysalis is also much like that of *eubule.* We need, however, fuller information than that which we possess, drawn, for the most part, from the pages of authors who wrote in the last century.

The species occurs in the hot parts of the Gulf States, and is common throughout tropical America.

Genus KRICOGONIA, Reakirt

Butterfly.—Medium sized, bright yellow on the upper and lower sides, with some dark markings, especially in the male. The primaries in the male are generally quite strongly falcate.

Early Stages.—Nothing has, as yet, been satisfactorily ascertained in relation to these.

The genus is not large, and is confined to the tropical regions of the New World, being represented in our fauna in the vicinity of the city of Brownsville, in Texas.

(1) **Kricogonia lyside,** Godart (form **terissa,** Lucas), Plate XXXIV, Fig. 20, ♂ ; Fig. 21, ♀ (Lyside).

FIG. 145.—Neuration of the genus *Kricogonia.*

Butterfly.—This insect, which may easily be distinguished from all its allies by its peculiar markings, is found in Florida and Texas, and is widely spread over the Antilles and tropical America. We know nothing of its life-history. A number of closely allied forms, reckoned as species, are known from the Antilles and Central America. They are so closely related to each other that it is believed that they are possibly only varieties or local races. We cannot, however, be sure of this until the test of breeding has been applied. Expanse, 1.90–2.10 inches.

Genus MEGANOSTOMA, Reakirt
(The Dog-face Butterflies)

" Let me smell the wild white rose,
Smell the woodbine and the may;
Mark, upon a sunny day,
Sated from their blossoms rise,
Honey-bees and butterflies."
JEAN INGELOW.

Butterfly.— Closely resembling those of the following genus, *Colias*, from which they may be readily distinguished by the more acutely pointed apex of the fore wings and by the remarkable coloration of these wings in the male sex, the dark outer borders being disposed upon the lighter ground-color so as to present the appearance of a rude outline of the head of a dog, whence these butterflies have sometimes been called the "dog-face butterflies."

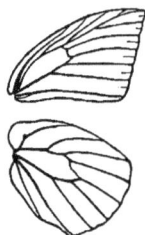

Egg.— Fusiform, strongly pointed at the apex, broader at the base, the sides marked with a few delicate ridges, between which are numerous cross-lines.

Caterpillar.—Elongate, cylindrical, the head relatively small, striped on either side by a whitish lateral line, each segment having a transverse darker line. They feed upon leguminous plants.

FIG. 146.—Neuration of the genus *Meganostoma.*

Chrysalis.—Pointed at the head, convex on the abdominal segments on the dorsal side, with a decided hump on the thorax. The wing-covers unite to form a moderately deep carinate, or keel-shaped, projection on the ventral side, not, however, nearly as large as in the genus *Catopsilia.*

But two species of the genus are found within our fauna, one widely distributed throughout the Southern and Southwestern States, the other confined to the Pacific coast.

(1) **Meganostoma eurydice**, Boisduval, Plate XXXVI, Fig. 1, ♂ ; Fig. 2, ♀ (The Californian Dog-face).

Butterfly.— The splendid purplish iridescence of the fore wings of the male is only faintly indicated in the plate. This beautiful

EXPLANATION OF PLATE XXXV

1. *Pieris monuste*, Linnæus, ♂.
2. *Pieris monuste*, Linnæus, ♀.
3. *Pieris rapæ*, Linnæus, ♀.
4. *Tachyris ilaire*, Godart, ♂.
5. *Tachyris ilaire*, Godart, ♀.
6. *Colias alexandra*, Edwards, ♂.
7. *Colias alexandra*, Edwards, ♀.
8. *Colias scudderi*, Reakirt, ♂.
9. *Colias scudderi*, Reakirt, ♀.
10. *Colias interior*, Scudder, ♂.
11. *Colias interior*, Scudder, ♀.
12. *Colias chrysomelas*, Henry Edwards, ♂.
13. *Colias chrysomelas*, Henry Edwards, ♀.
14. *Colias pelidne*, Boisduval, ♂.
15. *Colias eriphyle*, Edwards, ♂.

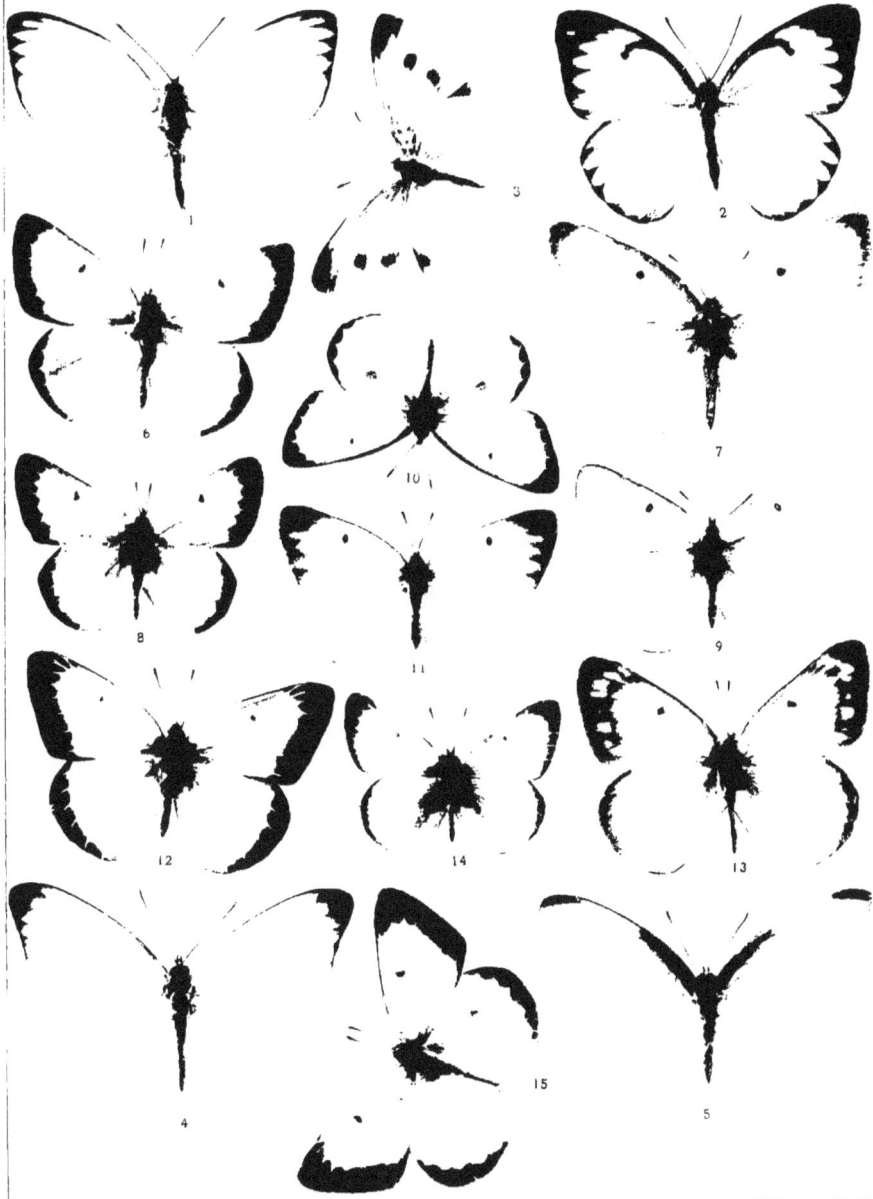

insect is peculiar to the Pacific coast, and there is a wide difference in appearance between the sexes. Expanse, 1.80–2.00 inches.

Early Stages.—The caterpillar feeds upon *Amorpha californica*. The life-history has been accurately described, and the various stages depicted, by Edwards.

(2) **Meganostoma cæsonia**, Stoll, Plate XXXVI, Fig. 3, ♂ ; Fig. 4, ♀ (The Southern Dog-face).

Butterfly.—The sexes are much alike in this species, which ranges widely over the Southern States, and is found even in southern Illinois and sometimes still farther north. Expanse, 2.25 inches.

Early Stages.—These have been fully described by various authors, most carefully by Edwards.

Genus COLIAS, Fabricius

(The Sulphurs)

" Above the arching jimson-weeds flare twos
And twos of sallow-yellow butterflies,
Like blooms of lorn primroses blowing loose,
When autumn winds arise."
JAMES WHITCOMB RILEY.

Butterfly.—Medium-sized butterflies, yellow or orange in color, with black borders upon the wings. In many species this border is heavier in the female than in the male.

Egg.—The egg is spindle-shaped, thickest at the middle, tapering at the apex and at the base, generally attached by an enlarged disk-like expansion to the point on which it is laid. The upper extremity is rounded; the sides are marked by small vertical ridges, between which are delicate cross-lines.

Caterpillar.—The caterpillars strongly resemble in appearance those of the preceding genus, from which, superficially, they cannot be distinguished by any anatomical peculiarities. They feed upon *Leguminosæ*, and especially upon clover (*Trifolium*).

Chrysalis.— The chrysalids do not generally differ in appearance from the chrysalids of the genus *Meganostoma*, though the wing-cases do

FIG. 147.—Neuration of the genus *Colias*.

289

not form as high a keel-shaped projection from the ventral side as in that genus.

This genus is very extensive, being represented throughout the temperate regions of both hemispheres, and also occurring in the cooler portions of South America, especially along the ranges of the Andes. One species is found in temperate South Africa. The brightly colored butterflies, which are sometimes found congregating in immense numbers in moist places, are familiar objects, and swarm upon the clover-fields and by the roadside in the summer months throughout the United States.

(1) **Colias meadi,** Edwards, Plate XXXVI, Fig. 5, ♂ ; Fig. 6, ♀ (Mead's Sulphur).

Butterfly.—The wings on the upper side are orange, greenish on the under side. The discal spot on the lower side is centered with green. Expanse, 1.75 inch.

Early Stages.—The life-history has been written by Edwards, and may be found in the pages of the " Canadian Entomologist," vol. xxi, p. 41. The larva feeds on clover.

The species is alpine in its habits, and is found in Colorado from nine to twelve thousand feet above sea-level.

(2) **Colias elis,** Strecker, Plate XXXVI, Fig. 13, ♂ ; Fig. 14, ♀ (Strecker's Sulphur).

Butterfly.—This species is discriminated from the preceding principally by the narrower black margins on the wings of the male and the more abundant yellow maculation of the borders in the female. Expanse, 1.55–1.90 inch.

Early Stages.—Closely resembling those of the preceding species, of which it may be only a varietal form.

The habitat of the species is on the lofty peaks of the Western Cordilleras.

(3) **Colias eurytheme,** Boisduval, Plate XLVIII, Fig. 18, ♀ ; Plate XXXIII, Fig. 5, ♂ , *under side;* Plate II, Fig. 1, *larva;* Plate V, Fig. 53, *chrysalis* (Eurytheme).

Butterfly.—This species has been made in recent years the subject of exhaustive study, and has been discovered to be strongly polymorphic — that is to say, liable to great variation. Not only does albinism assert itself in the production of white forms, but there are many seasonal and climatic forms. We are not yet through with our studies, and there is doubtless much more to be ascertained. The figures cited above represent the typical form of the species. We have given, in addition to these, the following forms:

(*a*) Winter form **ariadne,** Edwards, Plate XXXVI, Fig. 7, ♂ ; Fig. 8, ♀. This form, emerging from chrysalids which have overwintered, is like the type in having the fore wings tinged with orange. Expanse, 1.75 inch.

(*b*) Winter form **keewaydin,** Edwards, Plate XXXVI, Fig. 9, ♂ ; Fig. 10, ♀. This is a larger form, more deeply flushed with orange, though not quite as deeply as shown in the plate. Expanse, 1.85 inch.

(*c*) Summer form **eriphyle,** Edwards, Plate XXXV, Fig. 15, ♂ ; Plate XLIII, Fig. 3, ♂, *under side.* This summer form differs from typical *C. eurytheme* in being yellow and not laved with orange. Expanse, 2.00 inches.

Early Stages.—The caterpillar feeds on clover, as do most of the species of the genus.

The range of *eurytheme* is very wide. It extends from the Atlantic to the Pacific, and from Canada to the far South, though rare in the lower parts of Florida and Texas in the hot lands.

(4) **Colias philodice,** Godart, Plate I (Frontispiece), Fig. 4, ♂ ; Fig. 5, ♀ ; Plate II, Fig. 10, *larva;* Plate V, Figs. 54, 55, *chrysalis* (The Common Sulphur).

Butterfly.—We are all familiar with this species, the "puddle butterfly" of our childhood, which sits in swarms on moist places by the wayside, and makes the clover-fields gay with the flash of yellow wings in summer. There are many aberrational forms, albinos and negroes, white forms and dark forms, dwarfed forms and large forms, but in the main the species is remarkably constant, and seasonal forms and distinctly local races do not abound as in the case of the preceding species. Expanse, ♂, 1.25–1.80 inch; ♀, 1.60–2.25 inches.

Early Stages.—The food-plant is clover. The eggs are pale yellow, changing, after being laid, to crimson. The caterpillar is slender, green, striped longitudinally with paler green. The chrysalis is pale green.

The species ranges from New England to Florida, and westward to the Rocky Mountains.

(5) **Colias chrysomelas,** Henry Edwards, Plate XXXV, Fig. 12, ♂ ; Fig. 13, ♀ (The Gold-and-black Sulphur).

Butterfly.—Larger than *C. philodice.* The male on the upper side is bright lemon-yellow, with broad black margins on both wings. The female is paler, with the black margin of the hind wing lacking or very faintly indicated, and the margin of the fore wing much broken up by yellow spots. On the under side the wings of the male are dusky-orange, pale yellow on the disk of the primaries; the wings of the female on this side are pale yellow. Expanse, ♂, 2.00–2.10 inches; ♀, 2.25–2.30 inches.

Early Stages.—Undescribed.

The home of this species is on the Coast Range of northern California.

(6) **Colias alexandra,** Edwards, Plate XXXV, Fig. 6, ♂ ; Fig. 7, ♀ (The Alexandra Sulphur).

Butterfly.—Larger than *C. philodice.* The male is pale canary-yellow, with much narrower black borders than the preceding species. The female is pale yellow or white, without black borders, or, at most, faint traces of them at the apex of the primaries. On the under side the wings are silvery-gray, yellow only at the base and on the inner margin of the primaries. The discal spot on the hind wings is white. Expanse, ♂, 1.85 inch; ♀, 2.10–2.30 inches.

Early Stages.—The caterpillar is uniformly yellowish-green, with a white band on each side, broken with orange-red dashes running through it. The chrysalis, which resembles that of *C. philodice* in form, is yellowish-green, darkest on the dorsal side, and adorned with three small red dots on the ventral side of the abdomen near the wing-cases. The caterpillar eats *Astragalus, Thermopsis,* and white clover. Expanse, ♂, 1.90–2.15 inches; ♀, 2.00–2.30 inches.

The species is found in Colorado and the mountain regions to the north and west of that State.

(7) **Colias interior,** Scudder, Plate XXXV, Fig. 10, ♂ ; Fig. 11, ♀ (The Pink-edged Sulphur).

Butterfly.—The male on the upper side closely resembles *C. philodice,* but is smaller, the fringes of the wings rose-colored. The female is pale yellow above, more frequently white, with the tips of the fore wings lightly marked with blackish. On the under side the fore wings at the apex and the entire surface of the hind wings are rusty orange-yellow. The discal spot on the hind wings is silvery, bordered with rosy-red. Expanse, ♂, 1.30–1.75 inch; ♀, 1.60–2.00 inches.

Early Stages.—Little is as yet known of these.

The species was first found by Professor Louis Agassiz on the north shore of Lake Superior. It ranges through a rather narrow belt of country, through Quebec, Ontario, and westward to the Rocky Mountains north of the valley of the St. Lawrence and the Great Lakes.

(8) **Colias scudderi,** Reakirt, Plate XXXV, Fig. 8, ♂ ; Fig. 9, ♀ (Scudder's Sulphur).

Butterfly.—The male on the upper side is colored like *C. philodice,* but the black borders are much wider. The fringes are rosy. The female is generally white,—very rarely slightly yellow,—with very pale dark borders, or often without any trace of black on the outer margin of the wings. On the under side the apex of the fore wings and the entire surface of the hind wings are greenish-gray. The discal spot of the secondaries is well silvered and margined with pale red. Expanse, 1.80–2.00 inches.

Early Stages.—We know but little of these, except that the caterpillar feeds on the leaves of the huckleberry and the willow.

Scudder's Sulphur is found in Colorado, Utah, Montana, and British Columbia.

(9) **Colias pelidne,** Boisduval, Plate XXXV, Fig. 14, ♂ ; Plate XXXVI, Fig. 15, ♂ ; Fig. 16, ♀ (The Labrador Sulphur).

Butterfly.—The male on the upper side is pale yellow, with a greenish tinge on the hind wings; the black borders are narrow; the fringes are pink. The female on the upper side is white, with very little or no black on the outer borders, the black marking being confined to the apex of the fore wings. On the under side the wings are much as in *C. interior,* and it is possible that the two forms are varieties of one and the same species. Expanse, 1.60–1.85 inch.

Early Stages.—Little is known of these.

Pelidne is rather abundant in Labrador at the proper season, and ranges thence westward and northward in boreal America.

(10) **Colias nastes,** Boisduval, Plate XXXVI, Fig. 11, ♂ ; Fig. 12, ♀ (The Arctic Sulphur).

Butterfly.—Easily recognized in both sexes by the pale-greenish tint of the wings and the tendency of the outer border of the fore wings of the male to become divided, like those of the female, by a band of pale spots. Expanse, 1.50–1.65 inch.

Early Stages.—Unknown.

This is an arctic species, which is found in Labrador, Greenland, the far North in British America and Alaska, and on the summits of the Rocky Mountains in British Columbia.

(11) **Colias behri,** Edwards, Plate XXXVI, Fig. 17, ♂ (Behr's Sulphur).

Butterfly.—This very rare little species may be easily recognized by the dark-greenish tint of the upper side of the wings and the light spot on the upper side of the hind wings. The female has the outer borders dusky like the male, the dusky shade running inward on the lines of the veins and nervules. Expanse, 1.50 inch.

Early Stages.—We know little of these.

The insect has hitherto been taken only at considerable elevations among the Western Sierras, and the peaks and lofty meadows about the Yosemite Valley have been until recently the classic locality for the species.

There are a number of other species of the genus *Colias*, and numerous varieties which have been named and described from the western and northwestern portions of our region; but it requires almost as much skill to distinguish them as is required to discriminate between the different species of willows, asters, and goldenrods, among plants, and we do not think it worth while to burden the student with an account of these, and of the controversies which are being waged about them. If any reader of this book becomes entangled in perplexities concerning the species of *Colias*, the writer will be glad to try to aid him to correct conclusions by personal conference or correspondence.

Genus TERIAS, Swainson
(The Small Sulphurs)

"Hurt no living thing :
Ladybird, nor butterfly,
Nor moth with dusty wing,
Nor cricket chirping cheerily,
Nor grasshopper so light of leap,
Nor dancing gnat, nor beetle fat,
Nor harmless worms that creep."
CHRISTINA ROSSETTI.

Butterfly.—Small butterflies, bright orange or yellow, margined with black. They are more delicate in structure and have thinner wings than most of the genera belonging to the subfamily

294

EXPLANATION OF PLATE XXXVI

1. *Meganostoma eurydice*, Boisduval, ♂.
2. *Meganostoma eurydice*, Boisduval, ♀.
3. *Meganostoma carsonia*, Stoll, ♂.
4. *Meganostoma carsonia*, Stoll, ♀.
5. *Colias meadi*, Edwards, ♂.
6. *Colias meadi*, Edwards, ♀.
7. *Colias ariadne*, Edwards, ♂.
8. *Colias ariadne*, Edwards, ♀.
9. *Colias keewaydin*, Edwards, ♂.
10. *Colias keewaydin*, Edwards, ♀.
11. *Colias nastes*, Boisduval, ♂.
12. *Colias nastes*, Boisduval, ♀.
13. *Colias elis*, Strecker, ♂.
14. *Colias elis*, Strecker, ♀.
15. *Colias pelidne*, Boisduval (*labradorensis*, Scudder), ♂.
16. *Colias pelidne*, Boisduval (*labradorensis*, Scudder), ♀.
17. *Colias behri*, Edwards, ♂.

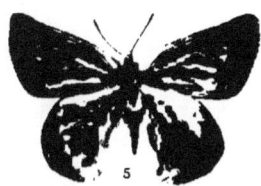

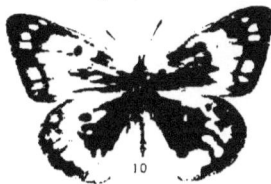

of the *Pierinæ*. The outer margin of the wings is generally straight or slightly rounded, though in a few species the apex is somewhat acuminate. The outer margin of the hind wings is generally rounded, though in a few species it is acuminate.

Egg.—Strongly spindle-shaped, pointed and rounded at the base and at the apex, much swollen at the middle, its sides marked by numerous broad but slightly raised vertical ridges.

Caterpillar.—The caterpillars are small, relatively long, cylindrical, with the head quite small, the thoracic segments somewhat larger than the others, giving the anterior portion of the body a slightly humped appearance. They feed upon leguminous plants.

Fig. 148.—Neuration of the genus *Terias*.

Chrysalis.—The chrysalis is compressed laterally, with the head pointed and the wing-cases forming a deep, keel-shaped projection on the ventral side, more pronounced than in any other genus except *Catopsilia*.

There are an immense number of species belonging to this genus scattered through the tropical and subtropical regions of both hemispheres. Many of the species are dimorphic or polymorphic, and much confusion has arisen, especially in relation to the Oriental species, on account of the great tendency to the production of seasonal varieties, many of which are strikingly different from one another.

(1) **Terias gundlachia**, Poey, Plate XXXVII, Fig. 1, ♂ (Gundlach's Sulphur).

Butterfly.—This species is easily recognized by the orange-yellow tint of the upper side of the wings and the sharply pointed hind wings. Expanse, 1.80 inch.

Early Stages.—We know nothing of these.

The species is found in Texas, Arizona, Mexico, and Cuba.

(2) **Terias proterpia**, Fabricius, Plate XXXVII, Fig. 2, ♂ (Proterpia).

Butterfly.—Even deeper orange than the preceding species. The hind wings are, however, less pointed; the veins and nervules are black at their ends, and the costal margin of the fore wings is evenly bordered with black, which does not run down on the outer margin as in *T. gundlachia*. Expanse, 1.50-1.75 inch.

Early Stages.—Unknown.

Proterpia is found in Texas, Arizona, and Mexico.

(3) **Terias nicippe,** Cramer, Plate XXXVII, Fig. 3, ♂ ; Fig. 4, ♀ ; Fig. 5, var. **flava,** ♂ ; Fig. 6, ♀, *under side;* Plate II, Fig. 6, *larva;* Plate V, Figs. 51, 52, *chrysalis* (Nicippe).

Butterfly.—The plate gives so full a presentation of this common species as to make a lengthy description unnecessary. It is subject to considerable variation. I have specimens of many varying shades of orange and yellow, and a few albino females. The orange form depicted in Plate XXXVII, Figs. 3 and 4, is typical. The form *flava* is not uncommon. Expanse, 1.50–2.00 inches.

Early Stages.—These are not as well known as they should be in view of the excessive abundance of the insect in long-settled parts of the country. The caterpillar feeds upon *Cassia* in preference to all other plants, but will eat other leguminosæ.

Nicippe is very rare in New England, but is common south of latitude 40° as far as the Rocky Mountains, and ranges over Cuba, Mexico, and Guatemala, into Venezuela and even Brazil. It fairly swarms at times in the Carolinas, Tennessee, Kentucky, and southern Indiana and Illinois. I have encountered clouds of it on the wing near Jeffersonville, Indiana, and thence north along the lines of the Pennsylvania Railroad as far as Seymour. It is not common in western Pennsylvania, but in former years was taken rather frequently about Pittsburgh.

(4) **Terias mexicana,** Plate XXXVII, Fig. 7, ♂ ; Fig. 8, ♀, *under side* (The Mexican Yellow).

Butterfly.—Easily distinguished from all other species in our fauna by the pointed hind wings, margined on the outer border with black, and by the heavy black border of the fore wings, deeply excised inwardly, recalling the fore wing of the species of the genus *Meganostoma*. Expanse, ♂, 1.75 inch; ♀, 1.85 inch.

Early Stages.—We do not, as yet, know much about these.

T. mexicana is very common in Arizona, and occurs also in Texas. It is abundant in Mexico.

(5) **Terias damaris,** Felder, Plate XXXVII, Fig. 9, ♂ ; Fig. 10, ♂, *under side* (Damaris).

Butterfly.—Allied to the preceding species, but readily distinguished from it by the less deeply excised outer border of the fore wing, by the fact that the black outer margin of the secondaries extends inwardly beyond the angulated point of the wing, and by the different color and style of the markings of the lower side. Expanse, 1.35–1.65 inch.

Early Stages.—Unknown.

Damaris occurs in Arizona, and thence ranges south into Venezuela.

(6) **Terias westwoodi,** Boisduval, Plate XXXVII, Fig. 11, ♂ (Westwood's Yellow).

Butterfly.—Pale yellow or orange-yellow, with a narrow black border on the fore wings, beginning on the costa beyond the middle, and not quite reaching the inner angle. On the under side the wings are pale yellow, immaculate, or at the apex of the fore wing and the outer angle of the hind wing broadly marked with very pale reddish-brown. Expanse, 1.75–2.00 inches.

Early Stages.—Unknown.

Westwood's Yellow occurs in Texas and Arizona, but is not common. It is abundant farther south.

(7) **Terias lisa,** Boisduval and Leconte, Plate XXXVII, Fig. 13, ♂ ; Plate II, Fig. 3, *larva;* Plate V, Fig. 56, *chrysalis* (The Little Sulphur).

Butterfly.—Allied to the three following species, from which it may at once be distinguished by the absence of the black bar on the inner margin of the fore wings and by the profusely mottled surface of the under side of the hind wings. It is subject to considerable variation, albino females and melanic males being sometimes found, as well as dwarfed specimens of very small size. Expanse, 1.25–1.60 inch.

Early Stages.—These have not been thoroughly studied and described, in spite of the fact that the insect is very common in many easily accessible localities. The caterpillar feeds on *Cassia* and on clover.

T. lisa ranges from New England south and west as far as the foot-hills of the Rocky Mountains, and into Mexico and Honduras. It is found in the Antilles and Bermuda. An interesting account of the appearance of a vast swarm of these butterflies in the Bermudas is given by Jones in " Psyche," vol. i, p. 121:

" Early in the morning of the first day of October last year (1874), several persons living on the north side of the main island perceived, as they thought, a cloud coming over from the northwest, which drew nearer and nearer to the shore, on reaching which it divided into two parts, one of which went eastward, and the other westward, gradually falling upon the land. They were not long in ascertaining that what they had taken for a cloud was an immense concourse of small yellow butterflies (*Terias lisa*, Boisduval), which flitted about all the open grassy patches and cultivated grounds in a lazy manner, as if fatigued after their long voyage over the deep. Fishermen out near the reefs, some few miles to the north of the island, very early that morning, stated that numbers of these insects fell upon their boats, literally covering them. They did not stay long upon the islands, however, only a few days, but during that time thousands must have fallen victims to the vigorous appetite of the bluebird (*Sialia sialis*, Baird) and blackbird (*Mimus carolinensis*, Gray), which were continually preying upon them."

As the nearest point of land is Cape Hatteras, about six hundred miles distant, it is seen that, weak and feeble as this little creature appears, it must possess, when aided by favoring winds, great power of sustained flight.

(8) **Terias elathea**, Cramer, Plate XXXVII, Fig. 12, ♂ (Elathea).

Butterfly.—Distinguished from its near ally, *T. delia*, by the fact that the ground-color of the hind wings is white. The female in this, as in the allied species, is without the black bar on the inner margin of the primaries. Expanse, 1.25–1.40 inch.

Early Stages.—Unknown.

Elathea is found in Florida, Mexico, and the Antilles.

(9) **Terias delia**, Cramer, Plate XXXVII, Fig. 14, ♂ (Delia).

Butterfly.—Almost exactly like the preceding species, but having the upper side of the hind wings yellow. On the under side the fore wing at the tip and the entire hind wing are red. Expanse, 1.25–1.50 inch.

Early Stages.—But little is known of them. The caterpillar feeds on *Cassia*.

Delia occurs commonly in the Gulf States.

(10) **Terias jucunda**, Boisduval and Leconte, Plate XXXVII, Fig. 15, ♂ ; Fig. 16, ♂ , *under side* (The Fairy Yellow).

Butterfly.—Distinguished from the preceding species by the dark marginal band surrounding the hind wing and the pale under surface. Expanse, 1.60–1.75 inch.

Early Stages.—Unknown.

This little species is found in the Gulf States.

Explanation of Plate XXXVII

1. *Terias gundlachia*, Poey, ♂.
2. *Terias proterpia*, Fabricius, ♂.
3. *Terias nicippe*, Cramer, ♂.
4. *Terias nicippe*, Cramer, ♀.
5. *Terias nicippe*, Cramer, var. *flava*, Strecker, ♂.
6. *Terias nicippe*, Cramer, ♀, under side.
7. *Terias mexicana*, Boisduval, ♂.
8. *Terias mexicana*, Boisduval, ♂, under side.
9. *Terias damaris*, Felder, ♂.
10. *Terias damaris*, Felder, ♂, under side.
11. *Terias westwoodi*, Boisduval, ♂.
12. *Terias elathea*, Cramer, ♂.
13. *Terias lisa*, Boisd.-Lec., ♂.
14. *Terias delia*, Cramer, ♂.
15. *Terias jucunda*, Boisd.-Lec., ♂.
16. *Terias jucunda*, Boisd.-Lec., ♂, under side.
17. *Dismorphia melite*, Linnæus, ♂.
18. *Dismorphia melite*, Linnæus, ♀.

1

2

3

4

5

6

7

8

9

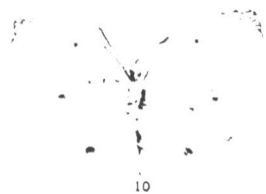
10

11

12

13

14

15

16

17

18

RED RAIN

"The lepidopterous insects in general, soon after they emerge from the pupa state, and commonly during their first flight, discharge some drops of a red-colored fluid, more or less intense in different species, which, in some instances, where their numbers have been considerable, have produced the appearance of a 'shower of blood,' as this natural phenomenon is sometimes called.

"Showers of blood have been recorded by historians and poets as preternatural—have been considered in the light of prodigies, and regarded, where they have happened, as fearful prognostics of impending evil.

"There are two passages in Homer, which, however poetical, are applicable to a rain of this kind; and among the prodigies which took place after the death of the great dictator, Ovid particularly mentions a shower of blood:

> " ' Sæpe faces visæ mediis ardere sub astris,
> Sæpe inter nimbos guttæ cecidere cruentæ.'

> " (' With threatening signs the lowering skies were fill'd,
> And sanguine drops from murky clouds distilled.')

"Among the numerous prodigies reported by Livy to have happened in the year 214 B. C., it is instanced that at Mantua a stagnating piece of water, caused by the overflowing of the river Mincius, appeared as of blood; and in the cattle-market at Rome a shower of blood fell in the Istrian Street. After mentioning several other remarkable phenomena that happened during that year, Livy concludes by saying that these prodigies were expiated, conformably to the answers of the aruspices, by victims of the greater kinds, and supplication was ordered to be performed to all the deities who had shrines at Rome. Again, it is stated by Livy that many alarming prodigies were seen at Rome in the year 181 B. C., and others reported from abroad; among which was a shower of blood which fell in the courts of the temples of Vulcan and Concord. After mentioning that the image of Juno Sospita shed tears, and that a pestilence broke out in the country, this writer adds that these prodigies, and the mortality which prevailed, alarmed the Senate so much that they ordered the consuls to sacrifice to such gods as their judgment should direct victims of the larger kinds, and that the decemvirs should consult their books. Pursuant to their direction, a supplication for one day was proclaimed to be performed at every shrine in Rome; and they advised, besides, and the Senate voted, and the consul proclaimed, that there should be a supplication and public worship for three days throughout all Italy. In the year 169 B. C., Livy also mentions that a shower of blood fell in the middle of the day. The decemvirs were again called upon to consult their books, and again were sacrifices offered to the deities. The account, also, of Livy, of the bloody sweat on some of the statues of the gods, must be referred to the same phenomenon, as the predilection of those ages to marvel, says Thomas Browne, and the want of accurate investigation in the cases recorded, as well as the rare occurrence of these atmospherical depositions in our own times, inclines us to include them among the blood-red drops deposited by insects.

"In Stow's 'Annales of England' we have two accounts of showers of blood, and from an edition printed in London in 1592, we make our quotations: 'Rivallus, sonne of Cunedagius, succeeded his father, in whose time (in the year 766 B. C.) it rained bloud three dayes: after which tempest ensued a great multitude of venemous flies, which slew much people, and then a great mortalitie throughout this lande, caused almost desolation of the same.' The second account is as follows: 'In the time of Brithricus (A. D. 786) it rayned blood, which falling on men's clothes, appeared like crosses.'

"Hollingshed, Grafton, and Fabyan have also recorded these instances in their respective chronicles of England.

"A remarkable instance of bloody rain is introduced into the very interesting Icelandic ghost-story of Thorgunna. It appears that in the year of our Lord 1009 a woman called Thorgunna came from the Hebrides to Iceland, where she stayed at the house of Thorodd; and during the hay season a shower of blood fell, but only, singularly, on that portion of the hay she had not piled up as her share, which so appalled her that she betook herself to her bed, and soon afterward died. She left, to finish the story, a remarkable will, which, from not being executed, was the cause of several violent deaths, the appearance of ghosts, and, finally, a legal action of ejectment against the ghosts, which, it need hardly be said, drove them effectually away.

"In 1017 a shower of blood fell in Aquitaine; and Sleidan relates that in the year 1553 a vast multitude of butterflies swarmed through a great part of Germany, and sprinkled plants, leaves, buildings, clothes, and men with bloody drops, as if it had rained blood. We learn also from Bateman's 'Doome' that these 'drops of bloude upon hearbes and trees' in 1553 were deemed among the forewarnings of the deaths of Charles and Philip, dukes of Brunswick.

"In Frankfort, in the year 1296, among other prodigies, some spots of blood led to a massacre of the Jews, in which ten thousand of these unhappy descendants of Abraham lost their lives.

"In the beginning of July, 1608, an extensive shower of blood took place at Aix, in France, which threw the people of that place into the utmost consternation, and, which is a much more important fact, led to the first satisfactory and philosophical explanation of this phenomenon, but too late, alas! to save the Jews of Frankfort. This explanation was given by M. Peiresc, a celebrated philosopher of that place, and is thus referred to by his biographer, Gassendi: 'Nothing in the whole year 1608 did more please him than that he observed and philosophized about, the *bloody rain*, which was commonly reported to have fallen about the beginning of July; great drops thereof were plainly to be seen, both in the city itself, upon the walls of the church-yard of the church, which is near the city wall, and upon the city walls themselves; also upon the walls of villages, hamlets, and towns, for some miles round about; for in the first place, he went himself to see those wherewith the stones were coloured, and did what he could to come to speak with those husbandmen, who, beyond Lambesk, were reported to have been affrighted at the falling of said rain, that they left their work, and ran as fast as their legs could carry them into the adjacent houses. Whereupon, he found that it was a fable that was reported, touching those husbandmen. Nor was he pleased that naturalists should refer this kind of rain to vapours drawn up out of red earth aloft in the air, which congealing afterwards into liquor, fall down in this form; because such vapours as are drawne aloft by heat, ascend without colour, as we may know by the alone example of red roses, out of which the vapours that arise by heat are congealed into transparent water. He was less pleased with the common people, and some divines, who judged that it was the work of the devils and witches who had killed innocent young children; for this he counted a mere conjecture, possibly also injurious to the goodness and providence of God.

" 'In the meanwhile an accident happened, out of which he conceived he had collected the true cause thereof. For, some months before, he shut up in a box a certain palmer-worm which he had found, rare for its bigness and form; which, when he had forgotten, he heard a buzzing in the box, and when he opened it, found the palmer-worm, having cast its coat, to be turned into a beautiful Butterfly, which presently flew away, leaving in the bottom of the box a red drop as broad as an ordinary sous or shilling; and because this happened about the beginning of the same month and about the same time an incredible multitude of Butterflies were observed flying in the air, he was therefore of opinion that such kind of Butterflies resting on the walls had there shed such like drops, and of the same bigness. Whereupon, he went the second time, and found, by experience, that those drops were not to be found on the house-tops, nor upon the round sides of the stones which stuck out, as it would have happened, if blood had fallen from the sky, but rather where the stones were somewhat hollowed, and in holes, where such small creatures might shroud and nestle themselves. Moreover, the walls which were so spotted, were not in the middle of towns, but they were such as bordered upon the fields, nor were they on the highest parts, but only so moderately high as Butterflies are commonly wont to fly.

" 'Thus, therefore, he interpreted that which Gregory of Tours relates touching a bloody rain seen at Paris in divers places, in the days of Childebert, and on a certain house in the territory of Senlis; also that which is storied, touching raining of blood about the end of June, in the days of King Robert; so that the blood which fell upon flesh, garments or stones could not be washed out, but that which fell on wood might; for it was the same season of Butterflies, and experience hath taught us, that no water will wash these spots out of the stones, while they are fresh and new. When he had said these and such like things to various, a great company of auditors being present, it was agreed that they should go together and search out the matter, and as they

!

EXPLANATION OF PLATE XXXVIII

1. *Papilio zolicaon*, Boisduval, ♂.   2. *Papilio daunus*, Boisduval, ♂.
3. *Papilio pilumnus*, Boisduval, ♂.

(The figures in this plate are reduced, being only two thirds of the natural size.)

PLATE XXXVIII.

went up and down, here and there, through the fields, they found many drops upon stones and rocks; but they were only on the hollow and under parts of the stones, but not upon those which lay most open to the skies.'

"This memorable shower of blood was produced by the *Vanessa urticæ* or *V. polychloros*, most probably, since these species of butterflies are said to have been uncommonly plentiful at the time when, and in the particular district where, the phenomenon was observèd."

FRANK COWAN, Curious History of Insects.

FOR A DESIGN OF A BUTTERFLY RESTING ON A SKULL

" Creature of air and light,
Emblem of that which may not fade or die,
　　Wilt thou not speed thy flight,
To chase the south wind through the glowing sky ?
　　What lures thee thus to stay,
　　With Silence and Decay,
Fix'd on the wreck of cold Mortality ?

"The thoughts once chamber'd there
Have gather'd up their treasures, and are gone —
　　Will the dust tell us where
They that have burst the prison-house are flown ?
　　Rise, nursling of the day,
　　If thou wouldst trace their way —
Earth hath no voice to make the secret known.

" Who seeks the vanish'd bird
By the forsaken nest and broken shell ?—
　　Far thence he sings unheard,
Yet free and joyous in the woods to dwell.
　　Thou of the sunshine born,
　　Take the bright wings of morn !
Thy hope calls heavenward from yon ruin'd cell."

MRS. HEMANS.

SUBFAMILY PAPILIONINÆ

Butterfly.—Generally large, and often with the hind wings adorned by tail-like projections. The most characteristic structural feature of the group is the absence of the internal vein of the hind wings. The submedian vein occupies the position usually held in other subfamilies by the internal.

Early Stages.—In that portion of the group which includes the genus *Parnassius* and its allies, the caterpillars are not, so far as is known, provided with scent-organs, and pupation takes place upon the ground, or among loosely scattered leaves, which are interwoven, at the time of pupation, with a few strands of silk. The genus *Papilio* and its allies have large, fleshy, more or less cylindrical caterpillars, possessed of osmateria, or offensive scent-organs, and a general resemblance runs through the chrysalids of all species, which are attached by a button of silk at the anal extremity and supported in the middle by a silk girdle.

Genus PARNASSIUS, Latreille

(The Parnassians)

" Some to the sun their insect wings unfold,
  Waft on the breeze, or sink in clouds of gold;
  Transparent forms, too fine for mortal sight,
  Their fluid bodies half dissolv'd in light."
                                                POPE.

Butterfly.—Of medium size, with more or less diaphanous wings, generally white or yellow in color, marked with black spots and round pink or yellow spots, margined with black. The head is relatively small, thickly clothed with hairs. The antennæ are short and straight, having a gradually thickened club. The palpi are very thin, straight, and clothed with long hairs.

The wings are generally translucent on the margin, with a rounded apex. The upper radial is lacking. The subcostal is five-branched, the third, fourth, and fifth nervules being emitted from a common stalk which springs from the upper outer angle of the cell. The first subcostal nervule rises well before the end of the cell; the second from the same point from which the stalk which bears the other three nervules springs. The cell of the hind wing is evenly rounded at its outer extremity. The inner margin of the hind wing is more or less excavated.

Early Stages.—The egg is turban-shaped, flattened, profusely covered with small elevations, giving it a sha-greened appearance. The caterpillars have very small heads. They are flat-

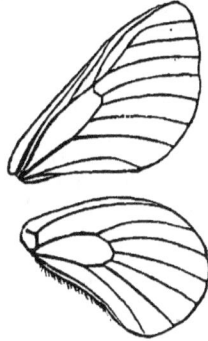

Fig. 149.—Neuration of the genus *Parnassius.*

tened, having a somewhat leech-like appearance; they are black or dark brown in color, marked with numerous light spots. The chrysalis is short, rounded at the head, and pupation takes place on the surface of the ground, among leaves and litter, a few loose threads of silk being spun about the spot in which transformation occurs.

The butterflies of this genus are classified with the *Papilionina*, because of the fact that the internal vein of the hind wings is always wanting, a characteristic of all papilionine genera.

(1) **Parnassius clodius,** Ménétries, Plate XXXIX, Figs. 7, ♂, ♂ ; Figs. 8, 10, ♀ (Clodius).

Butterfly.—The species may be distinguished from the following by the uniformly larger size and the more translucent outer margins of the fore wings in the male. Expanse, ♂, 2.50–2.75 inches; ♀, 2.50–3.00 inches.

Early Stages.—These await study. The egg and young larva were described by W. H. Edwards in the "Canadian Entomologist," vol. xi, p. 142, but we have no account of the later stages. The caterpillar feeds on *Sedum* and *Saxifraga*.

Clodius is found upon the mountains of California in spring and early summer. It is, like all its congeners, an alpine or boreal species.

(2) **Parnassius smintheus,** Doubleday and Hewitson, Plate XXXIX, Fig. 3, ♂ ; Fig. 4, ♀ ; var. **behri,** Edwards, Fig. 1, ♂ ; Fig. 2, ♀ ; var. **hermodur,** Henry Edwards, Fig. 6, ♀ ; *mate of hermodur,* Fig. 5, ♂ (Smintheus).

Butterfly.—This very beautiful insect is greatly subject to variation, and the plate shows a few of the more striking forms, of which the dark female, named *hermodur* by the late Henry Edwards, is one of the most beautiful. Expanse, ♂, 2.00–2.50 inches; ♀, 2.25–3.00 inches.

Smintheus is found at proper elevations upon the mountains from Colorado to California, and from New Mexico to Montana. The life-history is most exquisitely delineated by Edwards in "The Butterflies of North America," vol. iii.

The caterpillar feeds on *Sedum* and *Saxifraga.*

Genus PAPILIO, Linnæus

(The Swallowtails)

" The butterfly the ancient Grecians made
The soul's fair emblem, and its only name —
But of the soul, escaped the slavish trade
Of mortal life! For in this earthly frame
Ours is the reptile's lot — much toil, much blame,—
Manifold motions making little speed,
And to deform and kill the things whereon we feed."

COLERIDGE.

Butterfly.—Generally large, frequently with the hind wings tailed. A figure of the neuration characteristic of this genus is given on p. 20, Fig. 38. From this it will be seen that the internal vein of the hind wing is lacking, the submedian vein occupying the space which is commonly occupied by the internal vein. The median vein of the fore wing is connected with the submedian by a short vein, from the point of union of which with the submedian there proceeds a short internal vein in this wing. There is great diversity of form in the wings of this genus, some species even mimicking the species of the *Euplœinœ* and *Heliconidœ* very closely, and being entirely without tails. In all cases, however, in spite of obvious diversities in color and in form, there is substantial anatomical agreement in the structure of the wings; and the caterpillars and chrysalids reveal very strongly

306

Explanation of Plate XXXIX

1. *Parnassius smintheus*, Dbl.-Hew., var. *behri*, Edwards, ♂.
2. *Parnassius smintheus*, Dbl.-Hew., var. *behri*, Edwards, ♀.
3. *Parnassius smintheus*, Dbl.-Hew., ♂.
4. *Parnassius smintheus*, Dbl.-Hew., ♀.
5. *Parnassius smintheus*, Dbl.-Hew., ♂, mate of 4 *hermodur*.
6. *Parnassius smintheus*, Dbl.-Hew., var. *hermodur*, ♀, Henry Edwards.
7. *Parnassius clodius*, Ménétries, ♂ (*baldur*, Edwards).
8. *Parnassius clodius*, Ménétries, ♀ (*baldur*, Edwards).
9. *Parnassius clodius*, Ménétries, ♂.
10. *Parnassius clodius*, Ménétries, ♀.

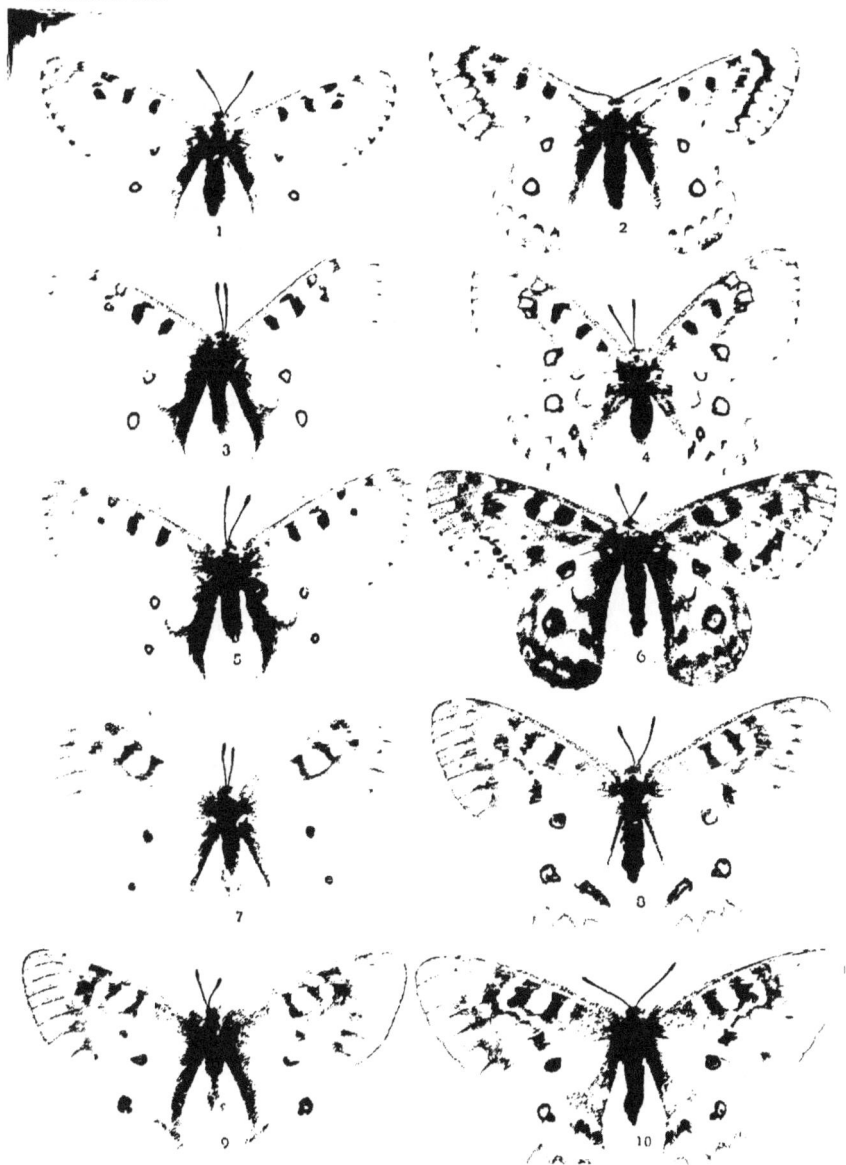

marked affinities throughout the whole vast assemblage of species, which at the present time includes about five hundred distinct forms.

Early Stages.—The eggs are somewhat globular, flattened at the base, and smooth. The caterpillars are cylindrical, smooth, fleshy, thicker in the anterior portion of the body than in the posterior portion, and are always provided with osmateria, or protrusive scent-organs, which, when the larva is alarmed, are thrust forth, and emit a musky odor, not highly disagreeable to the human nostrils, but evidently intended to deter other creatures from attacking them. The chrysalids are always attached by a button of silk at the anal extremity, and held in place by a girdle of silk about the middle. The chrysalids are, however, never closely appressed to the surface upon which pupation takes place.

There are about twenty-seven species of this genus found within the limits of boreal America. Our fauna is therefore much richer in these magnificently colored and showy butterflies than is the fauna of all Europe, in which but three species are known from the Dardanelles to the North Cape and Gibraltar. The genus is wonderfully developed in the tropics both of the New and the Old World, and has always been a favorite with collectors, containing many of the largest as well as the handsomest insects of the order.

(1) **Papilio ajax**, Linnæus, Plate II, Fig. 14, *larva;* Plate VI, Figs. 11, 12, *chrysalis* (Ajax).

Butterfly.—This insect, which is one of the most beautiful in our fauna, has been the subject of attentive study in recent years, and is now known to be seasonally polymorphic. We have given in Plate XLIV figures of several of the forms.

(*a*) Winter form **walshi;** Edwards, Plate XLIV, Fig. 4, ♂. In this form, which emerges from chrysalids which have been exposed to the cold of the winter, the black bands of the wings are narrower and a trifle paler than in the other forms, the tails of the hind wing tipped with white, and the crimson spot on the inner margin near the anal angle forming a conspicuous bent bar. A variety of this form, with a more or less distinct crimson line parallel to the inner margin on the upper side of the hind wing, has been named *Papilio ajax*, var. *abbotti*, by Edwards.

Another winter form, for which I propose the name **floriden-**

sis, is represented in Plate XLIV, Fig. 2, by a male specimen. It is characterized by the great breadth and intensity of the black bands on the upper side of the wings, which are quite as broad as in the summer form *marcellus*. I find this form prevalent in the spring of the year on the St. Johns River, in Florida. Expanse, 2.50–2.75 inches.

(*b*) Winter form **telamonides,** Felder, Plate XLIV, Fig. 1, ♂. In this form the tails of the hind wings are somewhat longer than in *walshi*, and are not simply tipped, but bordered on either side for half their length with white, and the red spots near the anal angle do not coalesce to form a crimson bar, but are separate. The black transverse bands on the upper side are wider than in *walshi*. Expanse, 2.75–3.00 inches.

(*c*) Summer form **marcellus,** Boisduval, Plate XLIV, Fig. 3, ♂. In this form, which represents the second generation emerging in the summer and fall from chrysalids produced from eggs of *walshi, floridensis,* and *telamonides,* the tails of the hind wings are greatly lengthened, being fully twice as long as in *walshi,* the black bands are greatly widened, and there is but a single small spot of crimson (sometimes none) above the anal angle of the secondaries. Expanse, 3.00–3.25 inches.

Early Stages.—These are well known. The caterpillar feeds on the leaves of the papaw (*Asimina triloba*), and wherever this plant is found the butterfly is generally common.

Ajax ranges from southern New England, where it is very rare, west and south over the entire country to the foot-hills of the Rocky Mountains. It is very common in the lower Appalachian region, and in southern Ohio, Indiana, Kentucky, and Tennessee is especially abundant.

(2) **Papilio eurymedon,** Boisduval, Plate XLIV, Fig. 5, ♂ (Eurymedon).

Butterfly. — This beautiful insect belongs to the same group as the four succeeding species. In the style of the markings it recalls *P. turnus,* but the ground-color is always pale whitish-yellow or white, the tails of the hind wings are more slender, and the white marginal spots on the under side of the fore wings are fused together, forming a continuous band. There are other differences, but these, with the help of the plate, will suffice for the ready identification of the species. Expanse, 3.50–4.00 inches.

Early Stages. — The caterpillar resembles that of *P. turnus*, but may be distinguished by its paler color and the much smaller spots composing the longitudinal series on the back and sides, and by the different color of the head. It feeds upon a variety of plants, and is especially partial to *Rhamnus californicus.*

The species ranges from Mexico to Alaska, and eastward as far as Colorado. It is abundant in the valleys of the Coast Range, and I have found it very common in the cañon of the Fraser River, in British Columbia, in the month of June.

(3) **Papilio rutulus,** Boisduval, Plate XLV, Fig. 1, ♂ (Rutulus).

Butterfly.—The insect very closely resembles the following species in color and markings, but the female is never dimorphic as in *P. turnus*, and the marginal spots on the under side of the fore wings run together, forming a continuous band, as in *eurymedon*, and are not separate as in *P. turnus.* By these marks it may always be distinguished. Expanse, ♂, 3.50–4.00 inches; ♀, 3.75–4.25 inches.

Early Stages.—These have been described with accuracy by W. H. Edwards in the second volume of his great work. The caterpillar differs from that of *P. turnus* in many minute particulars. It feeds on alder and willow. It is the representative on the Pacific coast of its Eastern congener, the common Tiger Swallowtail.

(4) **Papilio turnus,** Linnæus, Plate XLIII, Fig. 1, ♂ ; Fig. 2, dimorphic form **glaucus,** Linnæus, ♀ ; Plate II, Figs. 15, 26, 28, *larva;* Plate VI, Figs. 1–4, *chrysalis* (The Tiger Swallowtail).

Butterfly.—The "lordly Turnus" is one of the most beautiful insects of the Carolinian fauna. The plate shows the figures about one third smaller than in life, but they are sufficient for the immediate identification of the species. The species is dimorphic in the female sex in the southern portions of the territory which it occupies. The black form of the female was regarded for a long while as a distinct species, until by the test of breeding it was ascertained that some eggs laid by yellow females produced black females, and that, conversely, eggs laid by black females often produced yellow females. In Canada and northward and westward in northern latitudes the dark dimorphic female does not occur. A small yellow dwarfed form is common about Sitka, whence I have obtained numerous specimens. Expanse, ♂, 3.00–4.00 inches; ♀, 3.50–5.00 inches.

Early Stages.—The egg is outlined on p. 4, Fig. 3. It is green or bluish-green, quite smooth, with a few reddish spots in some specimens. The caterpillar feeds on a great variety of plants, but has a peculiar preference for the leaves of various species of wild cherry (*Cerasus*). The chrysalis is accurately portrayed in Plate VI, Figs. 1–4.

The metropolis of this species seems to be the wooded forests of the Appalachian ranges at comparatively low levels. It abounds in southwestern Pennsylvania, the Virginias, the Carolinas, Kentucky, and Tennessee. I have often found as many as a dozen of these magnificent butterflies congregated on a moist spot on the banks of the Monongahela. At Berkeley Springs, in West Virginia, I counted, one summer day, forty specimens hovering over the weeds and flowers in a small deserted field. The movements of the butterfly on the wing are bold and rapid. Its flight is dashing. Now aloft to the tops of the highest trees, now down in the shadows of the undergrowth, hither and thither it goes, often settling for a moment on some attractive flower, or staying its flight to quench its thirst on the sandy edge of a brook, and then away again over the fields and into the forests. In New England it is not very abundant, and in the Gulf States, while numerous, is still less common than about the head waters of the Ohio.

(5) **Papilio daunus**, Boisduval, Plate XXXVIII, Fig. 2, ♂ (Daunus).

Butterfly.—This magnificent species, which is even larger than *turnus* (the figures in the plate are greatly reduced), resembles the preceding species in color and markings, but may at once be distinguished by the two tails on the hind wing and the projection of the lobe at the anal angle of this wing. It is found among the eastern valleys of the Rocky Mountain ranges, and descends into Mexico. In Arizona it is quite common. Expanse, 4.00–5.25 inches.

Early Stages.—These have not yet been thoroughly studied, but what we know of them shows that the species is allied very closely to its immediate congeners, and the caterpillar feeds upon the same plants, principally *Rosaceæ*.

(6) **Papilio pilumnus**, Boisduval, Plate XXXVIII, **Fig.** 3, ♂ (Pilumnus).

Butterfly.—Resembling the preceding species, but smaller,

EXPLANATION OF PLATE XL

1. *Papilio asterias*, Cramer, ♂.   3. *Papilio hollandi*, Edwards, ♂.
2. *Papilio bairdi*, Edwards, ♂   4. *Papilio brucei*, Edwards, ♂.
                    5. *Papilio brevicauda*, Saunders, ♀.

having the bands and black margins of the wings decidedly broader, and the lobe of the anal angle of the hind wing so much lengthened as to give the wing the appearance of being furnished with three tails. Expanse, 3.80–4.30 inches.

Early Stages.—All we know of these is derived from the brief account given by Schaus in "Papilio," vol. iv, p. 100. Mr. Schaus says that the larva "feeds on laurel."

The insect is Mexican, and only occasionally occurs in Arizona.

(7) **Papilio thoas**, Linnæus, Plate XLII, Fig. 4, ♂ (Thoas).

Butterfly.—This species is readily distinguished from its near ally, *P. cresphontes*, by the greater and more uniform breadth of the median band of yellow spots traversing both the fore and the hind wing, and by the almost total absence of the curved submarginal series of spots on the primaries. There are other points of difference, but these are so marked as to make the determination of the species easy.

Early Stages.—These have never been fully described, but we know that the caterpillar feeds upon the leaves of the lemon, the orange, and other plants of the citrus group.

P. thoas is not common within the limits of the United States, where it is generally replaced by the following species; but it occasionally occurs in the hot lands of the extreme southern portion of Texas.

(8) **Papilio cresphontes**, Cramer, Plate XLII, Fig. 3, ♂ ; Plate II, Fig. 16, *larva;* Plate VI, Figs. 8–10, *chrysalis* (The Giant Swallowtail).

Butterfly.—The principal points of difference between this and the preceding species, its closest ally, have already been pointed out, and are brought into view upon the plate.

Early Stages.—These are quite well known. The caterpillar feeds upon *Ptelea, Xanthoxylon,* and various species of *Citrus.* It is very common in the orange-groves of Florida, where the people call the caterpillar the "orange-puppy," and complain at times of the ravages perpetrated by it upon their trees. It appears to have been gradually spreading northward, and in quite recent years has appeared at points in the Northern States where before it had never been observed. It has been recently taken in Ontario. It has become rather abundant in the vicinity of the city of Pittsburgh, where no observer had seen it prior to the year

1894. It is one of the largest and most showy species of the genus found within our faunal limits.

(9) **Papilio aliaska,** Scudder, Plate XLI, Fig. 1, ♂ (The Alaskan Swallowtail).

Butterfly.—This interesting form of the species, known to entomologists as *Papilio machaon,* Linnæus, and to every English school-boy as "the Swallowtail," represents a colonization from the Asiatic mainland of this insect, which is the sole representative of the genus on English soil. It differs from the English butterfly by having more yellow on the upper side of the wings, and by having the tails of the secondaries much shorter.

Early Stages.—Undoubtedly these are very much like those of the forms found in Europe and Asia, and the caterpillar must be sought upon umbelliferous plants.

Thus far this insect has been received only from Alaska, and is still rare in collections.

(10) **Papilio zolicaon,** Boisduval, Plate XXXVIII, Fig. 1, ♂ (Zolicaon).

Butterfly.—This species is somewhat nearly related to the preceding, but may at once be distinguished from it by the broader black borders of the wings, the deeper black on the upper side, and the longer tails of the secondaries. The figure given in the plate is only two thirds of the natural size.

Early Stages.—These have been fully described by Edwards, and are shown to be much like those of *P. asterias.* The caterpillar, like that of the last-mentioned species, feeds upon the *Umbelliferæ.*

Zolicaon ranges southward from Vancouver's Island to Arizona, and eastward to Colorado. It is more abundant in the valleys and foot-hills than on the sierras.

(11) **Papilio nitra,** Edwards, Plate XLI, Fig. 2, ♂ (Nitra).

Butterfly.—This insect, which is still very rare in collections, is very nearly related to the preceding species, it having, no doubt, with the succeeding species, sprung from the same original stock as *zolicaon* and *aliaska.*

Early Stages.—Unknown.

The insect occurs in Montana and the portions of British America adjacent on the north.

(12) **Papilio indra,** Reakirt, Plate XLI, Fig. 3, ♀ (Indra).

Butterfly.—Easily distinguished by the short tails of the secondaries, and the narrow bands of yellow spots on the wings closely resembling those found in the same location on the wings of *P. asterias*, ♂. Expanse, 2.50–2.75 inches.

Early Stages.—These still await description.

Indra occurs on the mountains of Colorado, Nevada, and California.

(13) **Papilio brevicauda,** Saunders, Plate XL, Fig. 5, ♀ (The Newfoundland Swallowtail).

Butterfly.—There are two varieties of this species—one with bright-yellow spots, one with the spots more or less deeply marked with orange-yellow on the upper sides of the wing. The latter variety is represented in the plate. The form with the yellow spots is common on the island of Anticosti; the other occurs quite abundantly in Newfoundland. Expanse, 2.75–3.00 inches.

Early Stages.—Both the caterpillar and the chrysalis show a very strong likeness to those of *P. asterias*. The larva feeds on umbelliferous plants.

The range of the species is confined to the extreme northeastern part of our faunal territory.

(14) **Papilio bairdi,** Edwards, Plate XL, Fig. 2, ♂ (Baird's Butterfly).

Butterfly.—This form, the male of which is represented in the plate, is the Western representative of *P. asterias*, and is characterized in general by the fact that the size is larger than that of *asterias*, and the postmedian band of yellow spots is broader. The female is generally darker and larger than that sex in *asterias*. Expanse, 3.25–3.50 inches.

Early Stages.—Not unlike those of *P. asterias*. The caterpillar feeds upon *Umbelliferæ*.

The seat of this species or form is Arizona, whence it ranges northward.

(15) **Papilio brucei,** Edwards, Plate XL, Fig. 4, ♂ (Bruce's Butterfly).

Butterfly.—This species, which is thought to be the result of a union between *P. oregonia* and *P. bairdi*, is found in Colorado. *Oregonia* is, unfortunately, not represented in our plates. It flies in Oregon and Washington, where *P. bairdi* is not found. In Colorado and adjacent regions meeting with the form *bairdi*, which ranges northward from Arizona, hybridization has occurred, and we have a fixed form breeding either toward *bairdi* or *oregonia*. To this form, characterized by more yellow on the bands of the wings than in *P. bairdi*, and less than in *oregonia*, Mr. Edwards has applied the name *P. brucei*, in honor of Mr. Bruce of Lockport, New York, who has done much to elucidate the problems connected with the species. Expanse, 3.25–3.60 inches.

Early Stages.—These have been fully described by Edwards. They are much like those of *asterias*, and the food-plants belong to the same class.

Bruce's Butterfly is found quite abundantly in Colorado.

(16) **Papilio hollandi,** Edwards, Plate XL, Fig. 3, ♂ (Holland's Butterfly).

Butterfly.—This species or form, which belongs to the Asterias-group, in the breadth of the yellow spots on the upper side of the wings holds a place intermediate between *P. bairdi* and *P. zolicaon*, between which it has been suggested that it may be a hybrid, which has become fixed, and therefore a species. It is characterized by the fact that the abdomen is always striped laterally with yellow or is wholly yellow. Expanse, 3.25–3.50 inches.

Early Stages.—We know as yet but little of these.

The insect occurs in Arizona and northward to Colorado.

(17) **Papilio asterias,** Fabricius, Plate XL, Fig. 1, ♂ ; Plate II, Figs. 17, 24, 27, *larva;* Plate VI, Figs. 13, 18, 19, *chrysalis* (The Common Eastern Swallowtail).

Butterfly.—The male is well represented in the plate. The female lacks the bright-yellow band of postmedian spots on the primaries, or they are but faintly indicated. The species is subject to considerable variation in size and the intensity of the markings. A very remarkable aberration in which the yellow spots cover almost the entire outer half of the wings has been found on several occasions, and was named *Papilio calverleyi* by Grote. The female of this form from the type in the author's collection is represented in Plate XLI, Fig. 6. Expanse, 2.75–3.25 inches.

Early Stages.—The caterpillar feeds on the *Umbelliferæ*, and is common on parsley and parsnips in gardens. In the South I have found that it had a special liking for fennel, and a few plants in the kitchen-garden always yielded me in my boyhood an abundant supply of the larvæ.

Explanation of Plate XLI

1. *Papilio machaon*, Linnæus, var. *aliaska*, Scudder, ♂.
2. *Papilio nitra*, Edwards, ♂.
3. *Papilio indra*, Reakirt, ♂.
4. *Papilio polydamas*, Linnæus, ♂.
5. *Papilio troilus*, Linnæus, ♂.
6. *Papilio asterias*, Cramer, var. *calverleyi*, Grote, ♀.

PLATE LI

P. asterias ranges all over the Atlantic States and the valley of the Mississippi.

(18) **Papilio troilus**, Linnæus, Plate XLI, Fig. 5, ♂ ; Plate II, Figs. 18, 19, 22, *larva;* Plate VI, Figs. 5–7, *chrysalis* (The Spice-bush Swallowtail).

Butterfly.—The upper side of the male is accurately depicted in the plate. The female has less bluish-green on the upper side of the hind wings. Expanse, 3.75–4.25 inches.

Early Stages.—The caterpillar lives upon the leaves of the common spicewood and sassafras, and draws the edges of a leaf together, thus forming a nest in which it lies hidden.

The insect is found throughout the Atlantic States and in the Mississippi Valley.

(19) **Papilio palamedes**, Drury, Plate XLII, Fig. 1, ♀ (Palamedes).

Butterfly.—The upper side of the wings is very accurately depicted in the figure just cited. On the under side the predominant tint is bright yellow. Expanse, 3.50–4.25 inches.

Early Stages.—These are described by Scudder in the third volume of his work on "The Butterflies of New England." The caterpillar feeds on *Magnolia glauca*, and on plants belonging to the order *Lauraceæ*.

The insect ranges from southern Virginia, near the coast, to the extreme southern end of Florida, and westward to southern Missouri and eastern Texas.

(20) **Papilio philenor**, Linnæus, Plate XLII, Fig. 2, ♂ ; Plate II, Figs. 13, 20, 21, *larva;* Plate VI, Figs. 14, 17, 20, *chrysalis* (The Pipe-vine Swallowtail).

Butterfly.—The figures in the plates obviate the necessity for describing this familiar but most beautiful insect, the glossy blue-green of which flashes all summer long in the sunlight about the verandas over which the *Aristolochia* spreads the shade of its great cordate leaves. Expanse, 3.75–4.25 inches.

Early Stages.—The caterpillar feeds upon the leaves of *Aristolochia sipho* (the Dutchman's-pipe) and *Aristolochia serpentaria*, which abound in the forest lands of the Appalachian region.

Philenor is always abundant during the summer months in the Middle Atlantic States, and ranges from Massachusetts to Arizona, into southern California and southward into Mexico. It is double-brooded in western Pennsylvania, and the writer has found females ovipositing as late as October. The caterpillars are familiar objects about houses on which the *Aristolochia* is grown as an ornamental vine.

(21) **Papilio polydamas**, Linnæus, Plate XLI, Fig. 4, ♂ (Polydamas).

Butterfly.— Easily distinguished by the absence of tails on the hind margin of the secondaries. The butterfly recalls the preceding species by the color of the wings on the upper side. On the under side the fore wings are marked as on the upper side; the hind wings have a marginal row of large red spots. Expanse, 3.00–3.50 inches.

Early Stages.—The caterpillar is dark brown, and in many points resembles that of *P. philenor* in outline, but the segments are spotted with ocellate yellow and red spots. It feeds on various species of *Aristolochia*. The chrysalis resembles that of *P. philenor*.

This lovely insect represents in the United States a great group of butterflies closely allied to it, which are natives of the tropics of the New World. It occurs in southern Florida and Texas, and thence ranges southward over Cuba, Mexico, and Central America.

THE CATERPILLAR AND THE ANT

" A pensy Ant, right trig and clean,
 Came ae day whidding o'er the green,
 Where, to advance her pride, she saw
 A Caterpillar, moving slaw.
 ' Good ev'n t' ye, Mistress Ant,' said he;
 ' How 's a' at hame ? I 'm blyth to s' ye.'
 The saucy Ant view'd him wi' scorn,
 Nor wad civilities return;
 But gecking up her head, quoth she,
 ' Poor animal ! I pity thee;
 Wha scarce can claim to be a creature,
 But some experiment o' Nature,
 Whase silly shape displeased her eye,
 And thus unfinish'd was flung bye.
 For me, I 'm made wi' better grace,
 Wi' active limbs and lively face;
 And cleverly can move wi' ease
 Frae place to place where'er I please;

316

Explanation of Plate XLII

1. *Papilio palamedes*, Drury, ♂.
2. *Papilio philenor*, Linnæus, ♂.

3. *Papilio cresphontes*, Cramer, ♂.
4. *Papilio thoas*, Linnæus, ♂.

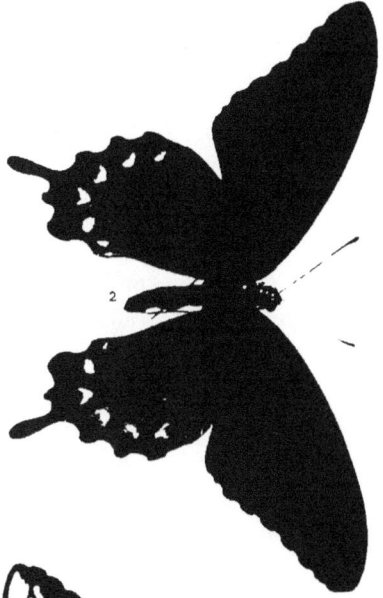
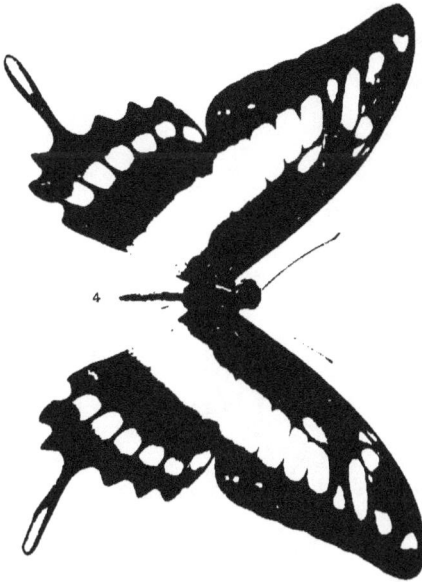

Can foot a minuet or jig,
And snoov 't like ony whirly-gig;
Which gars my jo aft grip my hand,
Till his heart pitty-pattys, and —
But laigh my qualities I bring,
To stand up clashing wi' a thing,
A creeping thing the like o' thee,
Not worthy o' a farewell t' ye.'
The airy Ant syne turned awa,
And left him wi' a proud gaffa.
The Caterpillar was struck dumb,
And never answered her a mum:
The humble reptile fand some pain,
Thus to be banter'd wi' disdain.
   But tent neist time the Ant came by,
The worm was grown a Butterfly;
Transparent were his wings and fair,
Which bare him flight'ring through the air.
Upon a flower he stapt his flight,
And thinking on his former slight,
Thus to the Ant himself addrest:
' Pray, Madam, will ye please to rest?
And notice what I now advise:
Inferiors ne'er too much despise,
For fortune may gie sic a turn,
To raise aboon ye what ye scorn:
For instance, now I spread my wing
In air, while you 're a creeping thing.' "
ALLAN RAMSAY.

JE Keeler del.

317

FAMILY V

HESPERIIDÆ (THE SKIPPERS)

Butterfly.—The butterflies belonging to this family are generally quite small, with stout bodies, the thorax strongly developed in order to accommodate the muscles of flight. They are exceedingly rapid in their movements. Both sexes possess six feet adapted to walking, and the tibiæ of the hind feet, with few exceptions, have spurs. The lower radial vein of the hind wing in many of the genera is lacking, or is merely indicated by a fold in the wing. There is great variety in the form as well as in the coloration of the wings.

Egg.—The eggs, so far as we are acquainted with them, may be said to be, almost without exception, more or less hemispherical, with the flat section of the hemisphere serving as the base. They are sometimes smooth, but not infrequently ornamented with raised longitudinal ridges and cross-lines, the ornamentation in some cases being very beautiful and curious.

Caterpillar.—The caterpillars are cylindrical, smooth, tapering forward and backward from the middle, and generally possess large globular heads. They commonly undergo transformation into chrysalids which have an anal hook, or cremaster, in a loose cocoon woven of a few strands of silk.

This family, the study of which presents more difficulties than are presented by any other family of butterflies, is not very well developed in the Palæarctic Region, but finds its most enormous development in the Nearctic and Neotropical Regions. It is also very strongly developed in the Indo-Malayan and Ethiopian Regions. There are, at the present time, in the neighborhood of two thousand species belonging to this family which have been named and described.

318

Explanation of Plate XLIII

1. **Papilio turnus**, Linnæus, ♂.
2. **Papilio turnus**, Linnæus, dimorphic
   ♀, glaucus, Linnæus.

3. Colias criphyle, Edwards. = Colias
   hageni, Edwards, ♂, under side.
4. Pyrameis atalanta, Linnæus, ♂.

5. Epargyreus tityrus, Fabricius, ♂.

(The figures in this plate are reduced, being only three fourths of the natural size.)

SUBFAMILY PYRRHOPYGINÆ

THIS subfamily is composed of closely related genera which are found only in the New World. They may be easily recognized by the large blunt club of the antennæ. The cell of the fore wing is always very long, being two thirds the length of the costa; the lower radial vein usually rises from the end of the cell, a little above the third median nervule, and at a considerable remove from the upper radial. They are said when at rest to extend all their wings horizontally.

But one genus belonging to this subfamily is represented within the limits of the United States.

FIG. 150.— Head and antenna of *Pyrrho-pyge*, magnified 2 diameters.

Genus PYRRHOPYGE, Hübner

FIG. 151.—Neuration of the genus *Pyrrhopyge*.

Butterfly.—The neuration is as represented in the cut, and need not, therefore, be described at length. The club of the antennæ is thickened, usually bluntly pointed and bent into a hook.

(1) **Pyrrhopyge araxes**, Hewitson, Plate XLV, Fig. 9, ♂ (Araxes).

Butterfly.—Easily recognized from the figure in the plate. The hind wings are prevalently yellow on the under side. It is wholly unlike any other species found within the faunal limits with which this book deals. The wings expand about two inches. We have no knowledge whatever of the life-history of the insect. It occurs in southern Texas occasionally, but is quite common in Mexico and more southern countries.

SUBFAMILY HESPERIINÆ (THE HESPERIDS)

" Twine ye in an airy round,
      Brush the dew and print the lea;
   Skip and gambol, hop and bound."
         DRAKE, *The Culprit Fay.*

THIS subfamily falls into two groups:

Group A.—In this group the cell of the fore wing is always more than two thirds the length of the costa; the lower radial vein lies approximately equidistant between the third median nervule and the upper radial. The hind wing is frequently produced at the extremity of the submedian vein into a long tail or tooth-like projection. The fore wing is usually furnished in the male sex with a costal fold, but is never marked with a discal stigma, or bunch of raised scales. The antennæ always terminate in a fine point and are usually bent into a hook. The butterflies when at rest, for the most part, hold their wings erect, though some of them hold them extended horizontally.

Group B.—In this group the cell of the fore wing is less than two thirds the length of the costa, and the lower radial is always emitted from the end of the cell near the upper angle, much nearer to the upper radial than to the third median. The hind wings are often somewhat lobed at the anal angle, but never produced as in the first group. The antennæ are very seldom hooked.

Genus EUDAMUS, Swainson

Butterfly.—The antennæ terminate in a fine point bent into a hook at the thickest part of the club. The cell of the fore wing is very long. The discocellulars are inwardly oblique and on the same straight line, the upper discocellulars being reduced to a mere point. The lower radial is equidistant between the upper radial

320

and the third median nervule. The hind wing is without the lower radial and is always produced into a long tail.

Egg.—The egg is more nearly globular than is true in most of the genera, but is strongly flattened at the base and is marked with a number of transverse longitudinal ridges, somewhat widely separated, between which are finer cross-lines. The micropyle at the summit is deeply depressed.

Caterpillar.—The caterpillar is cylindrical, tapering rapidly from the middle forward and backward. The head is much larger than the neck and is distinctly bilobed.

Chrysalis.—The chrysalis is provided with a somewhat hooked cremaster, is rounded at the head, humped over the thorax, and marked on the dorsal side of the abdominal segments with a few small conical projections. The chrysalis is formed between leaves loosely drawn together with a few strands of silk.

FIG. 152.—Neuration of the genus *Eudamus*.

This genus is confined to the tropics of the New World, and is represented in the extreme southern portions of the United States by the species figured in our plate—*E. proteus*.

(1) **Eudamus proteus**, Linnæus, Plate XLV, Fig. 6, ♀ ; Plate II, Fig. 34, *larva;* Plate VI, Fig. 23, *chrysalis* (The Long-tailed Skipper).

Butterfly.—The upper side of the wings is brown, glossed with green at the base of both wings. The spots on the primaries of both sexes are alike, and are well represented in the plate. On the under side the wings are pale brown; the primaries are marked as on the upper side; the secondaries have the anal portion and the tail dark brown; in addition they are crossed by a short dark band at the end of the cell, and another similar but longer postmedian band, which does not quite reach the costa and loses itself below in the dark shade which covers the anal portion of the wing. About the middle of the costa of the hind wings are two small subquadrate black spots. Expanse, 1.60–1.75 inch.

Early Stages.—The plates give us representations based upon Abbot's drawings of the mature caterpillar and the chrysalis. The student who desires to know more may consult the pages of Scudder's "Butterflies of New England." The caterpillar feeds

upon leguminous plants, especially upon the *Wistaria* and various species of *Clitoria* (Butterfly-pea). It makes a rude nest for itself by drawing two of the leaves together with strands of silk.

The species is tropical and is found all over the tropics and subtropical regions of the New World, but ranges northward along the Atlantic sea-coast, being occasionally found as far north as New York City, where it has been taken in Central Park.

Genus PLESTIA, Mabille

Butterfly.—The club of the antennæ is flattened, sickle-shaped, terminating in a fine point. The male has a costal fold upon the fore wing. The lower radial is nearer to the upper radial than to the third median nervule. The hind wing is produced into a short tail. The fifth vein is wanting.

Early Stages.—Unknown.

This genus is peculiar to Mexico and Central America. But one species is found within our limits, and is confined to Arizona.

Fig. 153.—Genus *Plestia*. Antenna, magnified 2 diameters. Neuration.

(1) **Plestia dorus**, Edwards, Plate XLV, Fig. 11, ♂ (The Short-tailed Arizona Skipper).

Butterfly.—The upper side is accurately depicted in the plate. On the under side the wings are hoary. The spots of the upper side reappear, the lower spots of the primaries being partially lost in the broad honey-yellow tint which covers the inner margin of that wing. The secondaries are crossed by obscure dark-brown basal, median, and postmedian bands, portions of which are annular, or composed of ring-like spots. The anal angle is clouded with dark brown. Expanse, 1.50–1.60 inch.

Early Stages.—Unknown.

The species has been taken in considerable numbers in Arizona, and ranges thence southward into Mexico.

Genus EPARGYREUS, Hübner

Butterfly.—The antennæ have the club stout, gradually thickened, tapering to a fine point, and abruptly bent into a hook.

322

1. *Papilio ajax*, Linnæus, var. *telamonides*, Felder. ♂.

2. *Papilio ajax*, Linnæus, var. *floridensis*, Holland. ♂. (This is the dark form found in Florida in the early spring.)

3. *Papilio ajax*, Linnæus, var. *marcellus*, Boisduval, ♂.

4. *Papilio ajax* Linnæus, var. *walski*, ♂.

5. *Papilio eurymedon*, Boisduval, ♂.

The palpi are profusely covered with thick scales, in which the third joint is almost entirely concealed. The fore wing of the male is furnished with a costal fold; the hind wing is prominently toothed at the extremity of the submedian vein.

Egg.—The egg is elevated, hemispherical; that is to say, it is flattened at the base, rounded above, its height being almost equal to the width. It is marked by about ten narrow, greatly elevated longitudinal ridges, which sometimes fork below the summit, and between which are a multitude of fine cross-lines. The micropyle is greatly depressed.

Caterpillar.—The caterpillar closely resembles the caterpillar of the genus *Eudamus*, but the head is not as strongly bilobed.

Fig. 154.—Neuration of the genus *Epargyreus.*

Chrysalis.—The chrysalis likewise resembles the chrysalis of the genus *Eudamus ;* the cremaster, however, is not as strongly hooked as in that genus.

(1) **Epargyreus tityrus**, Fabricius, Plate XLIII, Fig. 5, ♂ ; Plate II, Figs. 30, 31, 33, *larva ;* Plate VI, Figs. 22, 25, 26, *chrysalis* (The Silver-spotted Skipper).

Butterfly.—This very common and beautiful insect may easily be recognized from the figure in the plate. The broad, irregular silvery spot on the under side of the hind wings distinguishes it at a glance from all other related species in our fauna. Expanse, 1.75–2.00 inches.

Early Stages.—These have been accurately described by several authors, and a very full account of them is contained in "The Butterflies of New England." The caterpillar feeds upon leguminous plants, and is especially common upon the *Wistaria*, which is grown about verandas, and on the common locust (*Robinia pseudacacia*). The caterpillar makes a nest for itself in the same manner as *Eudamus proteus.* Pupation generally takes place among fallen leaves or rubbish at the foot of the trees upon which the caterpillar has fed.

This butterfly has a wide range, extending to the Gulf, south of a line passing from Quebec to Vancouver, and ranging still farther south as far as the Isthmus of Panama. It is single-brooded in the North, and double- or triple-brooded in the South.

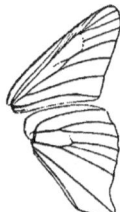

Genus THORYBES, Scudder
(The Dusky-wings)

Butterfly.—The club of the antennæ is not very heavy, hooked, the hooked portion about as long as the rest of the club. The palpi are directed forward, with the second joint heavily scaled, and the third joint very small. The fore wing may be with or without the costal fold in the male sex. The cut gives a correct idea of the neuration. The hind wing is evenly rounded on the outer margin, sometimes slightly angled at the extremity of the submedian vein.

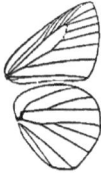

Egg.—The egg is subglobular, somewhat flattened at the base and on top, marked with numerous fine and not much elevated longitudinal ridges. The micropyle covers the upper surface of the egg and is not depressed.

FIG. 155.—
Neuration of the
genus *Thorybes*.

Caterpillar.—The caterpillar somewhat resembles that of the genus *Epargyreus*, but is relatively shorter, the head proportionately larger and more globular. The neck is greatly strangulated.

Chrysalis.—The chrysalis is somewhat curved in outline, with a strongly hooked cremaster and a prominent projection on the back of the thoracic region.

(1) **Thorybes pylades,** Scudder, Plate XLVIII, Fig. 6, ♀ ; Plate II, Figs. 25, 29, *larva;* Plate VI, Fig. 28, *chrysalis* (The Northern Dusky-wing).

Butterfly.—The upper side is represented correctly in Plate XLVIII. On the under side the wings are dark brown, shading into hoary-gray on the outer margins. The hind wings are crossed by irregular basal, median, and postmedian brown bands of darker spots, shaded with deeper brown internally. The translucent spots of the upper side reappear on the lower side of the fore wings. Expanse, 1.60 inch.

Early Stages.—These are elaborately described in the pages of Dr. Scudder's great work. The caterpillar feeds on clover, *Lespedeza*, and *Desmodium*.

This insect is found throughout the United States and Canada, but is not as yet reported from the central masses of the Rocky

Mountain region. It probably, however, occurs there also in suitable locations. It is very common in New England.

(2) **Thorybes bathyllus**, Smith and Abbot, Plate XLVIII, Fig. 5, ♀ ; Plate II, Fig. *32, larva;* Plate VI, Fig. 24, *chrysalis* (The Southern Dusky-wing).

Butterfly.—Easily distinguished from the preceding species by the much larger size of the translucent spots on the fore wings. Expanse, 1.40-1.50 inch.

Early Stages.—The habits of the larva are very similar to those of the preceding species, and the caterpillar feeds on herbaceous leguminosæ.

It ranges from the Connecticut Valley, where it is rare, southward along the coast and through the Mississippi Valley as far south and west as Texas.

(3) **Thorybes æmilia**, Skinner, Plate XLVI, Fig. *39,* ♂ (Mrs. Owen's Dusky-wing).

Butterfly.—This little species, which may readily be identified by the figure of the type given in the plate, is as yet quite rare in collections. We know nothing of the early stages. The types were taken at Fort Klamath, in Oregon. Dr. Skinner named it in honor of the estimable wife of Professor Owen of the University of Wisconsin, the discoverer of the species. Expanse, 1.20 inch.

(4) **Thorybes epigena**, Butler, Plate XLVIII, Fig. 13, ♂ (Butler's Dusky-wing).

Butterfly.—Readily distinguished by its large size, the conspicuous white fringes of the hind wings on the upper side, and the broad white marginal band of these wings on the under side. Expanse, 2.00 inches.

Early Stages.—Unknown.

This insect is common in Arizona and Mexico.

Genus ACHALARUS, Scudder

Butterfly.—The antennæ and palpi are as in the preceding genus. The neuration is represented in the cut. The hind wing is slightly lobed at the anal angle; the fore wing may or may not be provided with a costal fold.

(1) **Achalarus lycidas**, Smith and Abbot, Plate XLV, Fig. 10,

♀, *under side;* Plate II, Fig. 23, *larva;* Plate VI, Fig. 21, *chrysalis* (The Hoary-edge).

Butterfly.—The general appearance of the upper side of the wings strongly recalls *E. tityrus*, but the hoary edge of the secondaries and the absence of the broad median silvery spot found in *tityrus* at once serve to discriminate the two forms. Expanse, 1.65–1.95 inch.

Early Stages.—What is known of them may be ascertained by consulting the pages of "The Butterflies of New England." The caterpillar is found on the leaves of *Desmodium* (Beggar's-lice).

The insect is rare in southern New England, and ranges thence southward and westward to Texas, being scarce in the Mississippi Valley north of Kentucky, and apparently not ranging west of Missouri.

Fig. 156.—Neuration of the genus *Achalarus.*

(2) **Achalarus cellus**, Boisduval and Leconte, Plate XLV, Fig. 12, ♂ (The Golden-banded Skipper).

Butterfly.—The figure in the plate will enable the instant identification of this beautiful species, which, on the under side, has the hind wings banded much as in *E. proteus.* Expanse, 2.00 inches.

Early Stages.—What little we know of these is based mainly upon the observations of Abbot, and there is an opportunity here for some young naturalist to render a good service to science by rearing the insect through all stages from the egg. The habits of the larva are not greatly different from those of allied species.

A. cellus is found in the Virginias, and thence southward and westward to Arizona and Mexico. It is common in the Carolinas.

Genus HESPERIA, Fabricius

Butterfly.—The antennæ are relatively short; the club is stout and blunt at the tip. The palpi are bent upward, with the third joint buried in the scales covering the second joint. The hind wing is usually evenly rounded. In all the American species the male is provided with a costal fold. The neuration is represented in the cut.

Egg.—Hemispherical, ribbed.

326

Caterpillar.—The caterpillar is much like those which have been previously described, but is relatively much smaller.

Chrysalis.—The chrysalis has a somewhat blunt and not very distinctly developed cremaster.

(1) **Hesperia domicella,** Erichson, Plate XLVII, Fig. 19, ♂ (Erichson's Skipper).

Butterfly.—Allied to the following species, from which it is easily discriminated by the broad, solid white bands on both the fore and the hind wings. Expanse, 1.25 inch.

Early Stages.—Unknown.

H. domicella is found in Arizona, Mexico, and southward.

(2) **Hesperia montivaga,** Reakirt, Plate XLVII, Fig. 18, ♂ ; Plate VI, Fig. 35, *chrysalis* (The Checkered Skipper).

Fig. 157.—Genus *Hesperia.* Neuration. Antenna, magnified 3 diameters.

Butterfly.—The upper side is correctly delineated in the plate. The under side of the fore wings is much paler than the upper side, but with all the spots and markings of that side reproduced. The hind wings are creamy-white, crossed by median, postmedian, and marginal irregular bands of ochreous, somewhat annular spots. There is a triangular black spot at the anal angle of the secondaries. Expanse, 1.15 inch.

Early Stages.—We know little of these. The caterpillar probably feeds on malvaceous plants, as do most of the species of the genus.

The insect ranges from the Middle States to Arizona, and westward to the Rocky Mountains.

(3) **Hesperia centaureæ,** Rambur, Plate XLVII, Fig. 13, ♂ (The Grizzled Skipper).

Butterfly.—The upper side may easily be recognized by the help of the figure in the plate. On the under side the wings are darker than in the preceding species; the spots of the primaries reappear on this side, the submarginal curved row of spots coalescing to form a narrow white band, the white spot at the end of the cell flowing around the dark spot, which it only partly incloses on the upper side, and forming an eye-like spot. The hind wings are brown, scaled with green, and crossed by basal, median, and marginal bands of quadrate spots. The fringes are whitish, checkered with gray. Expanse, 1.15 inch.

Early Stages.—These await description.

This species, which originally was believed to be confined to Scandinavia and Lapland in Europe, and to eastern Labrador in this country, is now known to have a wide range in North America, extending from Labrador to the Carolinas on the Appalachian ranges, and occurring on the Rocky Mountains from British Columbia to southern Colorado.

(4) **Hesperia cæspitalis**, Boisduval, Plate XLVII, Fig. 14, ♀ (The Two-banded Skipper).

Butterfly.—On the upper side strongly resembling the preceding species, but the inner row of white spots on the hind wings is more complete. On the under side the fore wings are black, crossed by a double row of white spots, as on the upper side, these spots standing out conspicuously on the dark ground. The hind wings on the under side are more or less ferruginous, with the white spots more or less conspicuous. The fringes are checkered white and gray. Expanse, 1.00 inch.

Early Stages.—But little is known concerning these.

The species occurs in California, Oregon, and Nevada.

(5) **Hesperia xanthus**, Edwards, Plate XLVII, Fig. 15, ♂ (The Xanthus Skipper).

Butterfly.—Resembling the preceding species, but easily distinguished by the larger size of all the spots on the upper side of the wing and the paler under side, the secondaries being marked somewhat as in *H. montivaga*. Expanse, 1.00 inch.

Early Stages.—Hitherto undescribed.

The species has thus far been received only from southern Colorado, but undoubtedly will be found elsewhere in that portion of the land.

(6) **Hesperia scriptura**, Boisduval, Plate XLVII, Fig. 12, ♀ (The Small Checkered Skipper).

Butterfly.—Quite small. The hind wings on the upper side are almost entirely dark gray, the only white mark being a spot or two at the end of the cell. The fore wings are marked on this side as in the two foregoing species. On the under side the fore wings are blackish toward the base, with the costa, the apex, and the outer margin narrowly whitish. The hind wings below are pale, with an incomplete median band of white spots and broad white fringes, which are not checkered with darker color as in the preceding species. Expanse, .85 inch.

Early Stages.—Unknown.

The habitat of this species is southern Colorado, New Mexico, and Arizona.

(7) **Hesperia nessus**, Edwards, Plate XLVII, Fig. 17, ♂ (Nessus).

Butterfly.—This singularly marked little species, which probably might be separated from this genus on account of the slender and prolonged palpi, and no doubt would be by some of the hair-splitting makers of genera, I am content to leave where it has been placed by recent writers. It can be readily recognized by the figure in the plate, as there is nothing else like it in our fauna. Expanse, .80 inch.

Early Stages.—Unknown.

Nessus occurs in Texas and Arizona.

There are a few other species of this genus found within the limits of the United States, but enough have been represented to give a clear conception of the characteristics of the group, which is widely distributed throughout the world.

Genus SYSTASEA, Butler

Butterfly.—The palpi are porrect, the third joint projecting forward, the second joint densely scaled below. The antennæ are slender, the club moderately stout, somewhat bluntly pointed, bent, not hooked. The hind wings are somewhat crenulate, and deeply excised opposite the end of the cell. The fifth vein is lacking. In the fore wing the lower radial arises from a point nearer the upper radial than the third median nervule. The fore wings are crossed about the middle by translucent spots or bands.

Early Stages.—The early stages are unknown.

(1) **Systasea zampa**, Edwards, Plate XLVI, Fig. 1, ♂ (Zampa).

Fig. 158.—
Neuration of
the genus *Sys-
tasea*.

Butterfly.—The wings on the upper side are ochreous, mottled and clouded with dark brown. The primaries are marked about the middle and before the apex by translucent transverse linear spots. In addition there are a number of pale opaque spots on the primaries. The secondaries are traversed by

329

a pale submarginal whitish line. The under side of the wings is pale, with the light markings of the upper side indistinctly separated. Expanse, 1.10–1.25 inch.

Early Stages.—Unknown.

This interesting little species occurs in Arizona and northern Mexico.

Genus PHOLISORA, Scudder

Butterfly.—The palpi are porrect, the second joint loosely scaled, the third joint slender and conspicuous. The antennæ have the club gradually thickened, the tip blunt. The fore wing is relatively narrow, provided with a costal fold in the case of the male. The cut gives a correct idea of the neuration.

Egg.—The egg is curiously formed, much flattened at the base, marked on the side with longitudinal ridges and cross-lines, these ridges developing alternately at their apical extremities into thickened, more or less rugose elevations, the ridges pointing inwardly and surrounding the deeply depressed micropyle.

Fig. 150.—Neuration of the genus *Pholisora.*

Caterpillar.—Slender, with the head broad, rounded; the body stout, thickest in the middle, tapering toward either end, and somewhat flattened below.

Chrysalis.—The chrysalis is slender, very slightly convex on the ventral side, somewhat concave on the dorsal side behind the thorax. The wing-cases are relatively smaller than in the preceding genera.

(1) **Pholisora catullus,** Fabricius, Plate XLV, Fig. 4, ♂ ; Plate VI, Figs. 29, 36, 41, *chrysalis* (The Sooty-wing).

Butterfly.—Black on both sides of the wings, with a faint marginal series and a conspicuous submarginal series of light spots on the primaries in the male sex on the upper side, and, in addition to these, in the female sex, a faint marginal series on the secondaries. On the under side only the upper spots of the submarginal series of the primaries reappear. Expanse, .80–1.15 inch.

Early Stages.—The caterpillar feeds on "lamb's-quarter" (*Chenopodium album*) and the *Amarantaceæ*. It forms a case for itself by folding the leaf along the midrib and stitching the edges

Explanation of Plate XLV

1. *Papilio rutulus*, Boisduval, ♂.
2. *Pholisora alpheus*, Edwards, ♂.
3. *Calpodes ethlius*, Cramer, ♀.
4. *Pholisora catullus*, Fabricius, ♂.
5. *Thanaos afranius*, Lintner, ♂.
6. *Eudamus proteus*, Linnæus, ♂.
7. *Thanaos brizo*, Boisd.-Lec., ♀.
8. *Thanaos clitus*, Edwards, ♀.
9. *Pyrrhopyge araxes*, Hewitson, ♂.
10. *Achalarus lycidas*, Smith and Abbot, ♀, under side.
11. *Plestia dorus*, Edwards, ♀.
12. *Achalarus cellus*, Boisd.-Lec., ♂.

together with a few threaos of silk. It lies concealed during the day and feeds at night. A minute account of all its peculiarities is given by Scudder in "The Butterflies of New England," vol. ii, p. 1519.

The insect ranges over the whole of temperate North America.

(2) **Pholisora hayhursti**, Edwards, Plate XLVIII, Fig. 16, ♀ (Hayhurst's Skipper).

Butterfly.—Easily distinguished from the preceding species by the somewhat crenulate shape of the outer margin of the hind wings, the white color of the under side of the abdomen, and the different arrangement of the white spots on the fore wings, as well as by the dark bands which cross both the fore and the hind wings on the upper side. Expanse, .90–1.15 inch.

Early Stages.—Our information as to these is incomplete.

The species ranges from the latitude of southern Pennsylvania westward and southward to the Gulf, as far as the Rocky Mountains.

(3) **Pholisora libya**, Scudder, Plate XLVIII, Fig. 14, ♂ (The Mohave Sooty-wing).

Butterfly.—Easily distinguished from the two preceding species by the white fringes of the wings and by the markings of the under side. The primaries on the lower side are dark, tipped at the apex with light gray, and in the female having the costa and the outer margin broadly edged with light gray. The hind wings are pale gray of varying shades, marked with a number of large circular white spots on the disk and a marginal series of small white spots. Expanse, ♂, .80–1.25 inch; ♀, 1.15–1.40 inch.

Early Stages.—These await full description.

This species is found from Nevada to Arizona, and is apparently very common in the Mohave Desert.

(4) **Pholisora alpheus**. Edwards, Plate XLV, Fig. 2, ♂ (Alpheus).

Butterfly.—This little species is nearer *P. hayhursti* than any of the others we have described, but may at once be recognized and discriminated by the checkered margins and white tip of the fore wing and the linear shape of the spots composing the submarginal and median bands on the upper side of this wing. The hind wings on the under side are marked with a number of light spots arranged in marginal and median bands.

Early Stages.—Unknown.

Alpheus occurs in Nevada, Arizona, and New Mexico.

There are four other species of the genus found in our fauna.

Genus THANAOS, Boisduval
(The Dusky-wings)

Butterfly.—The antennæ have a moderately large club, curved, bluntly pointed. The palpi are porrect, the third joint almost concealed in the dense hairy vestiture of the second joint. The neuration of the wings is represented in the cut. The fore wing in the case of the male always has a costal fold. The butterflies comprised in this genus are all, without exception, dark in color, in a few species having bright spots upon the hind wings.

The genus reaches its largest development in North America. The discrimination of the various species is somewhat difficult.

Fig. 100.—
Neuration of the
genus *Thanaos.*

Egg.—The egg is somewhat like the egg in the genus *Achalarus*, but the micropyle at the upper end of the egg is relatively larger and not as deeply depressed below the surface. The sides are ornamented, as in *Achalarus*, by raised vertical ridges, between which are numerous cross-ridges; in a few cases the vertical ridges are beaded, or marked by a series of minute globose prominences, upon the edge.

Caterpillar.—The caterpillars are cylindrical, tapering from the middle forward and backward, marked with lateral and dorsal stripes, with the neck less strangulated than in the preceding genera.

Chrysalis.—Not greatly differing in outline from the chrysalis of the preceding genera, in most species having the outline of the dorsum straight on the abdominal segments, with the thoracic segments forming a slight hump or elevation; convex on the ventral side, the cremaster being usually well developed.

(1) **Thanaos brizo**, Boisduval and Leconte, Plate XLV, Fig. 7, ♀ ; Plate VI, Fig. 38, *chrysalis* (The Sleepy Dusky-wing).

Butterfly.—The band of postmedian spots on the fore wing is composed of annular dark markings, is regular, crosses the wing from the costa to the hind margin, and is reproduced on

332

the under side as a series of pale-yellowish spots more or less distinct. The hind wings have a double series of faint yellow spots; these as well as the marginal spots of the primaries are very distinct on the under side. Expanse, 1.25–1.60 inch.

Early Stages.—The caterpillar feeds on oaks, *Galactia*, and possibly *Baptisia*. The life-history has been only partially ascertained, in spite of the fact that the insect has a wide range and is not uncommon.

Brizo occurs from the Atlantic to the Pacific, ranging from the latitude of New England to that of Arizona.

(2) **Thanaos icelus**, Lintner, Plate XLVIII, Fig. 17, ♂ ; Plate VI, Fig. 27, *chrysalis* (The Dreamy Dusky-wing).

Butterfly.—Prevalently smaller in size than the preceding species. The under side of the wings is paler than the upper side, and the outer third of both the primaries and secondaries is marked with a profusion of small indistinct yellow spots, which do not form well-defined bands as in the preceding species. On the upper side of the fore wing the median area is generally marked by a broad band of pale gray, but this is not invariably the case. Expanse, 1.00–1.20 inch.

Early Stages.—These have been described by Scudder. The caterpillar feeds on a variety of plants, as the aspen, oaks, and witch-hazel.

Icelus ranges across the continent from Nova Scotia to Oregon, and south to Florida and Arizona.

(3) **Thanaos somnus**, Lintner, Plate XLVIII, Fig. 2, ♂ (The Dark Dusky-wing).

Butterfly.—A little larger than the preceding species, especially in the female sex. The male is generally quite dark, the banding of the fore wing on the upper side obscured. The hind wings have a row of light marginal and submarginal spots, more distinct on the under side than on the upper. The female generally is light gray on the upper side of the wings, with broad median and submarginal bands of dark brown, tending to fuse or coalesce at a point near the origin of the first median nervule. Expanse, ♂, 1.25 inch; ♀, 1.50 inch.

Early Stages.—But little is known of these.

All of the specimens I have ever seen came from southern Florida.

(4) **Thanaos lucilius**, Lintner, Plate XLVIII, Fig. 10, ♂ ; Plate VI, Figs. 30–32, *chrysalis* (Lucilius' Dusky-wing).

Butterfly.—This species may be distinguished from *T. pacuvius*, a near ally, by the more mottled surface of the secondaries, which in *pacuvius* are almost solidly black; and from *T. martialis,* another close ally, by the absence of the purplish-gray cast peculiar to both sides of the wings of the latter species, and the less regular arrangement of the bands of spots on the upper side of the fore wings. The plate does not show these delicate but constant marks of difference as well as might be desired. Expanse, 1.20–1.40 inch.

Early Stages.—Dr. Scudder has fully described these. The caterpillar feeds on the columbine (*Aquilegia canadensis*).

Lucilius ranges from New England to Georgia, is common in western Pennsylvania and West Virginia, and extends westward at least as far as the Rocky Mountains.

(5) **Thanaos persius,** Scudder, Plate XLVIII, Fig. 1, ♂; Plate VI, Fig. 34, *chrysalis* (Persius' Dusky-wing).

Butterfly.—This is a very variable species, some specimens being light and others dark in color. There is scarcely any positive clue to the specific identity of the insect except that which is derived from the study of the genital armature of the male, which is a microscopic research capable of being performed only by an expert in such matters. The student may be pardoned if, in attempting to classify the species of this genus, and the present species in particular, he should grow weary, and quote a few biblical expressions relating to Beelzebub, the " god of flies." Expanse, 1.20–1.45 inch.

Early Stages.—The caterpillar feeds on willows. Scudder has with patient care described its life-history.

The insect ranges from New England southward, and inland across the continent to the Pacific.

(6) **Thanaos afranius,** Lintner, Plate XLV, Fig. 5, ♂ (Afranius' Dusky-wing).

Butterfly.—Closely related to the preceding species. The hind wings on the upper side in the male sex are almost solid black, the fringes paler. On the under side there is a double row of light spots along the margin of the hind wing in both sexes. The female is generally paler in color on the upper side than the male.

Early Stages.—Unknown.

All the specimens I have seen come from Arizona, where the thing is apparently common.

(7) **Thanaos martialis,** Scudder, Plate XLVIII, Fig. 4, ♂ ; Plate VI, Fig. 37, *chrysalis* (Martial's Dusky-wing).

Butterfly.—The upper side of the wings is paler than in most species, and has a distinctly purplish-gray cast. The fore wings are crossed by irregular bands of dark spots. The hind wings on the outer half are profusely mottled with small pale spots. All the light spots are repeated on the under side of both wings, and are more distinct on this side than on the upper. Expanse, 1.25–1.40 inch.

Early Stages.—These are partly known. The caterpillar feeds on *Indigofera* and *Amaranthus.*

The species ranges from Massachusetts to Georgia, and westward to Missouri and New Mexico.

(8) **Thanaos juvenalis,** Fabricius, Plate XLVIII, Fig. 11, ♀ ; Plate VI, Fig. 33, *chrysalis* (Juvenal's Dusky-wing).

Butterfly.—Larger than the preceding species. The wings have a number of translucent spots arranged as a transverse series beyond the middle of the wing. They are far more distinct and larger in the female than in the male. The under side of the wings is paler than the upper side, and profusely but indistinctly marked with light spots. Expanse, 1.35–1.60 inch.

Early Stages.—For a full knowledge of these the reader may consult the pages of "The Butterflies of New England." The caterpillar feeds on oaks and leguminous plants of various species.

This insect ranges from Quebec to Florida, and westward as far as Arizona, where it appears to be common.

(9) **Thanaos petronius,** Lintner, Plate XLVIII, Fig. 7, ♂ (Petronius' Dusky-wing).

Butterfly. —Allied in size to the preceding species, but the translucent spots of the transverse band are not, as in that species, continued toward the inner margin, but terminate at the first median nervule. The outer third of the primaries is pale, the inner two thirds very dark. The under side of the wings of the male is uniformly dusky, slightly, if at all, marked with lighter spots. The under side of the wings of the female is less distinctly marked with light spots than is the case in allied species. Expanse, 1.50–1.75 inch.

Early Stages.— Unknown.

The species has thus far been found only in Florida.

(10) **Thanaos horatius,** Scudder, Plate XLVIII, Fig. 15, ♂ (Horace's Dusky-wing).

Butterfly. — Smaller than *T. juvenalis,* which it resembles in the long transverse series of translucent spots. It is, however, paler on the upper side of the wings, and more profusely mottled on the hind wing both above and below, though there is considerable variation in this regard. Expanse, 1.65 inch.

Early Stages. — The caterpillar probably feeds on the *Leguminosœ.* We know very little about the life-history of the species.

The butterfly ranges from Massachusetts to Texas.

(11) **Thanaos nævius,** Lintner, Plate XLVIII, Fig. 3, ♀ (Nævius' Dusky-wing).

Butterfly. — This insect is closely allied to *T. petronius,* but the translucent spots on the fore wing are smaller, and there is generally a light spot near the costa before the three subapical translucent spots.

Early Stages. —Unknown.

The habitat of this species is the region of the Indian River, in Florida.

(12) **Thanaos pacuvius,** Lintner, Plate XLVIII, Fig. 9, ♀ (Pacuvius' Dusky-wing).

Butterfly. —Small, with the fore wings on the upper side rather regularly banded with dark brown upon a lighter ground. The hind wings are almost solid black above, with the fringes toward the anal angle pure white. Expanse, 1.15–1.30 inch.

Early Stages. —Unknown.

This species occurs in Colorado, Mexico, and Arizona.

(13) **Thanaos clitus,** Edwards, Plate XLV, Fig. 8, ♂ (Clitus).

Butterfly. —Larger than the preceding species. The hind wings are solidly deep black, fringed broadly with pure white. The fore wings of the male are dark, of the female lighter. Expanse, 1.60–1.75 inch.

Early Stages. —Unknown.

The habitat of this species is Arizona and New Mexico.

(14) **Thanaos funeralis,** Lintner, Plate XLVIII, Fig. 12, ♂ (The Funereal Dusky-wing).

Butterfly. —Closely allied to the preceding species, of which it may be only a smaller varietal form. Expanse, 1.35 inch.

Early Stages. —Unknown.

Funeralis occurs in western Texas and Arizona.

The genus *Thanaos* is one of the most difficult genera to work out in the present state of our knowledge of the subject. The species are not only obscurely marked, but they vary in the most extraordinary manner. Except by a microscopic examination of the genital armature, which can be carried on only when the student possesses considerable anatomical knowledge and an abundance of material, there is no way of reaching a satisfactory determination in many cases.

COLLECTIONS AND COLLECTORS

In almost every community there is to be found some one who is interested in insects, and who has formed a collection. The commonest form of a collection is exceedingly primitive and unscientific, in which a few local species are pinned together in a glass-covered box or receptacle, which is then framed and hung upon the wall. Almost every village bar-room contains some such monstrous assemblage of insects, skewered on pins, in more or less frightful attitudes. As evidencing an innate interest in the beauties of natural objects, these things are interesting, but show a want of information which, as has been already pointed out, is largely due to a lack of literature relating to the subject in this country. In many of the schools of the land small collections, arranged more scientifically, have been made, and some of the collections contained in the high schools of our larger towns and cities are creditable to the zeal of teachers and of pupils. There is no reason why every school of importance should not, in the lapse of time, secure large and accurately named collections, not only of the insects, but of the other animals, as well as the plants and minerals of the region in which it is located. Every high school should have a room set apart for the use of those students who are interested in the study of natural history, and they ought to be encouraged to bring together collections which should be properly arranged and preserved. The expense is not great, and the practical value of the training which such studies impart to the minds of young people is inestimable.

The great systematic collections in entomology in the United States are for the most part in the hands of the museums and universities of the country. The entomological collections of the

United States government at Washington are large and rich in interesting material. The collections possessed by Harvard College and the Boston Society of Natural History are extensive; so are also the collections of the American Museum of Natural History, the Academy of Natural Sciences in Philadelphia, and those of the Carnegie Museum in Pittsburgh. The collection in the latter institution is altogether the largest and most perfect collection of the butterflies of North America in existence, and covers also very largely the butterflies of the world, there being about twelve thousand species of butterflies represented, including representatives of all known genera.

The formation of great collections has always had a charm for those who have possessed the knowledge, the time, and the means to form them; and the ranks of those who are engaged in the study of butterflies include many of the most famous naturalists, among them not a few of noble rank. One of the most enthusiastic collectors in Europe at the present time is the Grand Duke Nicholas of Russia. The Nestor among German collectors is Dr. Staudinger of Dresden. In France M. Charles Oberthür of Rennes is the possessor of the largest and most perfect collection on French soil. In England there are a number of magnificent collections, aside from the great collection contained in the British Natural History Museum. These are in the possession of Lord Walsingham, the Hon. Walter Rothschild, Mr. F. D. Godman, Mr. Herbert Druce, Mr. H. J. Elwes, and others, all of whom hold high rank in the domain of scientific research.

There are many men who make the collecting of natural-history specimens a business. They are among the most intrepid and indefatigable explorers of the present time. The late Henry W. Bates and Mr. Alfred Russel Wallace were in early life leaders in this work, and we are indebted to their researches for a knowledge of thousands of species. Two of the most successful collectors who have followed in their footsteps are Mr. Herbert H. Smith and Mr. William Doherty, both of them Americans; Mr. Smith one of the most enthusiastic and successful explorers in South and Central America, Mr. Doherty the most diligent explorer of the Indo-Malayan Region. The story of the travels and adventures of these two men is a tale full of romantic interest, which, alas! has been by neither of them fully told.

EXPLANATION OF PLATE XLVI

1. *Sislasea zampa*, Edwards, ?.
2. *Erynnis manitoba*, Scudder, ♂.
3. *Erynnis manitoba*, Scudder, ♀.
4. *Atalopedes huron*, Edwards, ♂.
5. *Atalopedes huron*, Edwards, ♀.
6. *Atrytone vitellius*, Smith and Abbot, ♂.
7. *Atrytone melane*, Edwards, ♂.
8. *Atrytone melane*, Edwards, ♀.
9. *Lerema bianna*, Scudder, ♂.
10. *Lerema bianna*, Scudder, ♀.
11. *Erynnis ottoë*, Edwards, ♂.
12. *Erynnis ottoë*, Edwards, ♀.
13. *Erynnis sassacus*, Harris, ♂.
14. *Phycanassa viator*, Edwards, ♂.
15. *Phycanassa viator*, Edwards, ♀.
16. *Limochores pontiac*, Edwards, ♂.
17. *Limochores pontiac*, Edwards, ♀.
18. *Hylephila phylæus*, Drury, ♂.
19. *Hylephila phylæus*, Drury, ♀.
20. *Atrytone hyssus*, Edwards, ♀.
21. *Limochores palatka*, Edwards, ♂.
22. *Thymelicus mystic*, Scudder, ♂.
23. *Thymelicus mystic*, Scudder, ♀.
24. *Atrytone delaware*, Edwards, ♂.
25. *Atrytone delaware*, Edwards, ♀.
26. *Erynnis morrisoni*, Edwards, ♂.
27. *Erynnis morrisoni*, Edwards, ♀.
28. *Thymelicus ætna*, Boisduval, ♂.
29. *Thymelicus ætna*, Boisduval, ♀.
30. *Limochores manataaqua*, Scudder, ♀.
31. *Euphyes metacomet*, Harris, ♂.
32. *Euphyes verna*, Edwards, ♂.
33. *Lerodea eufala*, Edwards, ♂.
34. *Prenes ocola*, Edwards, ♂.
35. *Oligoria maculata*, Edwards, ♂.
36. *Lerema carolina*, Skinner, ♂.
37. *Phycanassa aaroni*, Skinner, ♂.
38. *Phycanassa bouardi*, Skinner, ♂.
39. *Thocybes armilla*, Skinner, ♂.
40. *Limochores yehl*, Skinner, ♂.

PLATE XLVI.

SUBFAMILY PAMPHILINÆ

"Into the sunshine,
Full of light,
Leaping and flashing
From morn till night."
Russell.

THE *Pamphilinæ* found in our fauna fall into two groups.

Group A.—The antennæ are not greatly hooked and generally sharply pointed; the palpi have the third joint short and inconspicuous; the cell of the fore wing is always less than two thirds the length of the costa; the lower radial is somewhat nearer to the third median nervule than to the upper radial. The hind wing is often lobed. The lower radial in the hind wing is generally lacking. The male never has a costal fold on the fore wings, and but rarely is provided with a discal stigma.

But three genera belonging to this section of this subfamily are found in our fauna, namely, the genera *Amblyscirtes*, *Pamphila*, and *Oarisma*.

Group B.—The antennæ are sometimes curved, but never hooked, the palpi having the third joint minute, sometimes horizontally porrected. The cell of the fore wing is less than two thirds the length of the costa. The lower radial arises much nearer to the third median nervule than to the upper radial. The hind wing is elongated, but never tailed. The male is never provided on the fore wing with a costal fold, but is in many genera furnished with a discal stigma on the fore wing. When in a state of rest the majority of the species elevate their fore wings and depress their hind wings, an attitude which is peculiar to the insects of this group.

Genus AMBLYSCIRTES, Scudder

Butterfly.—The antennæ are short, with a moderately thick club, crooked at the end; the third joint of the palpi is bluntly

339

conical, short, and erect. The costa of the fore wing is straight, slightly curved inwardly before the apex. The neuration is represented in the cut.

Egg.— Hemispherical.

Caterpillar.— Not differing materially in its characteristics from the caterpillars of other hesperid genera.

Chrysalis.— Somewhat slender, with the dorsal and ventral outlines straighter than in any of the preceding genera, and the dorsum very slightly elevated in the region of the thoracic segments.

FIG. 161. — Neuration of the genus *Amblyscirtes.*

(1) **Amblyscirtes vialis**, Edwards, Plate XLVII, Fig. 5, ♂ ; Plate VI, Fig. 40, *chrysalis* (The Roadside Skipper).

Butterfly.—This little species, an exceptionally bright example of which is represented in the plate, may be known by the dark color of the upper surface, almost uniformly brown, with a few subapical light spots at the costa. In the specimen that is figured these light spots are continued across the wing as a curved band, but this is not usual. The wings on the under side in both sexes are very much as on the upper side, save that both wings on the outer third are lightly laved with gray. Expanse, 1.00 inch.

Early Stages.—These have been described with minute accuracy by Dr. Scudder.

The Roadside Skipper ranges from Montreal to Florida, and westward as far as Nevada and Texas. It is not a common species in the valley of the Mississippi; it seems to be far more common in southern New England and in Colorado. At all events, I have obtained more specimens from these localities than from any others.

(2) **Amblyscirtes samoset**, Scudder, Plate XLVII, Fig. 6, ♂ ; Plate VI, Fig. 45, *chrysalis* (Pepper-and-salt Skipper).

Butterfly.— This little species on the upper side has the ground-color as in the preceding species; the fringes on both wings are pale gray. There are three small subapical spots on the fore wing, three somewhat larger spots, one on either side of the second median nervule and the third near the inner margin, and two very minute spots at the end of the cell. On the under side the wings are pale gray, the white spots of the upper side of the fore wing reappearing. The hind wing is in addition marked by a semicircular median band of white spots, a small spot at the end of the cell, and another conspicuous white spot about the middle of the costa. Expanse, 1.00–1.10 inch.

Early Stages.—The caterpillar apparently feeds upon grasses. We know as yet very little of the life-history of the insect.

It is found in Maine, New Hampshire, along the summits of the Appalachian mountain-ranges as far south as West Virginia, and is reported to be common in Wisconsin and Michigan.

(3) **Amblyscirtes ænus,** Edwards, Plate XLVII, Fig. 7, ♀ (The Bronze Skipper).

Butterfly.—This obscure little species has the upper side of the wings somewhat tawny. The markings, which are similar to those in *A. samoset,* are not white, but yellow. The wings on the under side are darker than in *samoset.* The spots of the fore wing are the same, but the spots on the under side of the hind wing are different, and form a zigzag postmedian transverse band, with a single small spot at the end of the cell, and another of the same size beyond the middle of the costa. Expanse, 1.00–1.20 inch.

Early Stages.—These are unknown.

The species occurs in western Texas and Arizona.

(4) **Amblyscirtes simius,** Edwards, Plate XLVII, Fig. 8, ♂ (Simius).

Butterfly.—The upper side of the male is correctly figured in the plate. The wings on the under side are quite pale; the spots of the fore wing reappear on the under side, and the fore wing is blackish at the base; the hind wing has the angle at the base broadly white, with a broad white blotch at the end of the cell, and a semicircular curved band of obscure spots traversing the middle of the wing. Expanse, ♂, .90 inch; ♀, 1.20 inch.

Early Stages.—Unknown.

The species was originally described from Colorado.

(5) **Amblyscirtes textor,** Edwards, Plate XLVII, Fig. 16, ♂, *under side* (The Woven-winged Skipper).

Butterfly.—This little species, the under side of which is accurately delineated in the plate, needs no description to characterize it, as its peculiar markings serve at once to distinguish it from all other species. Expanse, 1.25–1.45 inch.

Early Stages.—Unknown.

This little insect ranges from North Carolina southward to Florida, Louisiana, and Texas.

Genus PAMPHILA, Fabricius

Butterfly.—The antennæ are very short, less than half the length of the costa. The club is stout, elongate, and blunt at its extremity; the palpi are porrect, densely clothed with scales, concealing the third joint, which is minute, slender, and bluntly conical. The body is long, slender, and somewhat produced beyond the hind margin of the secondaries. The neuration of the wings is represented in the cut.

Egg.—Hemispherical, vertically ribbed, the interspaces uniformly marked with little pitted depressions.

Fig. 162.—Neuration of the genus *Pamphila*.

Caterpillar.—The body is cylindrical, slender, tapering forward and backward; the neck less strangulated than in many of the genera. The body is somewhat hairy; the spiracles on the sides open from minute subconical elevations.

Chrysalis.—Not materially differing in outline and structure from the chrysalids of other genera which have already been described.

Only a single species belonging to the genus is found in North America.

(1) **Pamphila mandan,** Edwards, Plate XLXII, Fig. 1, ♂ (The Arctic Skipper).

Butterfly.—No description of this interesting little insect is necessary, as the figure in the plate will enable the student at once to distinguish it. It is wholly unlike any other species. Expanse, 1.10 inch.

Early Stages.—These have been described by Dr. Scudder and Mr. Fletcher. The caterpillar feeds on grasses.

The insect ranges from southern Labrador as far south as the White Mountains and the Adirondacks, thence westward, following a line north of the Great Lakes to Vancouver's Island and Alaska. It ranges southward along the summits of the Western Cordilleras as far as northern California.

Genus OARISMA, Scudder

Butterfly.—Closely related to the preceding genus. The antennæ are very short; the club is long, cylindrical, bluntly rounded at the apex, not curved. The palpi are stout, the apical joint very slender, elongated, and porrect. The head is broad; the body is long and slender, projecting somewhat beyond the posterior margin of the secondaries. The neuration of the wings is represented in the cut.

Early Stages.—So far as known to me the life-history of no butterfly of this genus has yet been ascertained.

(1) **Oarisma garita,** Reakirt, Plate XLVII, Fig. 3, ♂ (Garita).

Butterfly.—This obscure little insect is light fulvous on the upper side, with the costa of the hind wing somewhat broadly marked with leaden gray; on the under side the fore wings are brighter fulvous, with the inner margin laved with dark gray. The hind wings are paler fulvous, inclining to gray, with the inner margin brighter fulvous. Expanse, .75–1.00 inch.

Fig. 163.—Neuration of the genus *Oarisma.*

Early Stages.—We know little of these. The species is found in southern Colorado, ranging thence westward and southward to Arizona.

(2) **Oarisma powesheik,** Parker, Plate XLVII, Fig. 4, ♂ (Powesheik).

Butterfly.—This species may be distinguished from its ally *garita* by its larger size, the darker color of the upper side of the wings, and the red markings on the costa of the fore wings. On the under side the fore wings are black, edged on the costa and outer margin for a short distance below the apex with light fulvous. The hind wings are dusky, with the veins and nervules white, standing forth conspicuously upon the darker ground-color. Expanse, 1.00–1.25 inch.

Early Stages.—Unknown.

Powesheik occurs in Wisconsin, and ranges thence westward to Nebraska, northward to Dakota, and southward as far as Colorado.

EXCHANGES

One of the best ways of adding to a collection is by the method known as exchanging. A collector in one part of the country may find species which are rare, or altogether unknown, in another part of the country. By a system of exchanges with other collectors he is able to supply the gaps which may exist in his collection. No one, however, cares to effect exchanges with collectors who are careless or slovenly in the preparation of their specimens, or inaccurate in naming them. A collector who contemplates making an exchange should, as the first step, prepare double lists, in one of which he gives the names and the number of specimens of either sex of the butterflies which he is able to offer in exchange; in the other he sets forth the things which he desires to obtain. The first list is said to be a list of " offerta "; the second is a list of " desiderata." As an illustration of the manner in which such lists may be conveniently arranged, I give the following:

OFFERTA

Papilio turnus,	♂ 3; ♀ 4.
Dimorphic var. glaucus,	♀ 6.
Colias alexandra,	♂ 4; ♀ 6.

DESIDERATA

Papilio nitra,   ♀.
Papilio brevicauda, orange-spotted var.

The collector who receives these lists of offerta and desiderata will be able to decide what his correspondent has which he desires, and what there may be in his own collection which the correspondent wishes that he can offer in exchange; and the process of exchange is thus immediately facilitated.

Persons who exchange insects with others should always be extremely careful as to the manner of packing the specimens, and the directions given in the introductory portion of this book should be very carefully followed. Too much care cannot be taken in preventing damage to specimens in transit.

Genus ANCYLOXYPHA, Felder

Butterfly. — Very small, the antennæ very short, the club straight, bluntly pointed. The palpi have the third joint long,

344

slender, and suberect. The neuration of the wings is shown in the cut. The abdomen is slender, extending beyond the hind margin of the secondaries. The fore wings are without a discal stigma.

Egg. — Hemispherical, marked with lozenge-shaped cells; yellow when laid, later marked with orange-red patches.

Caterpillar.—The entire life-history has not yet been ascertained. The caterpillars live upon marsh grasses; they construct for themselves a nest by drawing together the edges of a blade of grass with bands of silk. In form they do not differ from other hesperid larvæ.

FIG. 164. — Neuration of the genus *Ancyloxypha.*

Chrysalis. — Not as yet accurately known.

(1) **Ancyloxypha numitor**, Fabricius, Plate XLVII, Fig. 2, ♂ (Numitor).

Butterfly.—The upper side is correctly delineated in the plate. On the under side the fore wings are black, margined on the costa and on the outer margin with reddish-fulvous. The hind wings are pale fulvous. Expanse, .75–.95 inch.

Early Stages.—What has been said in reference to these in connection with the description of the genus must suffice for the species.

This pretty little insect is widely distributed, and abounds among grasses about watercourses. It ranges from the province of Quebec to eastern Florida, thence westward across the Mississippi Valley as far as the Rocky Mountains.

Genus COPÆODES, Speyer

Butterfly. — The antennæ are very short; the club is thick, straight, rounded at the tip; the palpi are as in the preceding genus. The neuration of the wings is represented in the cut. The abdomen is slender, extending beyond the hind margin of the secondaries. The male is provided in most species with a linear stigma.

Early Stages.—These have not as yet been described.

(1) **Copæodes procris**, Edwards, Plate XLVII, Fig. 9, ♂ (Procris).

Butterfly. — The plate gives an excellent idea of the upper side of this diminutive species. On the under side the wings are col-

345

ored as on the upper side, save that the fore wings at the base near the inner margin are blackish, and that the hind wings are a trifle paler than on the upper side. The sexes do not differ in color. Expanse, .75–1.00 inch.

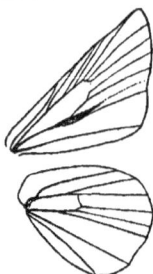

This pretty little butterfly is a Southern species, is found plentifully in Texas and Arizona, and occurs also very commonly in southern California.

(2) **Copæodes wrighti,** Edwards, Plate XLVII, Fig. 10, ♂ (Wright's Skipper).

Butterfly.—This species may be easily distinguished from the preceding by the dark

Fig. 165.—Neuration of the genus *Copærodes.*

fringes of both the fore and the hind wing and by the different arrangement of the discal stigma on the fore wing. On the under side it is colored very much as *procris.* Expanse, .75–1.10 inch.

Early Stages.— Unknown.

The species is found in the Mohave Desert and southern California.

(3) **Copæodes myrtis,** Edwards, Plate XLVII, Fig. 11, ♂ (Myrtis).

Butterfly. — This diminutive little species may be readily recognized by the plate. The fore wings are somewhat broadly margined with dusky at the apex and along the outer margin; the hind wings on the costa are broadly and on the outer edge narrowly margined with dusky. On the under side the fore wings are blackish at the base. Expanse, .75 inch.

The only specimens of this butterfly that I have ever seen came from Arizona. The type is figured in the plate.

Genus ERYNNIS, Schrank

Butterfly.—The antennæ are short, less than half the length of the costa; the club is robust, with a very minute terminal crook; the palpi have the third joint minute, suberect, and bluntly conical. There is a discal stigma on the fore wing of the male.

Egg.—Somewhat spherical.

Caterpillar.—Feeds upon grasses, and is stouter in form than most hesperid larvæ, and sluggish in proportion to its stoutness. It does not make a nest, but conceals itself between the leaves of grass at the point where they unite with the stem, and is not very difficult to discover.

Chrysalis.—The chrysalis is elongated, cylindrical. Our knowledge of this stage is not very accurate as yet.

(1) **Erynnis manitoba**, Scudder, Plate XLVI, Fig. 2, ♂ ; Fig. 3, ♀ (The Canadian Skipper).

Butterfly, ♂.—The upper side of the wings is depicted in the plate. On the under side the wings are paler, the fore wings fulvous on the cell, pale gray at the apex and on the outer margin. There

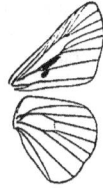

FIG. 166.—Neuration of the genus *Erynnis*, enlarged.

is a black shade at the base of the primaries, and a black streak corresponding in location to the discal stigma on the upper side. The hind wings are pale ferruginous, except a broad streak along the inner margin, which is whitish. All the light spots of the upper side of both wings reappear on the under side, but are more distinctly defined, and are pearly-white in color.

♀.—The female, on the under side of the fore wing, has the black discal streak replaced by a broad ferruginous shade. The hind wings are darker, and the light spots stand forth more conspicuously upon the darker ground. Expanse, ♂, 1.25 inch; ♀, 1.30 inch.

Early Stages.—These remain to be ascertained.

The Canadian Skipper is found across the entire continent north of a line roughly approximating the boundary between the United States and the Dominion of Canada. Along the Western Cordilleras it descends into the United States, as far south as Colorado and northern California.

(2) **Erynnis morrisoni**, Edwards, Plate XLVI, Fig. 26, ♂ ; Fig. 27, ♀ (Morrison's Skipper).

Butterfly.—The upper side of the wings in both sexes is sufficiently well delineated in the plate to obviate the necessity for description. On the under side the fore wings are pale fulvous, black at the base and ferruginous at the tip, the ferruginous shade interrupted by the subapical pale spots, which on this side of the wing are pearly-white. The hind wings are deep ferruginous, obscured on the inner margin by long pale-brown hairs. From the base to the end of the cell there is a broad silvery-white

347

ray. Beyond the cell the curved postmedian band of fulvous spots which appears above reappears as a band of pearly-white, which stands forth conspicuously on the dark ground. Expanse, ♂, 1.20 inch; ♀, 1.20–1.35 inch.

Early Stages.—Unknown.

The species ranges from southern Colorado to Arizona.

(3) **Erynnis sassacus**, Harris, Plate XLVI, Fig. 13, ♂ (The Indian Skipper).

Butterfly.—The upper side of the male is as shown in the plate. The female is larger, the fulvous ground-color paler, the outer marginal shades darker, and the discal stigma is replaced by a dark-brown shade. On the under side in both sexes the wings are pale fulvous, with the spots of the upper side feebly reproduced as faint lighter spots. The fore wings in both sexes are black at the base. Expanse, ♂, 1.10–1.25 inch; ♀, 1.25–1.35 inch.

Early Stages.—The caterpillar, which is plumper than most hesperid larvæ, feeds on grasses.

The insect ranges from New England to Georgia, and westward to Colorado.

(4) **Erynnis ottoë**, Edwards, Plate XLVI, Fig. 11, ♂ ; Fig. 12, ♀ (Ottoë).

Butterfly.—Considerably larger than the preceding species. The wings of the male on the upper side are pale fulvous, narrowly bordered with black. The discal stigma is dark and prominent. The female has the wings on the upper side more broadly but more faintly margined with dusky. The wings of both sexes on the under side are uniformly pale fulvous or buff, marked with dark brown or blackish at the base of the fore wings. Expanse, ♂, 1.35 inch; ♀, 1.45–1.50 inch.

Early Stages.—Unknown.

The habitat of this species is Kansas and Nebraska.

(5) **Erynnis metea**, Scudder, Plate XLVII, Fig. 33, ♂ ; Fig. 34, ♀ (The Cobweb Skipper).

Butterfly.—The upper side of the wings is fairly well represented in the plate, the male being a little too red, and the wings at the base and the discal stigma not being dark enough. On the under side the wings are brown, darker than on the upper side. The pale markings of the upper side are all repeated below as distinct pearly-white spots, and in addition on the hind wings near the base there is a curved band of similar white spots. Expanse, ♂, 1.20 inch; ♀, 1.25–1.30 inch.

Early Stages.—We know as yet but little of these.

The species occurs in New England, New York, and westward to Wisconsin.

(6) **Erynnis uncas,** Edwards, Plate XLVII, Fig. 27, ♂ ; Fig. 28, ♀ (Uncas).

Butterfly.—The upper side of the wings of both sexes is well represented in the plate. On the under side in both sexes the wings are beautifully marked with conspicuous pearly-white spots on a greenish-gray ground. The spots are defined inwardly and outwardly by dark olive shades and spots. Expanse, ♂, 1.30 inch; ♀, 1.55 inch.

Early Stages.—We know nothing of these.

The insect ranges from Pennsylvania to Colorado and Montana.

(7) **Erynnis attalus,** Edwards, Plate XLVII, Fig. 23, ♂ (Attalus).

Butterfly.—The male is fairly well depicted in the plate, but the light spots are too red. The female is larger and darker. On the under side the wings are dusky, with the light spots reproduced in faint gray. Expanse, ♂, 1.25 inch; ♀, 1.45 inch.

Early Stages.—Unknown.

The species occurs very rarely in New England, is found from New Jersey to Florida and Texas, and ranges westward to Wisconsin and Iowa.

(8) **Erynnis sylvanoides,** Boisduval, Plate XLVII, Fig. 44. ♂ (The Woodland Skipper).

Butterfly.—The upper side of the male is well shown in the plate. The female on the upper side has less fulvous, the wings being prevalently fuscous, and the red color reduced to a spot at the end of the cell. There is a median band of fulvous spots on both wings. On the under side in both sexes the wings are quite pale gray, with the costa near the base and the cell of the primaries reddish. The primaries at the base near the inner margin are black. The spots of the upper side reappear, but are pale and faint. Expanse, 1.25–1.35 inch.

Early Stages.—Unknown.

The species ranges along the Pacific coast from British Columbia to California, and eastward to Colorado.

(9) **Erynnis leonardus,** Harris, Plate XLVII, Fig. 35, ♂ ; Fig. 36, ♀ (Leonard's Skipper).

Butterfly.—Stouter and larger than the preceding species, and

349

notably darker in coloring. The upper side of the wings is shown in the plate. On the under side the wings are dark brick-red. The primaries are blackish on the outer half, interrupted by the spots of the median series, which on the under side are large, distinct, and shade from pale fulvous to white toward the inner margin. The secondaries have a round pale spot at the end of the cell, and a curved median band of similar spots, corresponding in location to those on the upper side. Expanse, ♂, 1.25 inch; ♀, 1.35 inch.

Early Stages.—These are only imperfectly known. The caterpillar feeds on grasses.

The butterfly, which haunts flowers and may easily be captured upon them, ranges from New England and Ontario southward to Florida, and westward to Iowa and Kansas.

(10) **Erynnis snowi,** Edwards, Plate XLVII, Fig. 29, ♂ ; Fig. 30, ♀ (Snow's Skipper).

Butterfly.—The upper side of the wings of both sexes is well represented in the plate. On the under side the wings are uniformly reddish-brown, with the primaries black at the base, and the median spots enlarged near the inner margin and whitish, as in the preceding species. The light spots of the upper side reappear below as pale spots, which are well defined on the dark ground-color. Expanse, 1.25–1.40 inch.

Early Stages.—Unknown.

The species ranges from southern Colorado to Arizona.

Genus THYMELICUS, Hübner

Butterfly.—The antennæ are short, less than half the length of the costa; the club is stout and short, somewhat crooked just at the end. The third joint of the palpi is conical, almost concealed in the thick vestiture of the second joint. The neuration is given in the cut.

Egg.— The egg is hemispherical, with the surface marked by irregular angular cells formed by slightly raised lines.

Caterpillar.—The caterpillars feed on grasses. They are long and slender, thicker behind than before, covered with short hair. They are generally dark in color, and not green as are the caterpillars in most of the hesperid genera.

EXPLANATION OF PLATE XLVII

1. *Pamphila mandan*, Edwards, ♀.
2. *Ancyloxypha numitor*, Fabricius, ♂.
3. *Oarisma garita*, Reakirt. ♂.
4. *Oarisma poweshiek*, Parker, ♂.
5. *Amblyscirtes vialis*, Edwards, ♂.
6. *Amblyscirtes samoset*, Scudder, ♂.
7. *Amblyscirtes œnus*, Edwards, ♀.
8. *Amblyscirtes simius*, Edwards, ♂.
9. *Copæodes procris*, Edwards, ♂.
10. *Copæodes wrighti*, Edwards, ♂.
11. *Copæodes myrtis*, Edwards, ♂.
12. *Hesperia scriptura*, Boisduval, ♀.
13. *Hesperia centaureæ*, Rambur, ♂.
14. *Hesperia cæspitalis*, Boisduval, ♀.
15. *Hesperia xanthus*, Edwards, ♂.
16. *Amblyscirtes textor*, Edwards, ♂,
    under side.
17. *Hesperia nessus*, Edwards, ♂.
18. *Hesperia montivaga*, Reakirt. ♂.
19. *Hesperia domicella*, Erichson, ♂.
20. *Lianocbores lannus*, Fabricius, ♂.
21. *Poanis massasoit*, Scudder, ♂.
22. *Poanis massasoit*, Scudder, ♀.
23. *Erynnis attalus*, Edwards, ♂.
24. *Polites peckius*, Kirby, ♂.
25. *Polites peckius*, Kirby, ♀.
26. *Polites mardon*, Edwards, ♂.
27. *Erynnis uncas*, Edwards, ♂.
28. *Erynnis uncas*, Edwards, ♀.
29. *Erynnis snowi*, Edwards, ♂.
30. *Erynnis snowi*, Edwards, ♀.
31. *Atrytone taxiles*, Edwards, ♂.
32. *Atrytone taxiles*, Edwards, ♀.
33. *Erynnis metea*, Scudder, ♂.
34. *Erynnis metea*, Scudder, ♀.
35. *Erynnis leonardus*, Harris, ♂.
36. *Erynnis leonardus*, Harris, ♀.
37. *Atrytone zabulon*, Boisd.-Lec., ♂.
38. *Atrytone zabulon*, Boisd.-Lec., ♀.
39. *Atrytone pocahontas*, Scudder, ♀
40. *Thymelicus brettus*, Boisd.-Lec., ♂.
41. *Thymelicus brettus*, Boisd.-Lec., ♀.
42. *Polites sabuleti*, Edwards, ♂.
43. *Polites sabuleti*, Edwards, ♀.
44. *Erynnis sylvanoides*, Boisduval, ♂.

PLATE XLVII.

Chrysalis.—I can discover no account of any observations made upon the chrysalids of this genus.

(1) **Thymelicus brettus,** Boisduval and Leconte, Plate XLVII, Fig. 40, ♂ ; Fig. 41, ♀ (The Whirlabout).

Butterfly.—The male on the upper side resembles *Hylephila phylœus,* but may be distinguished by the broader and darker spots on the under side of the wings. The costal and outer margins of the secondaries are also generally more broadly bordered with fuscous than in *phylœus,* a fact not shown in the specimen figured in the plate. The female is quite different from the female of *phylœus,* as will be seen by a comparison of the figures of the two sexes. Expanse, ♂ , 1.15 inch; ♀, 1.25 inch.

Early Stages.—These are only partially known. The caterpillar feeds on grasses.

FIG. 167.—Neuration of the genus *Thymelicus,* enlarged.

The insect is very rare in the North, a few specimens having been taken in New England and Wisconsin. It is found commonly in the Carolinas, and thence southward to the Gulf, and is abundant in the Antilles, Mexico, and Central America.

(2) **Thymelicus ætna,** Boisduval, Plate XLVI, Fig. 28, ♂ ; Fig. 29, ♀ ; Plate VI, Fig. 42, *chrysalis* (The Volcanic Skipper).

Butterfly.—Both sexes are well represented on the upper side in the plate. On the under side the wings are paler, with the light spots of the upper side repeated. Expanse, ♂ , 1.00 inch; ♀, 1.25 inch.

Early Stages.—What we know of these is well stated in the pages of Dr. Scudder's great work. The caterpillar usually feeds on grasses.

The species ranges from New England, Ontario, and Wisconsin on the north to the Gulf, and as far west as Iowa and Texas.

(3) **Thymelicus mystic,** Scudder, Plate XLVI, Fig. 22, ♂ ; Fig. 23, ♀ (The Long-dash).

Butterfly.—No description of the upper side is needed, the figures in the plate being sufficient to enable identification. On the under side the primaries are fulvous on the costa at the base. The remainder of the primaries and the secondaries are dark ferruginous, with the light spots of the upper side all repeated

351

greatly enlarged, pale, and standing out boldly upon the dark ground-color. The hind wings are pale brown on the inner margin. Expanse, ♂, 1.10 inch; ♀, 1.25 inch.

Early Stages.—These have been elaborately described by Scudder. The caterpillar feeds on grasses, making a tubular nest for itself among the leaves.

The insect ranges through southern Canada and New England to Pennsylvania, and westward to Wisconsin.

.

Genus ATALOPEDES, Scudder

Butterfly.—Antennæ short, less than half the length of the costa; club short, stout, crooked just at the end; the palpi as in the preceding genus. The cut shows the neuration. The only

Fig. 108.—Neuration of the genus *Atalopedes*, enlarged.

mark of distinction between this genus and the two genera that follow is found in the shape of the discal stigma on the wing of the male, which is described as follows by Dr. Scudder: "Discal stigma in male consisting of, first, a longitudinal streak at base of middle median interspace, of shining black, recurved rods; second, of a semilunar field of dead-black erect rods in the lowest median interspace, overhung above by long, curving scales; followed below by a short, small striga of shining black scales, and outside by a large field of erect, loosely compacted scales."

Egg.—Hemispherical, covered with a network of delicate raised lines describing small polygons over the surface; minutely punctate.

Caterpillar.—Cylindrical, tapering backward and forward; head large; the neck less constricted than in the genus *Eudamus* or in the genus *Thanaos;* dark in color.

Chrysalis.—The chrysalis is slender, cylindrical, a little humped upon the thorax, with the tongue-sheath free and projecting to the end of the fifth abdominal segment.

(1) **Atalopedes huron,** Edwards, Plate XLVI, Fig. 4, ♂ ; Fig. 5, ♀ ; Plate VI, Figs. 43, 47, *chrysalis* (The Sachem).

Butterfly.—The upper side of the wings in both sexes is well

352

represented in the plate. On the under side the wings are paler, with the light spots of the upper side faintly repeated. Expanse, ♂, 1.15 inch; ♀, 1.35 inch.

Early Stages.—These are described in full with painstaking accuracy by Scudder in " The Butterflies of New England." The caterpillar feeds on grasses.

The species ranges from southern New York to Florida, thence westward and southward into Mexico. '

Genus POLITES, Scudder

Butterfly. —The antennæ and the palpi are as in the preceding genus; the neuration of the wings is also very much the same. This is another genus founded by Dr. Scudder upon the shape of the discal stigma in the wing of the male. His description of this feature is as follows: " Discal stigma of male consisting of an interrupted, gently arcuate or sinuate streak of dead-black retrorse scales or rods, edged below, especially in the middle, by a border of similar, but dust-colored, erect rods, and followed beneath by an inconspicuous large area of loosely compacted, erect, dusky scales."

FIG. 160.—Neuration of the genus *Polites*, enlarged.

Egg.—Approximately hemispherical, the height, however, being greater than in the egg of the preceding genus; reticulated, the lines forming hexagonal figures upon the surface.

Caterpillar, etc.—Of the stages beyond the egg we know as yet comparatively little. The caterpillar feeds on grasses.

(1) **Polites peckius**, Kirby, Plate XLVII, Fig. 24, ♂ ; Fig. 25, ♀ (Peck's Skipper).

Butterfly. — This little species, the upper side of which in both sexes is correctly shown in the plate, has the under side of the wings dark brown, with the light spots of the upper side greatly enlarged, especially upon the disks of the wings, fused, and pale yellow, thus contrasting strongly with the rest of the wings. Expanse, ♂, 1.00 inch; ♀, 1.25 inch.

Early Stages.—These are not thoroughly known as yet. The larva feeds on grasses.

Peck's Skipper ranges from Canada southward to Virginia, and west to Kansas and Iowa.

(2) **Polites mardon,** Edwards, Plate XLVII, Fig. 26, ♂ (The Oregon Skipper).

Butterfly.— On the under side the wings are pale gray, with the light spots of the primaries and a curved median band of spots on the secondaries whitish. Expanse, ♂, 1.10 inch; ♀, 1.20 inch.

Early Stages.— Unknown.

The only specimens I have, including the types, were taken in Oregon and Washington.

(3) **Polites sabuleti,** Edwards, Plate XLVII, Fig. 42, ♂ ; Fig. 43, ♀ (The Sand-hill Skipper).

Butterfly.— Small, the male on the upper side looking like a diminutive and darkly bordered *phylæus.* On the under side the wings are paler than on the upper side; the still paler spots of the discal areas are defined outwardly and inwardly by elongated dark spots. Expanse, 1.00–1.10 inch.

Early Stages.—Unknown.

The habitat of this species is California.

Genus HYLEPHILA, Billberg

Butterfly.—The antennæ are very short, scarcely one third the length of the costa of the fore wing; the club is robust and short, with a very minute crook at the end; the palpi are as in the two preceding genera. The neuration of the wings is represented in the cut.

Early Stages.—As yet but partially known.

The larva feeds on grasses, and the mature form has been figured by Abbot, a copy of whose drawing is given by Dr. Scudder in Plate 77 of "The Butterflies of New England."

(1) **Hylephila phylæus,** Drury, Plate XLVI, Fig. 18, ♂ ; Fig. 19, ♀ ; Plate VI, Fig. 39, *chrysalis* (The Fiery Skipper).

Fig. 170.— Neuration of the genus *Hylephila,* enlarged.

Butterfly.—The upper side is correctly shown in the plate. On the under side the wings are pale yellow, with a few small, round spots on the margin and disk of the hind wings, a black patch at the base, large black marginal spots, and a central, interrupted, longitudinal black streak on the disk of the primaries. Expanse, 1.15–1.25 inch.

The insect ranges from Connecticut to Patagonia, over all the most habitable parts of the New World. I have taken it frequently in southern Indiana, where I often have collected in recent years.

Genus PRENES, Scudder

Butterfly.—The antennæ are short, not half the length of the costa. The head is broad, and the antennæ are inserted widely apart. The club is moderate, terminating in a fine point which is bent back at right angles, forming a distinct crook. The abdomen is long and slender, but does not project beyond the hind margin of the secondaries. The fore wings are pointed at the apex and are relatively longer and narrower than in the preceding genus. The neuration is illustrated in the cut.

Early Stages.—These have not yet been studied.

(1) **Prenes ocola**, Edwards, Plate XLVI, Fig. 34, ♂ (The Ocola Skipper).

Butterfly.—Accurately depicted in the plate. The under side is like the upper side, but a shade paler. The under side of the abdomen is whitish. Expanse, 1.45–1.60 inch.

Early Stages. — Unknown.

This is a Southern species, found commonly in the Gulf States, and ranging northward to Pennsylvania, southern Ohio, and Indiana.

FIG. 171.—Neuration of the genus *Prenes*, enlarged.

Genus CALPODES, Hübner

FIG. 172.—Neuration of the genus *Calpodes*, enlarged.

Butterfly.—Rather large, stout; head broad; antennæ as in the preceding genus, but stouter. The neuration, considerably enlarged, is accurately delineated in the cut.

Egg.—Hemispherical, ornamented with irregular, more or less pentagonal cells.

355

Caterpillar.—Cylindrical, slender, tapering forward and backward from the ninth segment, rapidly diminishing in size posteriorly; the head relatively small, the neck not much strangulated; spiracles surrounded by radiating blackish bristles.

Chrysalis.—The chrysalis is relatively slender, gently convex both on the ventral and dorsal aspects, with a curved delicate frontal tubercle. The tongue-case is long and projects for a considerable distance beyond the somewhat short cremaster.

(1) **Calpodes ethlius,** Cramer, Plate XLV, Fig. 3, ♀ ; Plate VI, Fig. 48, *chrysalis* (The Brazilian Skipper).

Butterfly.—There can be no mistaking this robust and thick-bodied species. The wings on the under side are dull olive, blackish at the base of the primaries, with all the spots of the upper side repeated. Expanse, 2.00–2.15 inches.

Early Stages.—The caterpillar feeds on the leaves of the canna.

It is common in the Gulf States, and ranges north to South Carolina. A stray specimen was once taken at West Farms, New York. Southward it ranges everywhere through the Antilles to Argentina, in South America.

Genus LERODEA, Scudder

Butterfly.—The antennæ are about half as long as the costa; the club is robust, slightly elongated, with a distinct crook at the extremity; the palpi have the third joint erect, minute, and bluntly conical. The neuration is represented in the cut.

Early Stages.—These are not known.

(1) **Lerodea eufala,** Edwards, Plate XLVI, Fig. 33, ♀ (Eufala).

Butterfly.—The plate shows the upper side of the female. The male is not different, except that the fore wings are a little more pointed at the apex. The under side is like the upper side, but a shade paler. The lower side of the abdomen is whitish. When seen on the wing the creature looks like a small *Prenes ocola.* Expanse, 1.10–1.20 inch.

Early Stages.—Unknown.

This butterfly is found in the Gulf States.

Fig. 173.—Neuration of the genus *Lerodea,* enlarged.

Genus LIMOCHORES, Scudder

Butterfly.—The antennæ are about half as long as the costa; the club is robust, elongate, with a very short terminal crook; the palpi have the third joint erect, short, bluntly conical. The male has a linear discal stigma on the upper side of the fore wing, as shown in the cut.

Egg.—Hemispherical, somewhat flattened on the top, the surface broken up by delicate raised lines into pentagonal cells.

Caterpillar.—Largest on the fourth and fifth abdominal segments, tapering to either end. The larvæ feed on grasses, and construct a tube-like nest of delicate films of silk between the blades.

FIG. 174.—Neuration of the genus *Limochores*, enlarged.

Chrysalis.—Comparatively slender, strongly convex on the thoracic segments and on the dorsal side of the last segments of the abdomen. On the ventral side the chrysalis is nearly straight. The cremaster, which is short, is bent upward at an oblique angle with the line of the ventral surface.

(1) **Limochores taumas,** Fabricius, Plate XLVII, Fig. 20, ♂ ; Plate VI, Fig. 44, *chrysalis* (The Fawny-edged Skipper).

Butterfly.—The upper side of the male is excellently portrayed in the plate. The female is without the tawny edge on the fore wing, the entire wing being olivaceous, with three small subapical spots and a median row of four spots beyond the end of the cell, increasing in size toward the inner margin. On the under side in both sexes the wings are uniformly dull olivaceous, with the spots of the upper side repeated. The costa of the male is edged with red on this side, as well as on the upper side. Expanse, ♂, 1.00 inch; ♀, 1.20 inch.

Early Stages.—The reader who wishes to know about them may consult the pages of "The Butterflies of New England." The caterpillar feeds on grasses.

The insect ranges from Canada to the Gulf, and westward to Texas, Colorado, and Montana.

(2) **Limochores manataaqua,** Scudder, Plate XLVI, Fig. 30, ♀ (The Cross-line Skipper).

Butterfly.—The male on the upper side is dusky-olive, with a black discal streak below the cell, which is slightly touched with

357

reddish, becoming deeper and clearer red on the costa at the base. The wings on the under side are more or less pale gray, with a transverse series of pale spots on the primaries, and a very faint curved discal series of similar spots on the secondaries. The female, the upper side of which is well shown in the plate, is marked below much like the male. Expanse, 1.10–1.20 inch.

Early Stages.—These have been described by Scudder.

The insect occurs in New England and Canada, and ranges westward to Nebraska.

(3) **Limochores pontiac,** Edwards, Plate XLVI, Fig. 16, ♂; Fig. 17, ♀ (Pontiac).

Butterfly.—This fine insect is so well represented in the plate as to require but little description. The wings are pale red, clouded with dusky on the under side, the spots of the upper side being indistinctly repeated. Expanse, ♂, 1.15 inch; ♀, 1.25 inch.

Early Stages.—Little is known of these.

The insect ranges from Massachusetts to Iowa and Nebraska, and seems to have its metropolis about the southern end of Lake Michigan.

(4) **Limochores palatka,** Edwards, Plate XLVI, Fig. 21, ♂ (The Palatka Skipper).

Butterfly.—The upper side of the male needs no description. The female closely resembles the female of *L. byssus,* which is shown in the plate at Fig. 20, but differs from the female of that species in having the median spots on the primaries much reduced in size, the band of spots being greatly interrupted beyond the end of the cell. On the hind wing the female has the entire surface of the secondaries inside of the broad outer band fulvous, as shown in the figure of the male, and not simply marked by a transverse narrow band of spots. On the under side the fore wings are bright fulvous, clouded with black at the base and near the outer angle. The hind wings are uniformly dull reddish-brown. This species has been identified by Dr. Scudder with a species named *dion* by Edwards, but which is a very different thing. Expanse, ♂, 1.50–1.65 inch; ♀, 1.90–2.00 inches.

Early Stages.—We know nothing of these.

The insect is confined to Florida, all the specimens which I have seen coming from the region of the Indian River.

(5) **Limochores byssus,** Edwards, Plate XLVI, Fig. 20, ♀ (Byssus).

Butterfly.—Allied to the preceding species. The discal stigma of the male upon the fore wings is much longer than in *L. palatka.* The outer margin of the secondaries is not as sharply defined as in that species, but shades insensibly into the lighter greenish-fulvous of the basal part of the wing. The female on the upper side is distinguished from the female of the preceding species by the restriction of the discal band of spots on the hind wing to a few small light-colored spaces beyond the end of the cell, and by the regular continuation of the band of yellow spots across the primaries from the subapical spots to the submedian nervule near the middle of the inner margin. On the under side the primaries and the secondaries are very bright, clear orange-red, with the base and inner margin of the primaries brightly laved with blackish. The median series of spots in the male are very faintly indicated on the fore wings, but are more strongly indicated on those of the female. Expanse, ♂, 1.45 inch; ♀, 1.65 inch.

Early Stages.—We know little of these.

The insect is found in Florida.

(6) **Limochores yehl**, Skinner, Plate XLVI, Fig. 40, ♂ (Skinner's Skipper).

Butterfly.—The upper side of the male is shown in the plate. On the under side the wings are lighter, the secondaries uniformly pale cinnamon-brown, marked with a semicircle of four yellowish round spots, with a small spot on the cell toward the base. Expanse, 1.25–1.35 inch.

Early Stages.—Unknown.

The species has been taken in Florida, and is as yet not common in collections. The figure is that of the type.

Genus EUPHYES, Scudder

Butterfly.— The antennæ have the club stout, elongate, furnished with a short crook at the end; the palpi are densely scaled; the third joint is slender, bluntly conical, projecting beyond the vestiture of the second joint. The neuration is shown in the cut.

Egg.—Hemispherical.

Caterpillar.—The head small, body cylindrical, tapering forward and backward from the middle, the body profusely covered with minute tapering hairs arising from small, wart-like protuberances.

Chrysalis.—Thus far undescribed.

(1) **Euphyes verna**, Edwards, Plate XLVI, Fig. 32, ♂ (The Little Glass-wing).

Butterfly.—The upper side of the male is correctly delineated in the plate. On the under side the wings are paler, inclining to purplish-red. The spots of the upper side are repeated, but in addition about the middle of the hind wings there is a semicircle of pale spots. Expanse, ♂, 1.15 inch; ♀, 1.35 inch.

Early Stages.—We do not know much of these; what little we do know may be found recorded in the pages of "The Butterflies of New England." The caterpillar feeds on grasses.

It ranges from southern New England to Virginia, westward to Kansas, and northward to the province of Alberta. It is quite common in Ohio, Indiana, and Illinois.

Fig. 175. — Neuration of the genus *Euphyes*, enlarged.

(2) **Euphyes metacomet**, Harris, Plate XLVI, Fig. 31, ♂ (The Dun Skipper).

Butterfly.—The male is dark in color on the upper side, and on the under side the wings are a shade lighter, the lower side of the abdomen being generally paler. The female has some faint traces of translucent apical spots near the costa, and two minute translucent spots on either side of the second median nervule near its origin. On the under side the spots of the upper side reappear. There is a faint trace of a semicircle of pale spots about the middle of the hind wing. The female specimens vary on the under side from pale brown to a distinctly purplish-brown. Expanse, ♂, 1.15 inch; ♀, 1.30 inch.

Early stages.—Next to nothing is known of these.

It ranges from Quebec to the Carolinas, and westward to Texas, New Mexico, and the British possessions east of the Rocky Mountains, as far north as the latitude of the northern shores of Lake Superior.

t

Explanation of Plate XLVIII

1. *Thanaos persius*, Scudder, ♂.
2. *Thanaos somnus*, Lintner, ♂.
3. *Thanaos nævius*, Lintner. ♂.
4. *Thanaos martialis*, Scudder, ♂.
5. *Thorybes bathyllus*, Smith and Abbot, ♀.
6. *Thorybes pylades*, Scudder, ♂.
7. *Thanaos petronius*, Lintner. ♂.
8. *Lerema accius*, Smith and Abbot, ♂.
9. *Thanaos pacuvius*, Lintner. ♀.
10. *Thanaos lucilius*, Lintner, ♂.
11. *Thanaos juvenalis*, Fabricius, ♂.
12. *Thanaos funeralis*, Lintner, ♂.
13. *Thorybes epigena*, Butler, ♂.
14. *Pholisora libya*, Scudder, ♂.
15. *Thanaos horatius*, Scudder, ♂.
16. *Pholisora hayhurstі*, Edwards, ♀.
17. *Thanaos icelus*, Lintner, ♂.
18. *Colias eurytheme*, Boisduval, ♀, albino.

PLATE XLVIII.

Genus OLIGORIA, Scudder

Butterfly.—The antennæ are as in the preceding genus; the palpi have the third joint minute and almost entirely concealed in the thick vestiture of the second joint. The neuration is represented in the cut.

Early Stages.—We know very little of these, and there is here a field for investigation.

(1) **Oligoria maculata,** Edwards, Plate XLVI, Fig. 35, ♂ (The Twin-spot).

Butterfly.— The upper side of the male is as shown in the plate. The female closely resembles the male, but the spots on the fore wing are larger. On the under side the wings are brown, almost as dark as on the upper side. The primaries are whitish near the outer angle. The spots of the upper side of the primaries are reproduced on the lower side. The hind wings have three conspicuous pearly-white spots about the middle, two located one on either side of the second median nervule, and one removed from these, located between the upper radial and the subcostal nervule. Expanse, ♂, 1.40 inch; ♀, 1.50 inch.

Fig. 176.—Neuration of the genus *Oligoria*, enlarged.

Early Stages.— But little is known of these.

This is a Southern species, found abundantly in Florida, and ranging northward into Georgia and the Carolinas. A specimen is reported to have been taken near Albany, New York, and diligent collecting may show that it has a far more northern range than has heretofore been supposed.

Genus POANES, Scudder

Butterfly.—The antennæ are short; the club is stout, bent, acuminate at the tip. The third joint of the palpi is slender, cylindrical, short. The neuration of the genus is shown in the cut.

Early Stages.—Nothing is known of these, and they await investigation.

(1) **Poanes massasoit,** Scudder, Plate XLVII, Fig. 21, ♂ ; Fig. 22, ♀ (The Mulberry-wing).

Butterfly.—The upper side of the wings in both sexes is correctly shown in the plate. On the under side the fore wings are black, with the costa and the outer margin bordered with reddish, with three small subapical light spots and two or three median spots. On the under side the hind wings are bright yellow, bordered on the costa and on the outer margin for part of their distance with reddish-brown. The female on the under side is more obscurely marked than the male, and the hind wings are more or less gray in many specimens, lacking the bright yellow which appears upon the wings of the male. There is considerable variation on the under side of the wings. Expanse, ♂, 1.15 inch; ♀, 1.20 inch.

Fig. 177.— Neuration of the genus *Poanes*, enlarged.

Early Stages.—Not known.

The species ranges from New England westward as far as Nebraska, and its range does not appear to extend south of Pennsylvania, though it has been reported from Colorado, and even from northern Texas, in the West.

Genus PHYCANASSA, Scudder

Butterfly.—Antennæ short; club straight, with a small crook at the end. The palpi are as in the preceding genus, but a trifle longer. The neuration is shown in the cut, and is very much like that of the preceding genus.

Early Stages.—These are wholly unknown.

(1) **Phycanassa viator**, Edwards, Plate XLVI, Fig. 14, ♂; Fig. 15, ♀ (The Broad-winged Skipper).

Butterfly.—Accurately delineated in the plate. On the under side the wings are as on the upper side, but paler, and the secondaries are traversed from the base to the middle of the outer margin by a pale light-colored longitudinal ray, which is more or less obscured in

Fig. 178.— Neuration of the genus *Phycanassa*, enlarged.

some specimens, especially of the female. The light spots of the upper side appear indistinctly on the under side. Expanse, ♂, 1.45 inch; ♀, 1.60 inch.

Early Stages.—Unknown.

It is not uncommon in the Gulf States, and has been found as far north as New Jersey, northern Illinois, and Wisconsin.

(2) **Phycanassa howardi**, Skinner, Plate XLVI, Fig. 38, ♂ (Howard's Skipper).

Butterfly.—The figure in the plate gives the upper side of the male, in which the discal streak is composed of light-colored scales of the same tint as the rest of the wing, in this respect resembling the allied *P. aaroni*. The under side of the wings is described by Dr. Skinner as follows: "Superiors with tawny central area and border same as upper side. There is a large triangular spot extending into the wing from the base. The tawny color above this spot is of a darker hue than that below and outside of it. Inferiors very light brown, generally with four or five very faint tawny spots in the central area." Expanse, ♂, 1.50 inch; ♀, 1.60 inch.

Early Stages.—Unknown.

The home of this species is Florida.

(3) **Phycanassa aaroni**, Skinner, Plate XLVI, Fig. 37, ♂ (Aaron's Skipper).

Butterfly.— This small species, the male of which is figured in the plate, may be easily recognized from the figure there given. On the under side the fore wings are black at the base; the middle area of the wing is tawny, paler than on the upper side, and bordered as above, but the border below is cinnamon-brown and not fuscous. The hind wings on the under side are uniformly light cinnamon-brown, without any spots. The female is like the male, but larger, the colors somewhat lighter and the markings not so well defined. Expanse, ♂, 1.00 inch; ♀, 1.25 inch.

Early Stages.—Unknown.

The specimens thus far contained in collections have all been taken about Cape May, in New Jersey, in the salt-marshes.

Genus ATRYTONE, Scudder

Butterfly.—The antennæ have a stout club, somewhat elongate, and furnished with a short crook at the end. The palpi are very much as in the preceding genus. The neuration is shown in the cut. There is no discal stigma on the fore wing of the male.

Egg.—The egg is hemispherical, somewhat broadly flattened at the apex, covered with small cells, the inner surface of which is marked with minute punctulations.

Caterpillar.—The caterpillar feeds upon common grasses, making a loose nest of silk for itself at the point where the leaf joins the stem. The head is small; the body is cylindrical, thick, tapering abruptly at either end.

Chrysalis.—Covered with delicate hair; the tongue-case free.

(1) **Atrytone vitellius,** Smith and Abbot, Plate XLVI, Fig. 6, ♂ (The Iowa Skipper).

Fig. 179.—
Neuration of the
genus *Atrytone,*
enlarged.

Butterfly.—The male on the upper side is as shown in the plate. The female on the upper side has the hind wings almost entirely fuscous, very slightly yellowish about the middle of the disk. The fore wings have the inner and outer margins more broadly bordered with fuscous than the male, and through the middle of the cell there runs a dark ray. On the under side the wings are bright pale yellow, with the inner margin of the primaries clouded with brown. Expanse, ♂, 1.25 inch; ♀, 1.45 inch.

Early Stages.—Very little is known of these.

The species ranges through the Gulf States, and northward in the valley of the Mississippi as far as Nebraska and Iowa. It seems to be quite common in Nebraska, and probably has a wider distribution than is reported.

(2) **Atrytone zabulon,** Boisduval and Leconte, Plate XLVII, Fig. 37, ♂ ; Fig. 38, ♀ (The Hobomok Skipper).

Butterfly.— The upper side of both sexes is shown in the plate. The color on the disk of the wings is, however, a little too red. On the under side the wings are bright yellow, with the bases and the outer margin bordered with dark brown. Expanse, ♂, 1.25 inch; ♀, 1.50 inch.

Early Stages.—The caterpillar feeds upon grasses. The life-history has been described with minute accuracy by Dr. Scudder.

The species ranges from New England to Georgia, and westward to the Great Plains. It is very common in Pennsylvania, Virginia, and the valley of the Ohio.

Dimorphic var. **pocahontas,** Scudder, Plate XLVII, Fig. 39, ♀. This is a melanic, or black, female variety of *zabulon,* which is

364

not uncommon. It is remarkable because of the white spots on the primaries and the dark color of the under side of the wings.

(3) **Atrytone taxiles,** Edwards, Plate XLVII, Fig. 31, ♂; Fig. 32, ♀ (Taxiles).

Butterfly.—The fore wings on the under side of the male are bright yellow, black at the base, slightly clouded on the outer margin with pale brown. The hind wings on the under side in this sex are still paler yellow, margined externally with pale brown, and crossed near the base and on the disk by irregular bands of pale brown. In the female sex the fore wings on the under side are fulvous, marked much as in the male, but darker, especially toward the apex, where the subapical spots and two small pale spots beyond the end of the cell near the outer margin interrupt the brown color. The hind wings on the under side are pale ferruginous, crossed by bands of lighter spots, and mottled with darker brown. Expanse, ♂, 1.45 inch; ♀, 1.50 inch.

Early Stages.—Unknown.

The range of this species is from Colorado and Nevada to Arizona.

(4) **Atrytone delaware,** Edwards, Plate XLVI, Fig. 24, ♂; Fig. 25, ♀ (The Delaware Skipper).

Butterfly.—No description of the upper side of the wings is necessary. On the under side the wings are bright orange-red, clouded with black at the base and on the outer angle of the fore wings. Expanse, ♂, 1.25–1.35 inch; ♀, 1.35–1.50 inch.

Early Stages.—Very little is known of these.

The butterfly is found from southern New England and northern New York as far south as Florida and Texas, ranging west to the Yellowstone and southern Colorado.

(5) **Atrytone melane,** Edwards, Plate XLVI, Fig. 7, ♂; Fig. 8, ♀ (The Umber Skipper).

Butterfly.—The male on the upper side somewhat resembles *A. zabulon,* var. *pocahontas;* the female likewise closely resembles specimens of this variety. The wings on the under side are ferruginous, clouded with blackish toward the base of the inner angle, the light spots of the upper side being repeated. The hind wings on the under side are reddish, with a broad irregular curved median band of pale-yellow spots. In the female the band of spots is far more obscure. Expanse, ♂, 1.30 inch; ♀, 1.50 inch.

Early Stages.—Unknown.

The insect is found in southern California.

Genus LEREMA, Scudder

Butterfly.—The antennæ are as in the preceding genus; the palpi have the third joint erect, short, conical. The neuration is represented in the cut. The male has a linear glandular streak on the upper side of the fore wing.

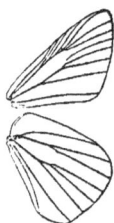

Egg.—Hemispherical, covered with more or less regularly pentagonal cells.

Caterpillar.— The caterpillar feeds upon grasses. The body is slender, tapering forward and backward; the head is small.

Chrysalis.—The chrysalis is slender, smooth, with a tapering conical projection at the head, and the tongue-case long and free, reaching almost to the end of the abdomen.

Fig. 180.—Neuration of the genus *Lerema*, enlarged.

(1) **Lerema accius,** Smith and Abbot, Plate XLVIII, Fig. 8, ♂ ; Plate VI, Fig. 46, *chrysalis* (Accius).

Butterfly.—The male on the upper side is dark blackish-brown, with three small subapical spots, and one small spot below these, near the origin of the third median nervule. The female is exactly like the male, except that it has two spots, the larger one being placed below the small spot corresponding to the one on the fore wing of the male. The wings on the under side are dark fuscous, somewhat clouded with darker brown, the spots of the upper side reappearing on the under side. Expanse, ♂, 1.40 inch; ♀, 1.50 inch.

Early Stages.—Very little has been written upon the early stages.

The butterfly ranges from southern Connecticut to Florida, thence westward to Texas, and along the Gulf coast in Mexico.

(2) **Lerema hianna,** Scudder, Plate XLVI, Fig. 9, ♂ ; Fig. 10, ♀ (The Dusted Skipper).

Butterfly.—The upper side is accurately represented in the plate. The wings on the lower side are as on the upper side, a trifle paler and somewhat grayer on the outer margin. Expanse, ♂, 1.15 inch; ♀, 1.25 inch.

Early Stages.—Unknown.

It ranges through southern New England, westward to Wisconsin, Iowa, and Nebraska, in a comparatively narrow strip of country.

(3) **Lerema carolina**, Skinner, Plate XLVI, Fig. 36, ♂ (The Carolina Skipper).

Butterfly.—On the upper side the butterfly is as represented in the plate. The spots are repeated on the under side of the fore wing, but less distinctly defined. The costa is edged with brownish-yellow. The hind wings on the under side are yellow, spotted with small dark-brown dots. Expanse, ♂, 1.00 inch. The female is unknown.

Early Stages.—Wholly unknown.

This species has thus far been found only in North Carolina, and is still extremely rare in collections. The figure in the plate represents the type. I have seen other specimens. I place it provisionally in the genus *Lerema*, though it undoubtedly does not belong here, and probably may represent a new genus. Lacking material for dissection, I content myself with this reference.

Genus MEGATHYMUS, Riley

This genus comprises butterflies having very stout bodies, broad wings, strongly clubbed antennæ, very minute palpi. The caterpillars are wood-boring in their habits, living in the pith and

Fig. 181.—*Megathymus yuccæ*, ♀.

underground roots of different species of *Yucca*. The life-history of the species represented in the cuts has been well described by the late Professor C. V. Riley, and the student who is curious to know more about this remarkable insect will do well to consult the "Eighth Annual Report of the State Entomologist of

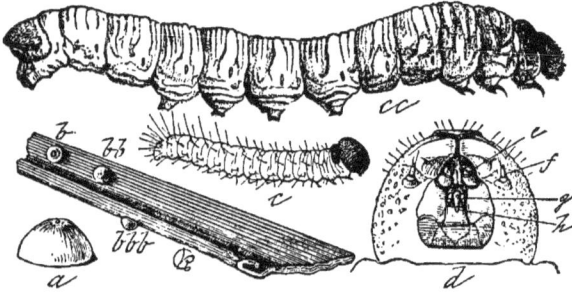

Fig. 182.—*Megathymus yucca:* a, egg, magnified; b, egg from which larva has escaped; bb, bbb, unhatched eggs, natural size; c, newly hatched larva, magnified; cc, larva, natural size; d, head, enlarged to show the mouth-parts; e, maxillary palpi; f, antenna; g, labial palpi; h, spinneret.

Missouri" (p. 169), or the "Transactions of the St. Louis Academy of Science" (vol. iii, p. 323), in which, with great learning, the author has patiently set forth what is known in reference to the insect.

The genus *Megathymus* is referred by some writers to the *Castniidæ*, a genus of day-flying moths, which seem to connect the moths with the butterflies; but the consideration of the anatomical structure of this insect makes such a reference impossible. The genus properly represents a subfamily of the *Hesperiidæ*, which might be named the *Megathyminæ*. The species represented in our cuts is *Megathymus yucca*, Boisduval and Leconte. There are a number of other species of *Megathymus* that are found in our Southern States, principally in Texas and Arizona. They are interesting insects, the life-history of which is, however, in many cases obscure, as yet.

Fig. 183.—Chrysalis of *Megathymus yucca.*

WE here bring to a conclusion our survey of the butterflies of North America. There are, in addition to the species that have been described and figured in the plates, about one hundred and twenty-five other species, principally *Hesperiidæ*, which have not been mentioned. The field of exploration has not by any means been exhausted, and there is no doubt that in the lapse of time a number of other species will be discovered to inhabit our faunal limits.

The writer of these pages would deem it a great privilege to aid those who are interested in the subject in naming and identifying any material which they may not be able to name and identify by the help of this book. In laying down his pen, at the end of what has been to him a pleasurable task, he again renews the hope that what he has written may tend to stimulate a deeper and more intelligent interest in the wonders of creative wisdom, and takes occasion to remind the reader that it is true, as was said by Fabricius, that nature is most to be admired in those works which are least—*"Natura maxime miranda in minimis."*

369

SUPPLEMENTARY NOTE TO THE SECOND EDITION

THE first edition of this book having been nearly exhausted in less than a month after publication, the author has not yet had opportunity to avail himself of the criticisms of scientific friends who are presumably looking for sins of omission and commission, of which it is sincerely hoped they will acquaint him when discovered. Thus far all criticisms have been of an approbatory character, and have only expressed pleasure.

The writer is indebted to Mr. Harrison G. Dyar, the Honorary Curator of the Department of Entomology in the United States National Museum, for reminding him of the fact, which he had carelessly overlooked, that the larva and chrysalis of *Eumæus atala* (see p. 237) have been fully described by Scudder, "Memoirs of the Boston Society of Natural History," vol. ii., p. 413, and by Schwartz, "Insect Life," vol. i., p. 39. The caterpillar is found abundantly upon the "coontie" (*Zamia integrifolia*, Willdenow), and the insect, according to Schwartz, fairly swarms in the pine-woods between the shores of Biscayne Bay and the Everglades.

INDEX

aaroni, Phycanassa, 363
Abbot, John, 70
abbotti, Papilio, 307
abdomen, 7, 17
aberrations, 24
acadica, Thecla, 242; Pieris, 280
acastus, Melitæa, 143
accius, Lerema, 366
Achalarus, genus, 325; cellus, 326; lycidas, 325
acis, Thecla, 240, 246
acmon, Lycæna, 266
Acræa, genus, 162
Acræinæ, subfamily, 162
Actinomeris, 157
Adelpha, genus, 187; californica, 187
adenostomatis, Thecla, 245
adiante, Argynnis, 123
Admiral, The Red, 170
Admirals, The White, 182; Hulst's, 185; Lorquin's, 185
æmilia, Thorybes, 325
ænus, Amblyscirtes, 341
ætna, Thymelicus, 351
affinis, Thecla, 249
afranius, Thanaos, 334
agarithe, Catopsilia, 287
Ageronia, genus, 193; feronia, 194; fornax, 194
Agraulis, genus, 96
Agrion, genus of dragon-flies, 86
ajax, Papilio, 307
alberta, Brenthis, 135
albinism, 24
albinos, 64
alcestis, Argynnis, 107; Thecla, 241
alexandra, Colias, 292
aliaska, Papilio, 312
alicia, Chlorippe, 190
alma, Melitæa, 147
alope, Satyrus, 215
alpheus, Pholisora, 331
Alpines, The, 208; Alaskan, 209; Colorado, 209; Common, 210; Red-streaked, 209
Althæa, 170
Amaranthaceæ, 330
Amarantus, 335
Amblyscirtes, genus, 339; ænus, 341; samoset, 340; simius, 341; textor, 341; vialis, 340
American Entomological Society, 73

ammon, Lycæna, 270
Amorpha californica, 289
ampelos, Cœnonympha, 207
amymone, Cystineura, 177
amyntula, Lycæna, 268
anal angle of wing, 19
Anartia, genus, 174; jatrophæ, 174
Anatomy of Butterflies, 14–25
Ancyloxypha, genus, 344; numitor, 345
andria, Pyrrhanæa, 9, 192
androconia, 18, 19
Angle-wings, The, 163; Colorado, 165; Graceful, 166
anicia, Melitæa, 140
Animal Kingdom, The Place of Butterflies in the, 58
annetta, Lycæna, 266
Anosia, genus, 81; berenice, 82, 84; plexippus, 4, 6, 7, 11, 12, 14, 15, 63, 82, 171; strigosa, 84
antennæ of caterpillar, 6; of butterfly, 14, 16, 23, 61
Antennaria, 170
Anthocharis, genus, 282
Anthrenus, a museum pest, 53
antiacis, Lycæna, 261
antiopa, Vanessa, 5, 7, 94, 169
Antirrhinum, 173
antonia, Chlorippe, 189
aortal chamber, 23
aphrodite, Argynnis, 107
apparatus, collecting, 26; for breeding butterflies, 34; for mounting butterflies, 39; for inflating caterpillars, 45; for preserving specimens, 48; pins, 56; forceps, 56
Aquilegia canadensis, 334
aquilo, Lycæna, 263
Arabis, 284
arachne, Melitæa, 148
Arachnida, 59
arctic butterflies, 171
Arctics, The, 218; Bruce's, 223; Greater, 220; Labrador, 223; Macoun's, 221; Mead's, 222; Uhler's, 222
Argynnis, genus, 96, 99, 101, 161, 172; adiante, 123; alcestis, 107; aphrodite, 18, 107; artonis, 123; atlantis, 108; atossa, 122; behrensi, 115; bischoffi, 124; brenneri, 113; callippe, 118; carpenteri, 106; chitone, 116; cipris, 107; clio, 124; columbia, 111; cornelia, 110; coronis, 117;

cybele, 106; diana, 103; edwardsi, 119;
egleis, 126; electa, 111; eurynome, 125;
halcyone, 116; hesperis, 112; hippolyta,
112; idalia, 103; inornata, 122; lais, 109;
laura, 120; leto, 105; liliana, 119; macaria, 121; meadi, 119; monticola, 114;
montivaga, 126; nausicaä, 108; nevadensis, 118; nitocris, 105; nokomis, 104;
opis, 124; oweni, 109; platina, 117; purpurescens, 114; rhodope, 115; rupestris,
120; semiramis, 121; snyderi, 118; zerene, 113
ariadne, Colias, 29
ariane, Satyrus, 216
Aristolochia, 315, 316
army-worm, 257
Arnold, Sir Edwin, quotations from, 214,
258
arota, Chrysophanus, 252
arrangement, of specimens, 52; of species,
62
arsace, Thecla, 248
arthemis, Basilarchia, 184
Arthropoda, definition of, 59; subdivisions
of, 59
artonis, Argynnis, 123
Asama-yama, volcano, 150
Asclepias, 81
Asimina triloba, 308
astarte, Brenthis, 135
aster, Lycæna, 266
asterias, Papilio, 314
Astragalus, 240
astyanax, Basilarchia, 184
atala, Eumæus, 237, 370
atalanta, Pyrameis, 170
Atalopedes, genus, 352; huron, 352
atlantis, Argynnis, 108
atossa, Argynnis, 122
Atrytone, genus, 363; delaware, 365; melane, 365; pocahontas, 364; taxiles, 365;
vitellius, 364; zabulon, 364
attalus, Erynnis, 349
augusta, Melitæa, 141
augustus, Thecla, 247
ausonides, Euchloë, 283
australis, Calephelis, 233
autolycus, Thecla, 241
Azalea occidentalis, 166

bachmanni, Libythea, 227
bairdi, Papilio, 313
baits for butterflies, 32
Banded Reds, The, 175
Baptisia, 333
Barbauld, Mrs., quotation from, 76
barnesi, Phyciodes, 155
baroni, Melitæa, 141; Satyrus, 216
base of wing, 19
Basilarchia, genus, 182; arthemis, 184;
astyanax, 183; disippus, 3, 8, 84, 185;
eros, 186; floridensis, 186; hulsti, 84,
185; lorquini, 185; proserpina, 184;
pseudodorippus, 185; weidemeyeri, 185
Bates, H. W., on study of butterflies, 3; as
a collector, 338
batesi, Phyciodes, 154
bathyllus, Thorybes, 325
battoides, Lycæna, 264
beani, Melitæa, 140

beating for lepidoptera, 33
beckeri, Pieris, 277
Beelzebub, the " god of flies," 334
behrensi, Argynnis, 115
behri, Colias, 294; Parnassius, 306; Thecla,
247
bellona, Brenthis, 134
Belt, " Naturalist in Nicaragua," 91
berenice, Anosia, 84
bischoffi, Argynnis, 124
Blake & Co., forceps, 56
bleaching wings of butterflies, 20
blenina, Thecla, 245
blow-fly, holding middle place in scale of
animal existence, 271
Blues, The, 236, 258; Arrow-head, 262;
Aster, 266; Behr's, 260, 264; Boisduval's,
260; Bright, 259; Colorado, 264; Common, 267; Couper's, 261; Dotted, 264;
Dwarf, 269; Eastern tailed, 268; Eyed,
261; Florida, 269; Gray, 263; Greenish,
260; Indian River, 270; Labrador, 263;
Marine, 270; Orange-margined, 265;
Pygmy, 269, 271; Reakirt's, 268; Rustic,
263; Scudder's, 265; Shasta, 265; Silvery,
262; Small, 262; Sonora, 263; Varied,
259; Western tailed, 268; West Indian,
270
Bœhmeria, 170
Boisduval, Dr. J. A., 70
Boisduval and Leconte, " Histoire Générale et Monographie des Lepidoptères et
des Chenilles de l'Amérique Septentrionale," 70
boisduvali, Brenthis, 132
Boisduval's Marble, 285
bolli, Melitæa, 147
Books about North American Butterflies,
69
boöpis, Satyrus, 216
borealis, Calephelis, 232
boxes for preserving collections, 48
brain, 22, 23
breeding butterflies, 34-37
breeding-cages, 35, 36
bremneri, Argynnis, 113
Brenthis, genus, 128, 224; alberta, 135;
astarte, 135; bellona, 134; boisduvali,
132; chariclea, 132; epithore, 135; freija,
132; frigga, 133; helena, 131; montinus,
131; myrina, 129; polaris, 133; triclaris,
130
brettus, Thymelicus, 351
brevicauda, Papilio, 313
British Museum, 338
brizo, Thanaos, 332
bronchial tubes, 22
Brongniart, M. Charles, 196
Brooklyn Entomological Society, 73
Brown, The Gemmed, 202; Henshaw's,
202
brucei, Œneis, 223; Papilio, 313
Brush-footed Butterflies. See Nymphalidæ
bryoniæ, Pieris, 279
Buckeye, The, 173
Buckland, Frank, story of, 68
"Bulletin Brooklyn Entomological Society,"
73
bumblebees in Australia, 256
Butterflies' Fad, The, 186

" Butterflies and Moths of North America,"
Strecker, 72
" Butterflies of New England, The," by
S. H. Scudder, 72; by C. J. Maynard, 72
" Butterflies of North America," by W. H.
Edwards, 71
Butterflies, Widely Distributed, 171
Butterfly, Baird's, 313; Bruce's, 313;
Chryxus, 221; Holland's, 314; Iduna,
220; Varuna, 222; White Mountain, 222
Butts, Mary, quotation from, 251
byssus, Limochores, 358

cabinets, 50
cænius, Calephelis, 232
cæsonia, Meganostoma, 289
cæspitalis, Hesperia, 328
calais, Œneis, 221
cnlanus, Thecla, 243
Calephelis, genus, 232; australis, 233; borealis, 232; cænius, 232; nemesis, 233
Calicoes, The, 193; Orange-skirted, 194;
White-skirted, 194
california, Cœnonympha, 205
californica, Adelpha, 187; Mechanitis, 87;
Vanessa, 168
callias, Erebia, 209
Callicore, genus, 178; clymena, 178
callippe, Argynnis, 118
Callosune, genus, 162
Calpodes, genus, 355; ethlius, 356
calverleyi, Papilio, 314
Camberwell Beauty, The, 169
camillus, Phyciodes, 155
canthus, Satyrodes, 200
Cardamine, 284
cardui, Pyrameis, 170, 171
Carduus, 170
carinenta, Libythea, 227
Carnegie Museum, The, 49, 50, 338
carolina, Lerema, 367
carpenteri, Argynnis, 106
Carryl, Charles Edward, quotation from,
208
caryæ, Pyrameis, 170
Cassia, 286
Castniidæ, family, 368
Caterpillar and the Ant, The, 316
caterpillars, structure, form, color, etc., 5-
11; social habits, 8; nests, 8; wood-
boring, 8; moulting, 9; manner of defense, 9; protected by color, 8; duration
of life of, 10; preservation of, 44-48;
carnivorous, 9. See Feniseca
Catopsilia, genus, 285; agarithe, 287;
eubule, 286; philea, 286
catullus, Pholisora, 330
Ceanothus thyrsiflorus, 168
cecrops, Thecla, 246
cellus, Achalarus, 326
celtis, Chlorippe, 189
Celtis, genus of plants, 188
centaureæ, Hesperia, 327
Cerasus (Wild Cherry), 310
Ceratinia, genus, 88; lycaste, 88; var.
negreta, 88
cethura, Euchloë, 284
chalcedon, Melitæa, 139
chalcis, Thecla, 244
chara, Melitæa, 146

chariclea, Brenthis, 132
charitonius, Heliconius, 92
charon, Satyrus, 217
Chenopodium album, 330
Chicken-thief, a supposed, 33
Chionobas, genus, 218
chiron, Timetes, 180
chitone, Argynnis, 116
Chlorippe, genus, 188; alicia, 190; antonia,
189; celtis, 189; clyton, 190; flora, 191;
leilia, 190; montis, 190
chrysalis, form of, 11; color, 12; duration
of life of, 13; preservation of, 43
chrysippus, Danais, 182
chrysomelas, Colias, 291
Chrysophanus, genus, 251; arota, 252;
editha, 253; epixanthe, 254; gorgon,
253; helloides, 254; hypophlæas, 254;
maripôsa, 254; rubidus, 255; sirius,
255; snowi, 255; thoë, 253; virginiensis,
252; xanthoïdes, 253
cipris, Argynnis, 107
citima, Thecla, 239
Citrus, 311
clara, Lycæna, 259
Clark, Willis G., quotation from, 250
Classification of Butterflies, 58
claudia, Euptoieta, 99
cleis, Lemonias, 232
Clerck, Charles, 69; " Icones," 69
clio, Argynnis, 124
Clitoria, 322
clitus, Thanaos, 336
clodius, Parnassius, 305
club-men, 176
clymena, Callicore, 178
clypeus, 14, 15
clytie, Thecla, 247
clyton, Chlorippe, 190
Cnicus, 170
Codling-moth, 257
cœnia, Junonia, 173
Cœnonympha, genus, 205; ampelos, 207;
california, 205; elko, 206; eryngii, 250;
galactinus, 205; haydeni, 207; inornata,
206; kodiak, 207; ochracea, 206; pamphiloides, 207; pamphilus, 207; typhon,
206
Colænis, genus, 94; delila, 95; julia, 95
Cold, In the Face of the, 224; effects of, on
butterflies, 225
Coleridge, S. T., quotation from, 306
Colias, genus, 161, 163, 289; alexandra,
292; ariadne, 291; behri, 294; chrysomelas, 291; elis, 290; eriphyle, 291; eurytheme, 290; interior, 292; keewaydin,
291; meadi, 290; nastes, 293; pelidne,
293; philodice, 17, 291; scudderi, 293
collecting apparatus, 26-34
collecting-jars, 28-30
Collections and Collectors, 337
colon, Melitæa, 140
colon, The, 22
color, of eggs of butterflies, 4; of caterpillars, 8
columbia, Argynnis, 111
comma, Grapta, 165
Comstock, John Henry, " A Manual for
the Study of Insects," 74
comyntas, Lycæna, 268

Index

Cook, Eliza, quotation from, 198
Copæodes, genus, 345; myrtis, 346; procris, 345; wrighti, 346
Coppers, The, 236, 251; American, 254;
  Bronze, 253; Great, 253; Least, 254; Nevada, 252; Purplish, 254; Reakirt's, 254;
  Ruddy, 255; Snow's, 255
coresia, Timetes, 180
cornelia, Argynnis, 110
coronis, Argynnis, 117
costal margin of wing, 19
costal vein, 20, 21
couperi, Lycæna, 261
Cowan, Frank, quotations from, 90, 299
Cowper, quotation from, 275
coxa, 17, 18
Cramer, Peter, 69; "Papillons Exotiques,"
  69
cremaster, 11
creola, Debis, 199
Creole, The, 199
Crescent-spots, The, 150; Pearl, 153
cresphontes, Papilio, 311
creusa, Euchloë, 283
Crimson-patch, The, 159
crocale, Synchloë, 160
Crustacea, 59
crysalus, Thecla, 239
cybele, Argynnis, 106
Cystineura, genus, 177; amymone, 177
cythera, Lemonias, 230

dædalus, Lycæna, 260
Dagger-wings, The, 179; Many-banded,
  180; Ruddy, 180
damaris, Terias, 296
damon, Thecla, 246
Danais chrysippus, 182
"darning-needles," 86
daunus, Papilio, 310
Debis, genus, 198; creola, 199; portlandia,
  199
delaware, Atrytone, 365
delia, Terias, 298
delila, Colænis, 95
Dermestes, a museum pest, 53
diana, Argynnis, 103, 127
Dichora, genus, 195
Diclippa, 157
dimorphism, 23
Dione, genus, 96; vanillæ, 97
dionysius, Neominois, 213
Dircenna, genus, 89; klugi, 89
disa, Erebia, 209
discal area of wing, 19
discocellular veins, 21
discoidalis, Erebia, 209; Thecla, 246
disippus, Basilarchia, 3, 8, 84, 185
Dismorphia, genus, 273; melite, 274
Distribution of Butterflies, 25
Dog-face Butterflies, 288; Californian, 288;
  Southern, 289
Doherty, William, 338
domicella, Hesperia, 327
dorsal vessel, 22
dorus, Plestia, 322
Drake, Joseph Rodman, quotation from,
  320
Druce, Herbert, 338
dryas, Grapta, 165

drying-boxes, 42
drying-ovens, 46, 47
dumetorum, Thecla, 249
duryi, Lemonias, 230
Dusky-wings, The, 324, 332; Afranius', 334;
  Butler's, 325; Dark, 333; Dreamy, 333;
  Funereal, 336; Horace's, 336; Juvenal's,
  335; Lucilius', 333; Martial's, 335; Nævius', 336; Northern, 324; Mrs. Owen's,
  325; Pacuvius', 336; Persius', 334; Petronius', 335; Sleepy, 332; Southern, 325
Dyar, Harrison G., 186
dymas, Melitæa, 145

eagle, white-headed, 63
editha, Chrysophanus, 253; Melitæa, 142
Edwards, W. H., Author of "Butterflies of
  North America," vi, 71; types of butterflies named by, vi
edwardsi, Argynnis, 119; Thecla, 243
eggs of butterflies, 3-5; how to secure, 34;
  preparation and preservation of, 43
egleis, Argynnis, 126
elada, Melitæa, 145
elathea, Terias, 298
electa, Argynnis, 111
Elfin, Banded, 249; Brown, 247; Hoary,
  248
elis, Colias, 290
elko, Cœnonympha, 206
Elwes, Henry J., 338
Emerson, Ralph Waldo, quotations from,
  197, 319
Emperor, The Mountain, 190; The Tawny,
  190
enoptes, Lycæna, 264
"Entomologica Americana," 73
"Entomologist, The Canadian," 73
entomology, definition of, 59; in high
  schools, 257
envelopes for butterflies, 37
Epargyreus, genus, 322; tityrus, 323
epigena, Thorybes, 325
epipsodea, Erebia, 210
epithore, Brenthis, 135
epixanthe, Chrysophanus, 254
Erebia, genus, 208, 224; callias, 209; disa,
  209; discoidalis, 209; epipsodea, 210;
  ethela, 210; magdalena, 211; mancinus,
  209; sofia, 210; tyndarus, 210
Eresia, genus, 157; frisia, 157; ianthe, 158;
  punctata, 158; texana, 158; tulcis, 158
Ericaceæ, 244
eriphyle, Colias, 291
eros, Basilarchia, 186
Erycininæ, subfamily, 228
eryngii, Cœnonympha, 205
Erynnis, genus, 346; attalus, 349; leonardus, 349; manitoba, 347; metea, 348;
  morrisoni, 347; ottoë, 348; sassacus,
  348; snowi, 350; sylvanoides, 349; uncas,
  349
eryphon, Thecla, 248
ethela, Erebia, 210
Ethiopian Faunal Region, 161
ethlius, Calpodes, 356
eubule, Catopsilia, 286
Euchloë, genus, 282; ausonides, 283; cethura, 284; creusa, 283; flora, 282; genutia, 4, 284; julia, 283; lanceolata, 285;

morrisoni, 284 ; pima, 284 ; reakirti, 282 ; rosa, 284 ; sara, 282 ; stella, 283
Eudamus, genus, 320; proteus, 321
eufala, Leroden, 356
Eumæus, genus, 237 ; atala, 237, 370 ; minyas, 237
Eunica, genus, 175 ; monima, 176 ; tatila, 176
Euphorbiaceæ, 192
Euphyes, genus, 359; metacomet, 360; verna, 360
Euplœinæ, subfamily, 78, 80 ; protected insects, 84 ; Indo-Malayan, 161
Euptoieta, genus, 98 ; claudia, 99 ; hegesia, 100
eurydice, Meganostoma, 288
eurymedon, Papilio, 308
eurynome, Argynnis, 125
eurytheme, Colias, 290
eurytus, Neonympha, 203
Exchanges, 344
exilis, Lycæna, 269
external angle of wing, 19
external margin of wing, 19
eyes, of caterpillars, 6 ; of butterflies, 14, 16

fabricii, Grapta, 164
Fabricius, Johann Christian, 69
Fad, The Butterflies', 186
Families of Butterflies, 64
Family names, 63
Faun, The, 165
Faunal Regions, 161
faunus, Grapta, 165
favonius, Thecla, 240
Fawcett, Edgar, quotation from, 228
felicia, Nathalis, 281
femur, 17, 18
Feniseca, genus, 250; tarquinius, 9, 251
feronia, Ageronia, 194
Field, Eugene, quotation from, 74
field-boxes, 30
flava, Terias, 296
Flint, Charles L., edition of Harris' Report, 71
flora, Chlorippe, 191 ; Euchloë, 282
floridensis, Basilarchia, 186 ; Papilio, 307
food of caterpillars, 10
food-plants, Selection of, by female butterfly, 5
Food-reservoir, 22
forceps, 56
fornax, Ageronia, 194
Fossil Insects, 195
freija, Brenthis, 132
French, Professor G. H., 72
frigga, Brenthis, 133
frisia, Eresia, 157
Fritillary, The Variegated, 99 ; Mexican, 100 ; Regal, 103; Great Spangled, 106; Miss Owen's, 110 ; Behr's, 114 ; Behrens', 115 ; Skinner's, 117 ; Snyder's, 118 ; Edwards'. 119 ; Cliff-dwelling, 120 ; Plain, 122 ; Bischoff's, 124 ; Silver-bordered, 129 ; Hübner's, 130 ; White Mountain, 131 ; Boisduval's, 132 ; Lapland, 132 ; Polar, 133 ; Meadow, 134
front, definition of, 14
fuliginosa, Lycæna, 258
fulla, Lycæna, 259
funeralis, Thanaos, 336

gabbi, Melitæa, 144 ; Satyrus, 216
Galactia, 333
galactinus, Cœnonympha, 205
ganglia, 22, 23
garita, Oarisma, 343
Geirocheilus, genus, 211 ; tritonia, 211
gemma, Neonympha, 202
genoveva, Junonia, 174
genus, definition of, 63
genutia, Euchloë, 284
Gerardia, 173
Geyer, Karl, 70
Gibson, William Hamilton, quotation from, 93
gigas, Œneis, 220
Glass-wing, The Little, 360
glaucon, Lycæna, 264
glaucus, Papilio, 309
Gnaphalium, 170
Goatweed Butterfly, The, 192 ; Morrison's, 193
Godman, F. D., 338
gorgon, Chrysophanus, 253
Gossamer-wing, The Sooty, 258
gracilis, Grapta, 166
Grapta, genus, 163 ; comma, 5, 165 ; dryas, 165 ; fabricii, 164 ; faunus, 165 ; gracilis, 166 ; harrisi, 165 ; hylas, 165 ; interrogationis, 164 ; marsyas, 165 ; progne, 166 ; satyrus, 165 ; silenus, 166 ; umbrosa, 164 ; zephyrus, 166
greasy specimens, 54
Grossulaceæ, 167
grunus, Thecla, 238
gundlachia, Terias, 295

Hackberry Butterflies, 188, 189
Hair-streaks, The, 236, 237 ; Acadian, 242; Banded, 243; Behr's, 247 ; Boisduval's, 238 ; Bronzed, 244 ; Colorado, 239 ; Common, 242 ; Coral, 250 ; Drury's, 246 ; Early, 249 ; Edwards', 243; Gray, 245 ; Great Purple, 239 ; Green-winged, 249 ; Green White-spotted, 249 ; Hedge-row, 244 ; Henry's, 248 ; Hewitson's, 245; Martial, 240 ; Nelson's, 245 ; Olive, 246 ; Southern, 240 ; Striped, 244 ; Texas, 241 ; Thicket, 245 ; White-M, 240 ; Wittfeld's, 241
halcyone, Argynnis, 116
halesus, Thecla, 239
Hamadryas, genus, 85
Hannington, Bishop, 172
hanno, Lycæna, 269
Harris, Dr. T. W., 70; " Report on the Insects of Massachusetts which are Injurious to Vegetation," 71
harrisi, Grapta, 165 ; Melitæa, 144
Harvester, The, 251
Haworth, quotation from, 236
haydeni, Cœnonympha, 207
hayhursti, Pholisora, 331
head, of butterfly, 14 ; of caterpillar, 6
heart, 22, 23
hegesia, Euptoieta, 100
Heine, quotation from, 281
helena, Brenthis, 131
Heliconiinæ, subfamily, 78, 91, 162
Heliconius, genus, 92, 162; charitonius, 92
helloides, Chrysophanus, 254

Hemans, Mrs. Felicia, quotation from, 303
henrici, Thecla, 248
henshawi, Neonympha, 202
hermodur, Parnassius, 306
Hesperia, genus, 326; cæspitalis, 328; centaureæ, 327; domicella, 327; montivaga, 327; nessus, 329; scriptura, 328; xanthus, 328
Hesperiidæ, family, 21, 66, 318; fossil, 196
Hesperiinæ, subfamily, 320
hesperis, Argynnis, 112
Heterocera, 62
Heterometabola, 59
heteronea, Lycæna, 259
hianna, Lerema, 366
hibernaculum of Basilarchia, 183
hibernation of caterpillars, 10, 37
hippolyta, Argynnis, 112
Hoary-edge, The, 326
hoffmanni, Melitæa, 143
Holland, Philemon, quotation from translation of Livy, 85
hollandi, Papilio, 314
Hood, Thomas, quotation from, 237
horatius, Thanaos, 336
Hornaday, W. T., vii
Hosackia argophylla, 249
howardi, Phycanassa, 363
Hübner, Jacob, 69; works of, 70
Hugo's "Flower to Butterfly," 74
hulsti, Basilarchia, 84, 185
humuli, Thecla, 242
Humulus, 170
huntera, Pyrameis, 170
Hunter's Butterfly, 170
huron, Atalopedes, 352
hylas, Grapta, 165
Hylephila, genus, 354; phylæus, 354
Hypanartia, genus, 175; lethe, 175
Hypolimnas, genus, 180; misippus, 171, 181
hypophlæas, Chrysophanus, 254

ianthe, Eresia, 158
icarioides, Lycæna, 260
icelus, Thanaos, 333
idalia, Argynnis, 103
iduna, Œneis, 220
ilaire, Tachyris, 276
imago, the, 13
Immortality, 57
Indigofera, 335
Indo-Malayan Faunal Region, 161
indra, Papilio, 312
ines, Thecla, 247
inflating larvæ, 44
Ingelow, Jean, quotation from, 150, 188
inner margin of wing, 19
inornata, Argynnis, 122; Cœnonympha, 206
Insect pests, 53
Insecta, 59
Insects, Fossil, 194
Instinct, 280
interior, Colias, 292
interrogationis, Grapta, 164
intestine, 22, 23
iole, Nathalis, 281
irus, Thecla, 248
ismeria, Phyciodes, 152

isola, Lycæna, 268
isophthalma, Lycæna, 269
isthmia, Mechanitis, 87
Ithomiinæ, subfamily, 78, 85, 162
itys, Thecla, 243
ivallda, Œneis, 222

Jackson, Helen Hunt (H. H.), quotation from, 318
j-album, Vanessa, 168
janais, Synchloë, 159
Japan, Collecting in, 149
jatrophæ, Anartia, 174
jucunda, Terias, 298
julia, Colænis, 95; Euchloë, 283
Juniperus virginiana, 246
Junonia, genus, 172; cœnia, 173; genoveva, 174; lavinia, 173
jutta, Œneis, 222
juvenalis, Thanaos, 335

Kansas grasshopper, 257
Karlsbader pins, 56
keewaydin, Colias, 291
Kenia, Mount, 172
Key to Subfamilies of Nymphalidæ, 79
Kilima-njaro, 172
Kirby, Beard, & Co.'s pins, 56
klugi, Dircenna, 89
kodiak, Cœnonympha, 207
Kricogonia, genus, 287; lyside, 287; terissa, 287

labels, 52
labial palpi. See Palpi
labium, of caterpillar, 6; of butterfly, 16
labrum, of caterpillar, 6; of butterfly, 14
lacinia, Synchloë, 159
Lady, The Painted, 170, 171; The West Coast, 170
læta, Thecla, 249
lais, Argynnis, 109
Lamb's-quarter, 330
lanceolata, Euchloë, 285
lappets, 17
Laria, genus of moths, 224; rossi, 224
larva. See Caterpillar
laura, Argynnis, 120
Lauraceæ, 192
Lavatera assurgentiflora, 171
lavinia, Junonia, 173
Leaf-wings, The, 191
leanira, Melitæa, 146
Leconte, Major John E., 70
legs, of caterpillars, 7; of butterflies, 17
leilia, Chlorippe, 190
Lemonias, genus, 229; cleis, 232; cythera, 230; duryi, 230; mormo, 229; nais, 230; palmeri, 231; virgulti, 230; zela, 231
Lemoniidæ, 65, 228
leonardus, Erynnis, 349
Leopard-spots, The, 178
Lepidoptera, 60; diurnal, 61
Lerema, genus, 366; accius, 366; carolina, 367; hianna, 366
Lerodea, genus, 356; eufala, 356
Lespedeza, 324
lethe, Hypanartia, 175
leto, 105
libya, Pholisora, 331

Libythea, genus, 226; bachmanni, 227; carinenta, 227; labdaca, 195
Libytheinæ, subfamily, 78, 226; fossil, 196
liliana, Argynnis, 119
limbal area of wing, 19
Limochores, genus, 357; byssus, 358; manataaqua, 357; palatka, 358; pontiac, 358; taumas, 357; yehl, 359
Linnæus, 58, 69
liparops, Thecla, 244
lisa, Terias, 297
Literature relating to North American butterflies, 69
Long-dash, The, 351
lorquini, Basilarchia, 185
lower discocellular vein, 21
lower radial vein, 20, 21
lucia, Lycæna, 267
lucilius, Thanaos, 333
Luther's Saddest Experience, 100
Lycæna, genus, 258; acmon, 266; ammon, 270; amyntula, 268; annetta, 266; antiacis, 261; aquilo, 263; aster, 266; battoides, 264; clara, 259; comyntas, 268; couperi, 261; dædalus, 260; enoptes, 264; exilis, 269; fuliginosa, 258; fulla, 259; glaucon, 264; hanno, 269; heteronea, 259; icarioides, 260; isola, 268; isophthalma, 269; lucia, 267; lycea, 259; lygdamas, 262; marginata, 267; marina, 270; melissa, 265; mintha, 260; neglecta, 267; nigra, 267; pheres, 261; piasus, 263; podarce, 263; pseudargiolus, 4, 267; rustica, 263; sæpiolus, 260; sagittigera, 262; scudderi, 265; shasta, 265; sonorensis, 263; speciosa, 262; theonus, 270; violacea, 267; xerxes, 261
Lycænidæ, 66, 161, 236
lycaste, Ceratinia, 88
lycea, Lycæna, 259
lycidas, Achalarus, 325
lygdamas, Lycæna, 262
lyside, Kricogonia, 287

macaria, Argynnis, 121
MacDonald, George, quotation from, 201
macglashani, Melitæa, 140
machaon, Papilio, 312
macouni, Œneis, 221
maculata, Oligoria, 361
magdalena, Erebia, 211
Malachites, The, 194; The Pearly, 195
Malacopoda, 59
m-album, Thecla, 240
Malpighian vessel, 22, 23
manataaqua, Limochores, 357
mancinus, Erebia, 209
mandan, Pamphila, 342
mandibles of caterpillar, 6
manitoba, Erynnis, 347
Many-banded Dagger-wing, The, 180
marcellus, Papilio, 308
marcia, Phyciodes, 153
mardon, Polites, 354
marginata, Lycæna, 267
marina, Lycæna, 270
mariposa, Chrysophanus, 254
maritima, Satyrus, 215
marsyas, Grapta, 165

martialis, Thecla, 240; Thanaos, 335
massasoit, Poanes, 361
Maxillæ, of caterpillars, 6; of butterflies, 14
Maynard, C. J., 72, 73
McDonald, quotation from, 177
meadi, Argynnis, 119; Satyrus, 216; Colias, 290
Mechanitis, genus, 86; californica, 87; isthmia, 87; polymnia, 88
median area of wing, 19
median nervules, 21
median vein, 20, 21
Meganostoma, genus, 288; cæsonia, 289; eurydice, 288
Megathyminæ, subfamily, 368
Megathymus, genus, 367; yuccæ, 368
melane, Atrytone, 365
melanism, 24
melinus, Thecla, 242
melissa, Lycæna, 265
Melitæa, genus, 137, 161, 163; acastus, 143; alma, 147; anicia, 140; arachne, 148; augusta, 141; baroni, 141; beani, 140; bolli, 147; chalcedon, 139; chara, 146; colon, 140; dymas, 145; editha, 142; elada, 145; gabbi, 144; harrisi, 144; hoffmanni, 143; leanira, 146; macglashani, 140; minuta, 148; nubigena, 141; nympha, 148; palla, 143; perse, 146; phaëton, 4, 138; rubicunda, 142; taylori, 142; thekla, 147; wheeleri, 141; whitneyi, 143; wrighti, 147
melite, Dismorphia, 274
menapia, Neophasia, 275
mesothorax, 17, 23
Metabola, 60
metacomet, Euphyes, 360
Metal-marks, The, 228
Metal-marks, The, 230; Behr's, 230; Dury's, 230; Dusky, 233; Little, 232; Northern, 232; Palmer's, 231; Southern, 233
metathorax, 17, 23
metea, Erynnis, 348
mexicana, Terias, 296
micropyle, 4
middle discocellular vein, 21
milberti, Vanessa, 169
mildew, 54
Milkweed Butterfly. See Anosia
Mime, The, 274
Mimic The, 181
Mimicry, 24, 235
mintha, Lycæna, 260
minuta, Melitæa, 148
minyas, Eumæus, 237
misippus, Hypolimnas, 171, 181
"Missouri Reports," The, by C. V. Riley, 73
Monarch, The, 82
monima, Eunica, 176
Monkey, story about, 68; butterflies distasteful to, 92
monstrosities, 24
montana, Phyciodes, 156
monticola, Argynnis, 114
montinus, Brenthis, 131
montis, Chlorippe, 190
montivaga, Argynnis, 126; Hesperia, 327
monuste, Pieris, 277
Moore, Thomas, quotation from, 58

Moravian Brethren, 127
mormo, Lemonias, 229
Mormon, The, 229
morpheus, Phyciodes, 154
Morris, Rev. John G., " Catalogue of the
Described Lepidoptera of North America," 71
morrisoni, Erynnis, 347; Euchloë, 284;
Pyrrhanæa, 193
moths, how to distinguish, from butterflies,
62
mould on specimens, 54
moulting of caterpillars, 9
mounting butterflies, 38 ; English method,
39 ; continental method, 39 ; on setting-
boards, 40; on setting-blocks, 42
Mount Washington, N. H., 220
Mourning-cloak, The, 169
Mulberry-wing, The, 361
Munkittrick, quotation from, 128
muscles of head of butterfly, 15, 16
mylitta, Phyciodes, 155
Myriapoda, 59
myrina, Brenthis, 129
myrtis, Copæodes, 346
mystic, Thymelicus, 351

nævius, Thanaos, 336
nais, Lemonias, 230
names, family, 63; generic, 63; specific,
63 ; scientific, 66; popular, 68; use of,
67
Naphthaline as a preventative of infection,
53
Naphthaline cones, 53
napi, Pieris, 279
nastes, Colias, 293
Nathalis, genus, 281 ; iole, 281; felicia, 281
nausicaä, Argynnis, 108
Nearctic Faunal Region, 161, 163
neglecta, Lycæna, 267
negreta, Ceratinia, 88
nelsoni, Thecla, 245
nemesis, Calephelis, 233
Neominois, genus, 212; dionysius, 213;
ridingsi, 213
Neonympha, genus, 201; eurytus, 18, 203;
gemma, 202 ; henshawi, 202; mitchelli,
203 ; phocion, 202; rubricata, 204; sosybius, 204
Neophasia, genus, 274
Neotropical Faunal Region, 161, 162
nephele, Satyrus, 215
nervous system of lepidoptera, 22, 23
nervules, 21
nessus, Hesperia, 329
nets, 26-28; the use of, 31
nevadensis, Argynnis, 118
" News, The Entomological," 73
" New York Entomological Society, Journal of the," 73
Nicholas, Grand Duke, 338
nicippe, Terias, 296
nigra, Lycæna, 267
niphon, Thecla, 249
nitocris, Argynnis, 105
nitra, Papilio, 312
nokomis, Argynnis, 104
Nova Scotian, The, 222
nubigena, Melitæa, 141

number of species of butterflies in the
United States, 25
numitor, Ancyloxypha, 345
nycteis, Phyciodes, 151
nympha, Melitæa, 148
Nymphalidæ, 65, 77; subfamilies of, 78;
fossil, 196
Nymphalinæ, subfamily, 78, 93; eggs of,
94; Indo-Malayan, 161
Nymphs, The (see Nymphalinæ) ; Eyed,
198 ; Common Grass, 200 ; Spangled, 201

Oarisma, genus, 343 ; garita, 343; powe-
shelk, 343
Oberland, Bernese, 172
Oberthür, M. Charles, 338
occidentalis, Pieris, 278
ochracea, Cœnonympha, 206
ocola, Prenes, 355
Œneis, genus, 218, 224; brucei, 223; calais, 221; chryxus, 221 ; gigas, 220; iduna,
220 ; ivallda, 222 ; jutta, 222 ; macouni,
221; semidea, 222; taygete, 223; uhleri,
222 ; varuna, 222
œsophagus, of butterfly, 15, 16, 23; of
caterpillar, 22
œtus, Satyrus, 218
oleracea-hiemalis, Pieris, 279
Oligoria, genus, 361; maculata, 361
olympus, Satyrus, 215
opis, Argynnis, 124
Orange-tips, The, 282 ; Falcate, 284 ; Pima,
284 ; Reakirt's, 282
oregonia, Papilio, 314
Ornithoptera, genus, 162, 272 ; paradisea,
162 ; victoria, 162
orseis, Phyciodes, 154
osmateria, 9
ottoë, Erynnis, 348
outer angle of wing, 19
oviduct, 23
oweni, Argynnis, 109

Packard, A. S., " Guide to the Study of
Insects," 74 ; " A Text-book of Entomology," 74
packing specimens, 55
pacuvius, Thanaos, 336
Palæarctic Faunal Region, 161
palamedes, Papilio, 315
palatka, Limochores, 358
palla, Melitæa, 143
pallida, Pieris, 297
palmeri, Lemonias, 231
palpi, of caterpillars, 6; of butterflies, 16,
23
Pamphila, genus, 342 ; mandan, 342
Pamphilinæ, subfamily, 339
pamphiloides, Cœnonympha, 207
pamphilus, Cœnonympha, 207
papering specimens, 37
Papilio, genus, 161, 162, 272, 306; abbotti,
307 ; ajax, 307; aliaska, 312; antimachus,
162; asterias, 6, 13, 314; bairdi, 313;
brevicauda, 313 ; brucei, 313; calverleyi,
314; cresphontes, 311; daunus, 310;
eurymedon, 308; floridensis, 307; glaucus, 309; hollandi, 314; indra, 312; machaon, 312; marcellus, 308; nitra, 312;
oregonia, 314; palamedes, 315; philenor,

6, 12, 315; pilumnus, 310; polydamas, 316; rutulus, 309; telamonides, 308; thoas, 311; troilus, 9, 315; turnus, 3, 23, 309; walshi, 307; zolicaon, 312
"Papilio," journal devoted to entomology, 73
Papilionidæ, 66, 272
Papilioninæ, subfamily, 304; fossil, 196
paradisea, Ornithoptera, 162
Parnassians, The, 304
Parnassius, genus, 304; behri, 306; clodius, 305; hermodur, 306; sminitheus, 306
Passiflora, 96
passion-flower, 92, 97, 98, 99
Patched Butterflies, The, 159
paulus, Satyrus, 217
Peacock Butterflies, 172
Peacock, The White, 174
Pearly-eye, The, 199
peckius, Polites, 353
pectus, 17
pegala, Satyrus, 215
pelidne, Colias, 293
Periodical literature of entomology, 73
perse, Melitæa, 146
persius, Thanaos, 334
petronius, Thanaos, 335
phaëton, Melitæa, 138
phaon, Phyciodes, 153
pheres, Lycæna, 261
philea, Catopsilia, 286
philenor, Papilio, 315
philodice, Colias, 291
phocion, Neonympha, 202
Pholisora, genus, 330; alpheus, 331; catullus, 330; hayhursti, 331; libya, 331
Phycanassa, genus, 362; aaroni, 363; howardi, 363; viator, 362
Phyciodes, genus, 150; barnesi, 155; batesi, 154; camillus, 155; ismeria, 152; marcia, 153; montana, 156; morpheus, 154; mylitta, 155; nycteis, 151; orseis, 154; phaon, 153; picta, 156; pratensis, 154; tharos, 153; vesta, 152
phylæus, Hylephila, 354
piasus, Lycæna, 268
picta, Phyciodes, 156
Pierinæ, subfamily, 272; fossil, 196
Pieris, genus, 276; acadica, 280; beckeri, 277; bryoniæ, 279; monuste, 277; napi, 279; occidentalis, 278; oleracea, 5, 13, 18; oleracea-hiemalis, 279; pallida, 279; protodice, 12, 278; rapæ, 280; sisymbri, 278; vernalis, 278; virginiensis, 279
pilumnus, Papilio, 310
pima, Euchloë, 284
pins, 56
Piperaceæ, 192
Plantago, 173
platina, Argynnis, 117
Plestia, genus, 322; dorus, 322
plexippus, Anosia, 82
Pliny, quotation from, 85
Poanes, genus, 361; massasoit, 361
pocahontas, Atrytone, 364
podarce, Lycæna, 263
Podostomata, 59
polaris, Brenthis, 133
Polites, genus, 353; mardon, 354; peckius, 353; sabuleti, 354

polydamas, Papilio, 316
polymnia, Mechanitis, 88
polymorphism, 23
pontine, Limochores, 358
Pope, Alexander, quotation from, 304
Populus, 169
portia, Pyrrhanæa, 193
portlandia, Debis, 199
potato-bug, 257
powesheik, Oarisma, 343
pratensis, Phyciodes, 154
precostal veins of Erycininæ, 228
Prenes, genus, 355; ocola, 355
proboscis of butterflies, 14–16, 23
procris, Copæodes, 345
progne, Grapta, 166
prolegs, of caterpillars, 7; anal, 8
proserpina, Basilarchia, 184
protective mimicry, 25
proterpia, Terias, 295
proteus, Eudamus, 321
prothorax, 17, 23
protodice, Pieris, 278
pseudargiolus, Lycæna, 267
pseudodorippus, Basilarchia, 185
"Psyche," journal devoted to entomology, 73
Ptelea, 311
punctata, Eresia, 158
pupa. See Chrysalis
Purple, The Banded, 184; The Red-spotted, 183
purpurescens, Argynnis, 114
pylades, Thorybes, 324
Pyrameis, genus, 169; atalanta, 170; cardui, 170, 171; caryæ, 170; huntera, 170; indica, 172
Pyrrhanæa, genus, 191; andria, 9, 192; morrisoni, 193; portia, 193
Pyrrhopyge, genus, 319; araxes, 319
Pyrrhopyginæ, subfamily, 319

Queen, The, 84
Queens, The Tropic, 180
Quercus, chrysolepis, 239
Question-sign, The, 164

Race after a Butterfly, 127
Ramsay, Allan, quotation from, 316
rapæ, Pieris, 280
Reakirt, 87–90
reakirti, Euchloë, 282
rectum, 22, 23
Red Rain, 299
Reds, The Banded, 175
Regions, Faunal, 161
relaxing specimens, 41
Repairing broken specimens, 55
Rhamnus californicus, 309
rhodope, Argynnis, 115
Rhopalocera, origin of term, 16; suborder of lepidoptera, 60, 62
Ribes, 252
ridingsi, Neominois, 213
Riley, James Whitcomb, quotation from, 276
Riley, Professor C. V., vii, 73, 80, 256
Ringlets, The, 205; Alaskan, 207; Californian, 205; Elko, 206; Hayden's, 207;

Ochre, 206; Plain, 206; Ringless, 207;
   Utah, 207
Robinia pseudacacia, 323
Rogers, quotation from, 294
rosa, Euchloë, 284
Ross, Commander James, 224
Rossetti, Christina, quotation from, 294
rossi, Laria, 224
Rothschild, Hon. Walter, 338
rubicunda, Melitæa, 142
rubidus, Chrysophanus, 255
rubricata, Neonympha, 204
Ruddy Dagger-wing, The, 180
Rumex, 253
rupestris, Argynnis, 120
Russell, quotation from, 339
rustica, Lycæna, 263
rutulus, Papilio, 309

sabuleti, Polites, 354
Sachem, The, 352
sæpiolus, Lycæna, 260
sæpium, Thecla, 244
sagittigera, Lycæna, 262
samoset, Amblyscirtes, 340
sara, Euchloë, 282
sassacus, Erynnis, 348
sassafras, 315
Satyr, The, 165
Satyrinæ, subfamily, 78, 197; fossil, 196
Satyrs, The: Baron's, 216; Boisduval's,
   218; Carolinian, 204; Gabb's, 216; Georgian, 202; Little Wood-, 203; Mead's,
   216; Mitchell's, 203; Red, 204; Ridings',
   213; Scudder's, 213
Satyrodes, genus, 200; canthus, 200
Satyrus, genus, 214; alope, 215; ariane,
   216; baroni, 216; boöpis, 216; charon,
   217; gabbi, 216; maritima, 215; meadi,
   216; nephele, 215; œtus, 218; olympus,
   215; paulus, 217; pegala, 215; sthenele,
   218; texana, 215
satyrus, Grapta, 165
sauer-kraut, 257
Saxifraga, 306
scales of wings, 18; how to remove, 19;
   arrangement on wing, 20
scale-insects, injurious to orange-trees, 256
Schaus, William, 160
scriptura, Hesperia, 328
Scudder, Dr. S. H., author of " The Butterflies of New England," vi, vii, 72, 73
Scudderi, Lycæna, 265; Colias, 293
Sedum, 306
segments constituting external skeleton of
   caterpillar, 6
semidea, Œneis, 222
semiramis, Argynnis, 121
setting-blocks, 39
setting-boards, 39
setting-needles, 40
sex, 64
sex-signs, 64
Shakespeare, quotations from, 91, 205,
   218, 273
shasta, Lycæna, 265
shellac, 55
Shelley, quotation from, 26
" Shingling " butterflies when packing for
   shipment, 55

Sigourney, Mrs., quotation from, 57
silenus, Grapta, 166
Silver-spot, Arizona, 108; Bremner's, 113;
   Columbian, 111; Mead's, 119; Mountain, 108; New Mexican, 107; Nevada,
   118; Northwestern, 109; Owen's, 119
simæthis, Thecla, 246
simius, Amblyscirtes, 341
sirius, Chrysophanus, 255
Sisters, The, 187; Californian, 187
sisymbri, Pieris, 278
Sisymbrium, 284
siva, Thecla, 246
size, 271
Skinner, Dr. Henry, 325, 363
Skippers, The, 318; Aaron's, 363; Arctic,
   342; Brazilian, 336; Brond-winged, 362;
   Bronze, 341; Canadian, 347; Carolina,
   367; Checkered, 327; Cobweb, 348;
   Cross-line, 357; Delaware, 365; Dun,
   360; Dusted, 366; Erichson's, 327;
   Fiery, 354; Golden-banded, 326; Grizzled, 327; Hayhurst's, 331; Hobomok,
   364; Howard's, 363; Indian, 348; Iowa,
   364; Leonard's, 349; Long-tailed, 321;
   Morrison's, 347; Ocola, 355; Oregon,
   354; Palatka, 358; Peck's, 353; Pepper-
   and-salt, 340; Roadside, 340; Sand-hill,
   354; Short-tailed, 322; Silver-spotted,
   323; Skinner's, 359; Small-checkered,
   328; Snow's, 350; Tawny-edged, 357;
   Two-banded, 328; Umber, 365; Volcanic, 351; Woodland, 340; Woven-
   winged, 341; Wright's, 346; Xanthus,
   328
Slosson, Mrs. Annie Trumbull, quotation
   from, 233
Small Sulphurs, 294; Gundlach's, 295
smintheus, Parnassius, 306
Smith, Herbert H., 338
Smith, Sir James Edward, 70
Smith and Abbot, " The Natural History
   of the Rarer Lepidopterous Insects of
   Georgia," 70
Snout-butterflies, 226, 227; Southern, 227
Snow, Chancellor F. H., 255
snowi, Chrysophanus, 255; Erynnis, 350
snyderi, Argynnis, 118
sofia, Erebia, 210
somnus, Thanaos, 333
sonorensis, Lycæna, 263
Sooty-wing, The, 330; Mohave, 331
sosybius, Neonympha, 204
species, definition of, 62
speciosa, Lycæna, 262
Spenser, Edmund, Quotation from, 226
spermatheca, 23
spicewood, 315
spinetorum, Thecla, 245
spinneret, 6, 22
spinning-vessel, 22
Staudinger, Dr. Otto, 338
stella, Euchloë, 283
steneles, Victorina, 195
sthenele, Satyrus, 218
stomach, 22, 23
Strecker, Herman, 72
strigosa, Anosia, 84
subcostal nervules, 21
subcostal vein, 20, 21

subfamily names, 63
submedian vein, 20, 21
subœsophageal ganglion, 22, 23
"sugaring," 32
Sulphurs, The, 272, 289; Alexandra, 292;
Arctic, 293; Behr's, 294; Cloudless, 286;
Common, 291; Gold-and-black, 291;
Great, 285; Labrador, 293; Large Orange,
287; Little, 297; Mead's, 290; Pink-
edged, 292; Red-barred, 286; Scudder's,
293; Strecker's, 290
Superstitions, 90
Suspicious Conduct, 136
Swallowtails, The, 272, 306; Alaskan, 312;
Common Eastern, 314; Giant, 311;
Newfoundland, 313; Pipe-vine, 315;
Spice-bush, 315; Tiger, 309
Swinburne, quotation from, 272
sylvanoides, Erynnis, 349
Synchloë, genus, 159; crocale, 160; janais,
159; lacinia, 159
Systasea, genus, 329; zampa, 329

Tachyris, genus, 275; ilaire, 276
tacita, Thecla, Plate XXIX, Fig. 30
tarquinius, Feniseca, 251
tarsi, 17, 18
tatila, Eunica, 176
taumas, Limochores, 357
taxiles, Atrytone, 365
taygete, Œneis, 223
taylori, Melitæa, 142
tegulæ, 17
telamonides, Papilio, 308
Tennyson, quotation from, 213
Terias, genus, 294; damaris, 296; delia,
298; elathea, 298; flava, 296; gundlachia,
295; jucunda, 298; lisa, 297; mexicana,
296; nicippe, 296; proterpia, 295; westwoodi, 297
terissa, Kricogonia, 287
testis, 22
texana, Eresia, 158; Satyrus, 215
textor, Amblyscirtes, 341
Thanaos, genus, 332; afranius, 334; brizo,
332; clitus, 336; funeralis, 336; horatius,
336; icelus, 333; juvenalis, 335; lucilius,
333; martialis, 335; nævius, 336; pacuvius, 336; persius, 334; petronius, 335;
somnus, 333
tharos, Phyciodes, 153
Thecla, genus, 237; acadica, 242; acis,
240, 246; adenostomatis, 245; affinis,
249; alcestis, 241; arsace, 248; augustus,
247; autolycus, 241; behri, 247; blenina,
245; calanus, 243; cecrops, 246; chalcis,
244; citima, 239; clytie, 247; crysalus,
239; damon, 246; discoidalis, 246; dumetorum, 249; edwardsi, 243; eryphon, 248;
favonius, 240; grunus, 238; halesus, 239;
henrici, 248; humuli, 242; ines, 247;
irus, 248; itys, 243; læta, 249; liparops,
244; m-album, 240; martialis, 240; melinus, 242; nelsoni, 245; niphon, 249;
sæpium, 244; simætnis, 246; siva, 246;
spinetorum, 245; tacita, Plate XXIX,
Fig. 30; titus, 250; wittfeldi, 241
thekla, Melitæa, 147
theonus, Lycæna, 270
Thibet, 172

thoas, Papilio, 311
thoë, Chrysophanus, 253
thorax, 7, 14, 17, 22, 23
Thoreau, Quotation from, 93
Thorybes, genus, 324; æmilia, 325; bathyllus, 325; epigena, 325; pylades, 324
Thymelicus, genus, 350; ætna, 351; brettus, 351; mystic, 351
tibia, 17, 18
tiger, 63
Timetes, genus, 179; chiron, 180; coresia,
180; petreus, 180
tip for inflating tube, 46
titus, Thecla, 250
tityrus, Epargyreus, 323
Tokyo, 149
Tongue. See Proboscis
Tortoise, The Compton, 168
Tortoise-shells, The, 167; the California,
168; Milbert's, 169
tracheæ, 15, 22
"Transactions of the American Entomological Society," 73
transformations, egg to caterpillar, 5;
caterpillar to chrysalis, 11; chrysalis to
butterfly, 13
triclaris, Brenthis, 130
tritonia, Geirocheilus, 211
trochanter, 17, 18
troilus, Papilio, 315
tulcis, Eresia, 158
turnus, Papilio, 309
Turritis, 285
Twin-spot, The, 361
tyndarus, Erebia, 210
types of butterflies named by W. H. Edwards, vi; used in preparation of this
book, vii
typhon, Cœnonympha, 206

uhleri, Œneis, 204
Umbelliferæ, 312, 313, 314
umbrosa, Grapta, 164
uncas, Erynnis, 349
Uncle Jotham's Boarder, 233
United States Department of Agriculture,
49, 73
United States National Museum, 73
upper discocellular vein, 21
upper radial vein, 20, 21
Urtica, 164, 169
urtica, Vanessa, 169
Urticaceæ, 164, 165
Utility of Entomology, The, 256

Vanessa, genus, 167; antiopa, 5, 7, 94, 169;
californica, 168; j-album, 168; milberti,
169; urticæ, 169; vau-album, 168; xanthomelas, 168
vanillæ, Dione, 97
varieties, 64; insular, 64
varuna, Œneis, 222
vau-album, Vanessa, 168
veins of wings, 20, 21
verna, Euphyes, 360
vernalis, Pieris, 278
Vertex, definition of, 14
vesta, Phyciodes, 152
vialis, Amblyscirtes, 340
viator, Phycanassa, 362

Viceroy, The, 185
victoria, Ornithoptera, 162
Victorina, genus, 194; steneles, 195
violacea, Lycæna, 267
violets, 98, 102
Violet-wings, The, 175; The Dingy, 176
virginiensis, Chrysophanus, 252; Pieris, 279
virgulti, Lemonias, 230
vitellius, Atrytone, 364

Wallace, Alfred Russel, 92, 338
walshi, Papilio, 307
Walsingham, Lord, 338
weidemeyeri, Basilarchia, 185
westwoodi, Terias, 297
wheeleri, Melitæa, 141
Whirlabout, The, 351
White Admirals, The, 182
White Peacock, The, 174
Whites, The, 272; Becker's, 277; Cabbage, 280; California, 278; Common, 278; Florida, 276; Great Southern, 277; Mustard, 279; Pine, 275; Western, 278
whitneyi, Melitæa, 143
Wilcox, Ella Wheeler, quotation from, 186
wings of butterflies, 18, 21
winter quarters of Basilarchia, 183

Wistaria, 322
wittfeldi, Thecla, 241
Wood-nymphs, The, 214; Clouded, 215; Common, 215; Dark, 217; Least, 218; Small, 217; Southern, 215
wrighti, Melitæa, 147; Copæodes, 346
writers, early, upon butterflies of North America, 69; later, 71

xanthoides, Chrysophanus, 253
xanthomelas, Vanessa, 168
xanthus, Hesperia, 328
xerxes, Lycæna, 261

"Yale Literary Magazine," 100
yehl, Limochores, 359
Yellow, The Dwarf, 281; The Fairy, 298
The Mexican, 296; Westwood's, 297
yuccæ, Megathymus, 368

zabulon, Atrytone, 364
zampa, Systasea, 329
Zebra, The. See Charitonius
zela, Lemonias, 231
zephyrus, Grapta, 166
zerene, Argynnis, 113
zolicaon, Papilio, 312